EXPORT RESTRICTIONS ON CRITICAL MINERALS AND METALS

Conventional wisdom on the insufficiency of existing WTO disciplines on export restrictions has triggered momentum on the issue. In this book, Ilaria Espa offers a comprehensive analysis of the scope and coverage of WTO disciplines on export restrictions in light of emerging case law. She investigates whether such rules still provide a credible and effective framework capable of preventing abuses in the use of export restrictive measures on critical minerals and metals during a period of economic crisis and change in international trade patterns. Giving a broad overview of the export restrictions applied to these materials, Espa identifies distinctive features in the proliferation of export barriers and analyses the existing WTO rules to reveal their gaps and inconsistencies. She goes on to present solutions based upon her findings with the aim of bringing more coherence and equity to WTO rules on the export side.

ILARIA ESPA is a Marie Curie senior research fellow at the World Trade Institute (WTI), University of Berne, and a member of the Work Package on 'Trade and Climate Change' of the Swiss National Centre of Competence in Research (NCCR) programme in Trade Regulation. She was a visiting scholar at Columbia Law School and she has served as a consultant for the Trade and Environment Division of the WTO and for the Swiss Federal Office of Energy.

CAMBRIDGE INTERNATIONAL TRADE AND ECONOMIC LAW

Series Editors

As the processes of regionalization and globalization have intensified, there have been accompanying increases in the regulations of international trade and economic law at the levels of international, regional and national laws.

The subject matter of this series is international economic law. Its core is the regulation of international trade, investment and cognate areas such as intellectual property and competition policy. The series publishes books on related regulatory areas, in particular human rights, labour, environment and culture, as well as sustainable development. These areas are vertically linked at the international, regional and national levels, and the series extends to the implementation of these rules at these different levels. The series also includes works on governance, dealing with the structure and operation of related international organizations in the field of international economic law, and the way they interact with other subjects of international and national law.

Books in the series

Optimal Regulation and the Law of International Trade: The Interface between Societal Values and WTO Law
Boris Rigod

The Social Foundations of World Trade: Norms, Community and Constitution
Sungjoon Cho

Public Participation and Legitimacy in the WTO
Yves Bonzon

The Challenge of Safeguards in the WTO
Fernando Piérola

General Interests of Host States in International Investment Law
Edited by Giorgio Sacerdoti, Pia Acconci, Mara Valenti and Anna De Luca

The Law of Development Cooperation: A Comparative Analysis of the World Bank, the EU and Germany
Philipp Dann

WTO Disciplines on Subsidies and Countervailing Measures: Balancing Policy Space and Legal Constraints
Dominic Coppens

Domestic Judicial Review of Trade Remedies: Experiences of the Most Active WTO Members
Müslüm Yilmaz

International Organizations in WTO Dispute Settlement: How Much Institutional Sensitivity?
Marina Foltea

Public Services and International Trade Liberalization: Human Rights and Gender Implications
Barnali Choudhury

The Law and Politics of WTO Waivers: Stability and Flexibility in Public International Law
Isabel Feichtner

African Regional Trade Agreements as Legal Regimes
James Thuo Gathii

Processes and Production Methods (PPMs) in WTO Law: Interfacing Trade and Social Goals
Christiane R. Conrad

Non-Discrimination in International Trade in Services: 'Likeness' in WTO/GATS
Nicolas Diebold

The Law, Economics and Politics of Retaliation in WTO Dispute Settlement
Edited by
Chad P. Bown and
Joost Pauwelyn

The Multilateralization of International Investment Law
Stephan W. Schill

Trade Policy Flexibility and Enforcement in the WTO: A Law and Economics Analysis
Simon A.B. Schropp

The WTO and International Investment Law: Converging Systems
Jürgen Kurtz

The Law, Economics and Politics of International Standardisation
Panagiotis Delimatsis

Optimal Regulation and the Law of International Trade: The Interface between the Right to Regulate and WTO Law
Boris Rigod

Export Restrictions on Critical Minerals and Metals: Testing the Adequacy of WTO Disciplines
Ilaria Espa

EXPORT RESTRICTIONS ON CRITICAL MINERALS AND METALS

Testing the Adequacy of WTO Disciplines

ILARIA ESPA
World Trade Institute, University of Berne

CAMBRIDGE
UNIVERSITY PRESS

CAMBRIDGE
UNIVERSITY PRESS

University Printing House, Cambridge CB2 8BS, United Kingdom

Cambridge University Press is part of the University of Cambridge.

It furthers the University's mission by disseminating knowledge in the pursuit of education, learning and research at the highest international levels of excellence.

www.cambridge.org
Information on this title: www.cambridge.org/9781107085961

First published 2015

A catalogue record for this publication is available from the British Library

Library of Congress Cataloging in Publication data
Espa, Ilaria, author.
Export restrictions on critical minerals and metals : testing the adequacy of WTO disciplines / Ilaria Espa ; foreword by Giorgio Sacerdoti.
pages cm. – (Cambridge international trade and economic law ; 19)
Based on author's thesis (doctoral – Università Commerciale Luigi Bocconi, Milan, 2013) issued under title: Emerging trends in critical raw materials trade and WTO regulation of export restrictions.
ISBN 978-1-107-08596-1 (hardback)
1. Export controls. 2. Mineral industries – Law and legislation. 3. World Trade Organization. I. Title.
K3952.M56E87 2015
382′.42–dc23
2015023764

ISBN 978-1-107-08596-1 Hardback

CONTENTS

Foreword by Giorgio Sacerdoti xiii
Acknowledgements xv
List of abbreviations xvii
WTO reports xix
GATT panel reports xxiii

Introduction 1

I.1 Historical trends in the use of export restrictions 1

I.2 The panoply of export restrictions on primary commodities in recent times 8

I.2.1 Agricultural export restrictions and food security concerns 11

I.2.2 Export restrictions on energy commodities between new and old concerns 22

I.2.3 Export restrictions on raw materials of mineral origin and the revival of industrial policies 28

I.3 'Critical' minerals and metals and the debate over the use of export restrictions 31

I.4 Goals, method and structure of this study 34

PART I **Minerals and metals critical to the world economy and the recent proliferation of export restrictions** 39

1 Evolving world trade patterns in critical minerals and metals 41

1.1 Introduction 41

1.2 Defining mineral raw materials 41

1.2.1 A definition of raw materials 41

1.2.2 Raw materials of mineral origin 44
1.2.3 A classification of minerals and metals 46

1.3 Identifying ‘critical’ minerals and metals 48
1.3.1 Demand and supply determinants for mineral raw materials 48
1.3.2 Specific factors of mineral criticality 51
1.3.3 A list of critical minerals and metals 54

1.4 Distinctive features of critical minerals and metals 56

1.5 Main patterns of international trade in critical minerals and metals 63

1.6 Conclusions 65

2 The landscape of export restrictions on critical minerals and metals 67

2.1 Introduction 67

2.2 A definition of export restrictions 67

2.3 The explosion of attention towards export restrictions 69

2.4 Recent trends in the use of export taxes 72
2.4.1 A definition of export taxes or duties 72
2.4.2 The proliferation of export duties or taxes 74
2.4.3 Export duties or taxes applied to critical mineral and metals 77
2.4.4 Other price or tax measures imposed on exportation 84

2.5 Recent trends in the use of QRs on exports 86
2.5.1 Export prohibitions or bans 86
2.5.2 Export quotas 89
2.5.3 Export licensing schemes or requirements 93
2.5.4 Minimum export prices and other types of export restrictions 96

2.6 Conclusions 98

3 Economic effects and policy goals of mineral export restraints 101

3.1 Introduction 101

3.2 The economic effects of export restrictions 102

3.2.1 Effects predicted by theoretical economic models 103

3.2.2 Changes in welfare and redistributional effects linked to export restrictions 105

3.2.3 Sector-specific factors influencing the effects of export restrictions on minerals and metals 108

3.3 Main policy goals of export restrictions in the minerals and metals sector 111

3.3.1 Export restrictions as developmental tools 111

3.3.2 Export restrictions as environmental tools 116

3.3.3 Other policy objectives associated with export restrictions 118

3.4 Empirical evidence on the effects of export restrictions on critical minerals and metals 119

3.5 Conclusions 122

PART II **Existing WTO disciplines on export restrictions** 125

4 WTO disciplines on export duties 127

4.1 Introduction 127

4.2 Asymmetry in the WTO treatment of import and export tariffs: origins and rationale 128

4.3 Export duties or taxes in the GATT 131

4.3.1 Enforceability of export duty concessions included in GATT schedules 131

4.3.2 Differences in the treatment of import and export duty concessions in the GATT 135

4.3.3 Admissibility of export duties under Article XI:1 GATT 140

4.3.4 Other GATT provisions applicable to export duties 143

4.4 Export duty commitments in post-1994 accession protocols: the emergence of WTO-plus obligations 145

4.4.1 An overview of export duty commitments assumed by newly acceded WTO Members 147

4.4.2 Limited flexibility of WTO-plus obligations on export duties 161

4.5 Transparency and notification requirements applicable to export duties 163

4.6 Conclusions 165

5 WTO disciplines on export quantitative restrictions 169

5.1 Introduction 169

5.2 GATT rules on export quantitative restrictions 169

5.2.1 The general elimination obligation under Article XI:1 GATT 169

5.2.2 Other GATT provisions relevant to quantitative restrictions on exports 175

5.3 GATT exceptions applicable to quantitative restrictions on exports 179

5.3.1 The critical shortage of supply clause under Article XI:2(a) GATT 180

5.3.2 Limited applicability of other Article XI-specific exceptions 184

5.3.3 Non-availability of other GATT exceptions applicable to quantitative restrictions 185

5.4 Transparency and notification requirements applicable to export quantitative restrictions 188

5.5 Conclusions 190

6 GATT general exceptions relevant and applicable to WTO-inconsistent export restrictions 193

6.1 Introduction 193

6.2 Availability of Article XX GATT to WTO-plus obligations on export duties assumed under post-1994 accession protocols 194

6.2.1 The existence of an 'objective link' as a precondition for the applicability of Article XX GATT to post-1994 accession protocol provisions 194

6.2.2 Implications for WTO Members assuming specific obligations on export duties under post-1994 accession protocols 202

6.2.3 Systemic imbalances produced by the Appellate Body's approach in light of the fragmentation of WTO disciplines on export duties 204

6.3 The relevance of Article XX GATT to justify export restrictions in light of recent WTO case law 208

6.3.1 Article XX(b) GATT 209

6.3.2 Article XX(g) GATT 214

6.3.3 Industrial policy exceptions under Article XX GATT 221

6.3.4 The chapeau of Article XX GATT 223

6.4 Conclusions 225

PART III **Regulatory prospects** 229

7 Understanding the multifaceted discourse on export restrictions in the WTO: a brief conceptual framework 231

7.1 Introduction 231

7.2 Access to supplies vs. development needs in the debate over export restrictions in the GATT era 232

7.3 Arguments in favour of export restrictions as developmental tools 234

7.3.1 The regained centrality of the principle of sovereignty over natural resources 235

7.3.2 Export restrictions as safeguard tools 241

7.3.3 Export restrictions as instruments to relieve commodity dependence and vulnerability to price volatility 243

7.4 Arguments against export restrictions as beggar-thy-neighbour instruments 245

7.4.1 Disruptions in global value chains caused by export restrictions 245

7.4.2 Subsidy-equivalent effects of export restrictions 247

7.4.3 Rent-seeking behaviours induced by export restrictions and absence of competition rules in the WTO 252

7.5 Conclusions 254

8 Export restrictions in the Doha Round negotiations 257

8.1 Introduction 257

8.2 The Doha mandate on export restrictions 257

8.3 The EU Proposals on export taxes 258

8.3.1 The proposal for a WTO Agreement on Export Taxes 258

8.3.2 The EU Revised Submission on Export Taxes 262

8.4 The proposed Protocol on Transparency in Export Licensing to the GATT 266

8.5 Conclusions 269

9 The way forward 273

9.1 Introduction 273

9.2 Building blocks 274

9.2.1 Enhanced transparency in the use of export restrictions 274

9.2.2 Negotiation of export duty concessions in WTO Members' GATT schedules 277

9.2.3 Expansion of GATT flexibilities 282

9.2.4 Envisaging an export safeguard mechanism 284

9.3 Implementing options 286

9.3.1 A deal on transparency 286

9.3.2 Launching negotiations on export duty concessions: options and modalities 287

9.3.3 A new WTO Agreement on Export Taxes 289

9.4 Conclusions 290

Appendix 1 Export taxes applied to critical minerals and metals (2008–2014) 293

Appendix 2 Export licensing schemes applied to critical minerals and metals (2008–2014) 329

Bibliography 335

Index 357

FOREWORD

This book by Dr Ilaria Espa is very timely in view of the fact that in recent years export restrictions have been increasing in terms of their economic impact and the range of countries resorting to them.

This in turn has raised the issue of their compatibility with multilateral trade regulation, which has inspired the objectives of liberalization and non-discrimination, and more broadly that of ensuring the adequacy of existing rules of the WTO to cope with trade-restrictive measures applied on the export side. In fact, the recent *China–Raw Materials* and *China–Rare Earths* disputes brought by a number of countries against the Chinese export regime have shown both the sensitivity of these restrictions and the lack of adequate provisions in the GATT to address them. This weakness has been exacerbated by the divergent regulation of these restrictions in the WTO accession protocols of mineral exporting members, leading to fragmentation of the trade rules in this respect, when they should be striving for uniformity and coherence.

Dr Espa has not only produced a timely work but also a high-quality one. This is due to her accurate analytical approach and investigation, and also to the breadth of her coverage, as she analyses the relevant legal aspects from a wide economic perspective. She observes both the reasons for exporting countries to resort to such restrictions in various forms and their effects on the economy (not always beneficial) and the negative impact that these obstacles to the freedom of commerce can have on trade flows, affecting the downstream industries in other countries along with their economies.

This leads the author to put forward her interpretation of the current legal regime, balancing the diverse concerns that are invoked by different groupings of countries and interests. On the one hand, there is the question of promotion of free trade and the protection of reasonable access to foreign natural resources (for which so many wars have been fought in the past!); on the other hand, there is the right to national sovereignty over natural resources and development. Between the conflicting national interests, there is a real risk that the safeguard of global commons, such as the environment and sustainable development for all, may be lost.

Dr Espa does not just content herself with a detailed analysis of the current situation. She goes further in considering how coherent and equitable disciplines might be developed multilaterally to redesign and complete the present regulatory framework to take care of those divergent interests, while remaining faithful to the overall WTO principles.

Thus, this is not just an informed and reliable book for those interested in such key aspects of today's international trade relations; it is also an enlightening reflection on the challenges ahead and the risks we all face, from shortages to conflicts, if they are not properly addressed in the next round of multilateral trade negotiations.

Giorgio Sacerdoti, Milan, 26 February 2015

ACKNOWLEDGEMENTS

This book is the result of a journey started in 2010 as a PhD student at Bocconi University, Milan, and concluded at the World Trade Institute of the University of Berne, where I have had the privilege of pursuing my postdoctoral studies as a Marie Curie (COFIT) senior research fellow. The completion of this work would not have been possible without the encouragement and guidance of the many people I have encountered during these years.

First of all, I would like to express my gratitude to Professor Giorgio Sacerdoti, who also kindly agreed to write the foreword to this book, for supporting my choice of subject for the PhD dissertation from the very beginning and for his continuous encouragement during the process of writing and updating the thesis. His insight, thorough comments and helpful suggestions have all guided me at different stages of this project and I am grateful for his personal and professional generosity throughout. I am indebted to my PhD examiners, Professor Petros C. Mavroidis and Professor Michele Vellano, whose advice and thought-provoking questions have assisted me in improving my reach and created the basis for many more fruitful and stimulating exchanges over the years. During the doctoral years, I also greatly benefitted from a visit to Columbia Law School in the 2012 Spring term, for which I am deeply grateful to Professor Merit E. Janow and Professor Karl Sauvant.

I was privileged to be able to upgrade my PhD thesis at the World Trade Institute. I am especially grateful to Professor Thomas Cottier for the enthusiasm he showed for this project since I first arrived at the institute. His constant encouragement and the patience with which he engaged in many enlightening discussions on the topic with me always brought fresh perspectives, enabling me to see the broader picture. This work has also greatly benefitted from exposure to the vibrant community of researchers participating in the Swiss National Centre for Excellence in Research (NCCR) programme in Trade Regulation at the institute, particularly the Work Package 5 team members, and the moral support I received from everybody working at the institute.

Many more people have contributed, in different ways, to the completion of this work with their encouragement, suggestions and questions. Particular thanks go to Giovanna Adinolfi, Elisa Baroncini, Claudio Dordi, Manfred Elsig, Marina Foltea, Joseph Francois, Christian Häberli, Gary Horlick, Kateryna Holzer, Susan Kaplan, Jane Korinek, David Luff, Gabrielle Marceau, Paola Mariani, Irene Musselli, Roberta Piermartini, Luca Rubini, Pierre Sauvé, Yulia Selivanova and Weiwei Zhang. I am very grateful to Rafael Tiago Juk Benke, Sofie Flensborg, Ricardo Meléndez-Ortiz and Scott Vaughan for asking me to join the Expert Group on 'Trade and Investment in Extractive Industries' of the E15 Initiative, jointly implemented by the International Centre for International Trade and Development and the World Economic Forum, and to all the members of the group, where I was able to sharpen some of the arguments on export restrictions presented in this book. I would like to thank Liz Obee for her editorial assistance on this work, and I am indebted to Kim Hughes, Fleur Jones, Gaia Poggiogalli, Ramya Ranganathan, Jonathan Ratcliffe and Richard Woodham of Cambridge University Press for their patience and excellent professional support.

Finally, I would like to thank my family and friends for always being there for me. I am indebted to Anna and Pino for their generosity and the warm environment they created around me in many critical circumstances. My deepest thanks go to Stefano and to my parents, Emma and Alberto, for their daily love, their patience and their faith in me throughout the ups and downs of these years. This book could not have come into being without their support and is dedicated to them.

28 February 2015

ABBREVIATIONS

AoA	WTO Agreement on Agriculture
AB	Appellate Body
APSEG	Asian Pacific School of Economics and Governance
ASCM	WTO Agreement on Subsidies and Countervailing Measures
ATC	WTO Agreement on Textiles and Clothing
BIS	US Bureau of Industry and Security
BoP	balance of payments
CCCMC	China's Chamber of Commerce of Metals, Minerals and Chemicals Importers and Exporters
CDDCs	commodity-dependent developing countries
CRS	US Congressional Research Service
DoE	US Department of Energy
DRC	Democratic Republic of the Congo
DSB	Dispute Settlement Body
DSU	Dispute Settlement Understanding
ECT	Energy Charter Treaty
EPI	Environmental Performance Indicator
ERSD	WTO Economic Research and Statistics Division
EU	European Union
FAO	Food and Agriculture Organization
FOB	free on board
GATS	General Agreement on Trade in Services
GATT	General Agreement on Tariffs and Trade
GDP	gross domestic product
GTAP	Global Trade Analysis Project
GVCs	global value chains
HLTF	High Level Task Force on the Global Food Security Crisis
HREE	heavy rare earth elements
HS	Harmonized System Nomenclature
ICTSD	International Centre for Trade and Sustainable Development
IFAD	International Fund for Agricultural Development
IFIs	International Financial Institutions
IFPRI	International Food Policy Research Institute

IMF	International Monetary Fund
IPC	International Food and Agricultural Trade Policy Council
ITA	Information Technology Agreement
ITAC	International Trade Administration Commission of South Africa
ITO	International Trade Organization
LDCs	least developed countries
LEDs	light-emitting diodes
LPG	liquefied petroleum gas
LREE	light rare earth elements
MEAs	multilateral environmental agreements
MEP	minimum export price
MFA	Multi-Fibre Agreement
MFN	most favoured nation clause
MOFCOM	Chinese Ministry of Commerce
NAMA	non-agricultural market access
NBER	US National Bureau of Economic Research
NFIDCs	net-food importing developing countries
NGMA	Negotiating Group on Non-Agricultural Market Access
NIEO	new international economic order
NTBs	non-tariff barriers
OECD	Organisation for Economic Co-operation and Development
OMA	orderly market arrangements
OPEC	Organization of the Petroleum Exporting Countries
QRs	quantitative restrictions
PGMs	platinum group metals
SSGs	special agricultural safeguards
STEs	state trading enterprises
STIC	Standard International Trade Classification
TPRB	Trade Policy Review Body
TPRs	Trade Policy Reviews
UN	United Nations
UNCTAD	United Nations Conference on Trade and Development
UNEP	United Nations Environment Programme
USGS	United States Geological Survey
VCLT	Vienna Convention on the Law of Treaties
VERs	voluntary export restraints
VRAs	voluntary restraint arrangements
WCO	World Customs Organization
WFP	World Food Programme
WGTCP	Working Group on the Interaction between Trade and Competition Policy
WTO	World Trade Organization

WTO REPORTS

Argentina – Footwear (EC) Appellate Body Report, *Argentina – Safeguard Measures on Imports of Footwear*, WT/DS121/AB/R, adopted 12 January 2000, DSR 2000:I, p. 515

Argentina – Hides and Leather Panel Report, *Argentina – Measures Affecting the Export of Bovine Hides and Import of Finished Leather*, WT/DS155/R and Corr.1, adopted 16 February 2001, DSR 2001:V, p. 1779

Brazil – Retreaded Tyres Appellate Body Report, *Brazil – Measures Affecting Imports of Retreaded Tyres*, WT/DS332/AB/R, adopted 17 December 2007, DSR 2007:IV, p. 1527

Brazil – Retreaded Tyres Panel Report, *Brazil – Measures Affecting Imports of Retreaded Tyres*, WT/DS332/R, adopted 17 December 2007, as modified by Appellate Body Report WT/DS332/AB/R, DSR 2007:V, p. 1649

Canada – Periodicals Panel Report, *Canada – Certain Measures Concerning Periodicals*, WT/DS31/R and Corr.1, adopted 30 July 1997, as modified by Appellate Body Report WT/DS31/AB/R, DSR 1997:I, p. 481

Canada – Wheat Appellate Body Report, *Canada – Measures Relating to Exports of Wheat and Treatment of Imported Grain*, WT/DS276/AB/R, adopted 27 September 2004, DSR 2004:VI, p. 2739

Canada – Wheat Panel Report, *Canada – Measures Relating to Exports of Wheat and Treatment of Imported Grain*, WT/DS276/R, adopted 27 September 2004, upheld by Appellate Body Report WT/DS276/AB/R, DSR 2004:VI, p. 2817

China – Audiovisuals Panel Report, *China – Measures Affecting Trading Rights and Distribution Services for Certain Publications and Audiovisual Entertainment Products*, WT/DS363/R and Corr.1, adopted 19 January 2010, as modified by Appellate Body Report WT/DS363/AB/R, DSR 2010:II, p. 261

China – Rare Earths Appellate Body Reports, *China – Measures Related to the Exportation of Rare Earths, Tungsten and Molybdenum*, WT/DS431/AB/R / WT/DS432/AB/R / WT/DS433/AB/R, 29 August 2014

China – Rare Earths Panel Reports, *China – Measures Related to the Exportation of Rare Earths, Tungsten and Molybdenum*, WT/DS431/R / WT/DS432/R / WT/DS433/R /Add.1, adopted 29 August 2014, as modified by Appellate Body Reports WT/DS431/AB/R / WT/DS432/AB/R / WT/DS433/AB/R

China – Raw Materials Appellate Body Reports, *China – Measures Related to the Exportation of Various Raw Materials*, WT/DS394/AB/R / WT/DS395/AB/R / WT/DS398/AB/R, adopted 22 February 2012, DSR 2012:VII, p. 3295

China – Raw Materials Panel Reports, *China – Measures Related to the Exportation of Various Raw Materials*, WT/DS394/R / WT/DS395/R / WT/DS398/R /Add.1 and Corr.1, adopted 22 February 2012, as modified by Appellate Body Reports WT/DS394/AB/R / WT/DS395/AB/R / WT/DS398/AB/R, DSR 2012:VII, p. 3501

Colombia – Ports of Entry Panel Report, *Colombia – Indicative Prices and Restrictions on Ports of Entry*, WT/DS366/R and Corr.1, adopted 20 May 2009, DSR 2009:VI, p. 2535

EC – Asbestos Appellate Body Report, *European Communities – Measures Affecting Asbestos and Asbestos-Containing Products*, WT/DS135/AB/R, adopted 5 April 2001, DSR 2001:VII, p. 3243

EC – Asbestos Panel Report, *European Communities – Measures Affecting Asbestos and Asbestos-Containing Products*, WT/DS135/R and Add.1, adopted 5 April 2001, as modified by Appellate Body Report WT/DS135/AB/R, DSR 2001:VIII, p. 3305

EC – Banana III (Ecuador) Panel Report, *European Communities – Regime for the Importation, Sale and Distribution of Bananas, Complaint by Ecuador*, WT/DS27/R/ECU, adopted 25 September 1997, as modified by Appellate Body Report WT/DS27/AB/R, DSR 1997:III, p. 1085

EC – Banana III (Guatemala and Honduras) Panel Report, *European Communities – Regime for the Importation, Sale and Distribution of Bananas, Complaint by Guatemala and Honduras*, WT/DS27/R/GTM, WT/DS27/R/HND, adopted 25 September 1997, as modified by Appellate Body Report WT/DS27/AB/R, DSR 1997:II, p. 695

EC – Banana III (Mexico) Panel Report, *European Communities – Regime for the Importation, Sale and Distribution of Bananas, Complaint by Mexico*, WT/DS27/R/MEX, adopted 25 September 1997, as modified by Appellate Body Report WT/DS27/AB/R, DSR 1997:II, p. 803

EC – Banana III (US) Panel Report, *European Communities – Regime for the Importation, Sale and Distribution of Bananas, Complaint by the United States*, WT/DS27/R/USA, adopted 25 September 1997, as modified by Appellate Body Report WT/DS27/AB/R, DSR 1997:II, p. 943

EC – Tariff Preferences Panel Report, *European Communities – Conditions for the Granting of Tariff Preferences to Developing Countries*, WT/DS246/R, adopted 20 April 2004, as modified by Appellate Body Report WT/DS246/AB/R, DSR 2004:III, p. 1009

EC-IT Products Panel Reports, *European Communities and its Member States – Tarif Treatment of Certain Information Technology Products*, WT/DS375/R / WT/DS376/R / WT/DS377/R, adopted 21 September 2010, DSR 2010:III, p. 933

India – Autos Panel Report, *India – Measures Affecting the Automotive Sector*, WT/DS146/R, WT/DS175/R and Corr.1, adopted 5 April 2002, DSR 2002:V, p. 1827

India – Quantitative Restrictions Panel Report, *India – Quantitative Restrictions on Imports of Agricultural, Textile and Industrial Products*, WT/DS90/R, adopted 22 September 1999, upheld by Appellate Body Report WT/DS90/AB/R, DSR 1999:V, p. 1799

Japan – Alcoholic Beverages II Appellate Body Report, *Japan – Taxes on Alcoholic Beverages*, WT/DS8/AB/R, WT/DS10/AB/R, WT/DS11/AB/R, adopted 1 November 1996, DSR 1996:I, p. 97

Korea – Alcoholic Beverages Appellate Body Report, *Korea – Taxes on Alcoholic Beverages*, WT/DS75/AB/R, WT/DS84/AB/R, adopted 17 February 1999, DSR 1999:I, p. 3

Korea – Various Measures on Beef Appellate Body Report, *Korea – Measures Affecting Imports of Fresh, Chilled and Frozen Beef*, WT/DS161/AB/R, WT/DS169/AB/R, adopted 10 January 2001, DSR 2001:I, p. 5

Korea – Various Measures on Beef Panel Report, *Korea – Measures Affecting Imports of Fresh, Chilled and Frozen Beef*, WT/DS161/R, WT/DS169/R, adopted 10 January 2001, as modified by Appellate Body Report WT/DS161/AB/R, WT/DS169/AB/R, DSR 2001:I, p. 59

Turkey – Textiles Panel Report, *Turkey – Restrictions on Imports of Textile and Clothing Products*, WT/DS34/R, adopted 19 November 1999, as modified by Appellate Body Report WT/DS34/AB/R, DSR 1999:VI, p. 2363

US – DRAMs Appellate Body Report, *United States – Countervailing Duty Investigation on Dynamic Random Access Memory Semiconductors (DRAMs) of One Megabit or Above from Korea*, WT/DS296/R, adopted 20 July 2005, DSR 2005:XVI, p. 8131

US – Export Restraints Panel Report, *United States – Measures Treating Export Restraints as Subsidies*, WT/DS194/R and Corr.2, adopted 23 August 2001, DS 2001:XI, p. 5767

US – Gambling Appellate Body Report, *United States – Measures Affecting the Cross-Border Supply of Gambling and Betting Services*, WT/DS285/AB/R, adopted 20 April 2005, DSR 2005:XII, p. 5663 (Corr.1, DSR 2006:XII, p. 5475)

US – Gasoline Appellate Body Report, *United States – Standards for Reformulated and Conventional Gasoline*, WT/DS2/AB/R, adopted 20 May 1996, DSR 1996:I, p. 3

US – Gasoline Panel Report, *United States – Standards for Reformulated and Conventional Gasoline*, WT/DS2/R, adopted 20 May 1996, as modified by Appellate Body Report WT/DS2/AB/R, DSR 1996:I, p. 29

US – Poultry (China) Panel Report, *United States – Certain Measures Affecting Imports of Poultry from China*, WT/DS392/R, adopted 25 October 2010, DSR 2010:V, p. 1909

US – Shrimp Appellate Body Report, *United States – Import Prohibition of Certain Shrimp and Shrimp Products*, WT/DS58/AB/R, adopted 6 November 1998, DSR 1998:VII, p. 2755

US – Shrimp Panel Report, *United States – Import Prohibition of Certain Shrimp and Shrimp Products*, WT/DS58/R and Corr.1, adopted 6 November 1998, as modified by Appellate Body Report WT/DS58/AB/R, DSR 1998:VII, p. 2821

US – Softwood Lumber IV Appellate Body Report, *United States – Final Countervailing Duty Determination with Respect to Certain Softwood Lumber from Canada*, WT/DS257/AB/R, adopted 17 February 2004, DSR 2004:II, p. 571

US – Softwood Lumber IV Panel Report, *United States – Final Countervailing Duty Determination with Respect to Certain Softwood Lumber from Canada*, WT/DS257/R and Corr.1, adopted 17 February 2004, as modified by Appellate Body Report WT/DS257/AB/R, DSR 2004:II, p. 641

US – Stainless Steel (Mexico) Appellate Body Report, *United States – Final Anti-Dumping Measures on Stainless Steel from Mexico*, WT/DS344/AB/R, adopted 20 May 2008, DSR 2008:II, p. 513

US – Wool Shirts and Blouses Appellate Body Report, *United States – Measure Affecting Imports of Woven Wool Shirts and Blouses from India*, WT/DS33/AB/R, adopted 23 May 1997 and Corr.1, DSR 1997:I, p. 323

GATT PANEL REPORTS

Canada - FIRA GATT Panel Report, *Canada Administration of the Foreign Investment Review Act*, L/5504, adopted 7 February 1984, BISD 30S, p. 140

Canada - Herring and Salmon GATT Panel Report, *Canada - Measures Affecting Exports of Unprocessed Herring and Salmon*, L/6268, adopted 22 March 1988, BISD 35S, p. 98

EEC - Dessert Apples GATT Panel Report, *European Economic Community - Restrictions on Imports of Dessert Apples - Complaint by Chile*, L/6491, adopted 22 June 1989, BISD 36S, p. 93

EEC - Minimum Import Price Panel Report, *EEC - Programme of Minimum Import Prices, Licences and Surety Deposits for Certain Processed Fruits and Vegetables*, L/4687, adopted 18 October 1978, BISD 25S, p. 68

France - Import Restrictions GATT Panel Report, *French Import Restrictions*, L/1921, adopted 14 November 1962, BISD 11S, p. 94

Japan - Agricultural Products I GATT Panel Report, *Japan - Restrictions on Imports of Certain Agricultural Products*, L/6253, adopted 2 March 1988, BISD 35S, p. 163

Japan–Semiconductors GATT Panel Report, *Japan - Trade in Semiconductors*, L/6309, adopted 4 May 1988, BISD 35S, p. 116

US - Customs User Fee GATT Panel Report, *United States - Customs User Fee*, L/6264, adopted 2 February 1988, BISD 35S, p. 245

US–Sugar GATT Panel Report, *United States - Restrictions on Imports of Sugar*, L/6514, adopted 22 June 1989, BISD 36S, p. 331

US - Sugar Waiver GATT Panel Report, *United States - Restrictions on the Importation of Sugar and Sugar-Containing Products Applied under the 1955 Waiver and under the Headnote to the Schedule of Tariff Concessions*, L/6631, adopted 7 November 1990, BISD 37S, p. 228

Introduction

I.1 Historical trends in the use of export restrictions

Today international commodity markets are being affected by the longest and most pervasive 'wave' of export restrictions since the Second World War.[1] The origin of this phenomenon can be traced back to the 2000s commodity boom,[2] and its endurance to an exceptional mix of concurring factors affecting patterns of world supply and demand for primary products.[3] The explosion of attention towards export restraints directly stems from the unique scope and length of the present wave. Yet, export restrictions are not a recent practice. Their use has varied in scale and intensity throughout the centuries. Governments have used export restrictions since at least the eleventh century, mainly as a tool of industrial policy.[4] The British Empire was, in particular, the first great

[1] The use of export restrictions since the Second World War has intensified at particular points in time, giving rise to what could conventionally be called 'waves'. In the most recent decades, these waves have coincided with periods of commodity boom. M. Radetzki, *A Handbook of Primary Commodities in the Global Economy* (Cambridge University Press, 2010), pp. 66–72.

[2] The third since the Second World War, the 2000s commodity boom – or, as it is often called, the 2000s 'super-cycle' – started in 2002–2003 and came to an end after more than a decade due to the effects of the global financial and economic crisis. Radetzki, *A Handbook of Primary Commodities*, pp. 70–2; C. A. Carter, G. C. Rausser and A. Smith, 'Commodity Booms and Busts', *Annual Review of Resource Economics* 3 (2011), 87, pp. 90–1; D. S. Jacks, 'From Boom to Bust: A Typology of Real Commodity Prices in the Long Run', US National Bureau of Economic Research (NBER) Working Paper No. 18874, 2013 (Cambridge, Massachusetts: National Bureau of Economic Research), http://www.nber.org/papers/w18874.pdf (accessed July 2014).

[3] These include a profound structural transformation of geopolitical and geoeconomic determinants of international trade in primary supplies and the macroeconomic downturn associated with the outbreak of the global financial crisis towards the end of the decade. For a detailed account see below, Section I.2.

[4] For an historical overview see J. Gorton, 'British Preferential Export Taxes', *American Economic Association* 14 (1924), 56; L. R. Edminster, 'Control of Exports of Raw Materials: An International Problem', *The Annals of the American Academy of Political and Social Science* 150 (1930), 89; R. Goode and G. Lent, 'Role of Export Taxes in Developing Countries', *Staff Papers (International Monetary Fund)* 13 (1966), 453; S. Devarajan,

power to make extensive use of export restrictions on selected agricultural raw materials such as wool, hides and skins in the early stages of its industrialization, and many analysts agree that in several cases these measures contributed significantly to the building of England's industrial greatness.[5] From the nineteenth century, export restrictions were progressively dismantled in Europe although never completely eliminated by countries aiming to accelerate their industrial development.[6] Other areas of the world including Canada and several countries of Latin America introduced restrictive measures on the exportation of various agricultural

D. Go, M. Schiff and S. Suthiwart-Narueput, 'The Whys and Why Nots of Export Taxes', The World Bank Policy Research Working Paper No. 1684, 1996, pp. 1–25; E. S. Reinert, *How Rich Countries Got Rich . . . and Why Poor Countries Stay Poor* (London: Constable & Robinson Ltd, 2008); and C. V. Prestowitz, 'Export Restraints: The Key to Getting Rich', *Foreign Policy Magazine* (7 July 2011). Although well-documented records trace export restrictions back to the eleventh century, these types of instruments were most probably utilized at the time of the Roman Empire as well, at least with respect to certain goods. Evidence of such measures, however, remains vague and poorly documented. See H. Meredith, 'Evaluating Movements of Open-Work Glassware', in M. Mundell Mango (ed.), *Byzantine Trade, 4th—12th Centuries: The Archaeology of Local, Regional and International Exchange*. Papers of the Thirty-eighth Spring Symposium of Byzantine Studies (Farnham and Burlington: Ashgate, 2004), p. 197.

5 England started to introduce export restrictions in the form of taxes on the exportation of wool and hides in 1275 in order to promote domestic textile processing, and maintained such measures for several centuries (at least until 1660) in order to achieve industrial development. Particularly during the Tudor era, Britain achieved a substantial wool fabric monopoly through export duties on raw wool applied during the realm of Henry VII and, even more, by means of the raw wool export ban subsequently introduced by Elizabeth I. For an analysis of the effects produced by the exploitation of the monopolistic position of Great Britain in the wool industry, particularly in terms of competition with the Florentines, see Reinert, *How Rich Countries*, pp. 80–1. During the same period, other European territories, such as Venice and Holland, used export restrictions as a tool of industrial and trade strategy in order to acquire 'the same triple rent situation: a strong industrial sector, a raw material monopoly, and overseas trade' (ibid., pp. 81-3). Another significant example of recourse to restrictive export measures as a means to achieving a dominant position in the international market, and thereby eliminating prospective competitors (in particular, the US), is the case of the export duty on tin ore introduced by Great Britain in the first decades of the twentieth century. In this case, England exploited its power of colonial dominion to apply a preferential export duty on tin ore shipped from British colonies such as the Federated Malay States and West Africa, i.e. a duty applied only on the exportation of tin ore to states other than the United Kingdom. See Gorton, 'British Preferential Export Taxes', pp. 57 ff. Preferential export duties were frequently applied by Great Britain (e.g. a ban on untanned hides and skins coming from India was introduced in 1919; an export duty on crude rubber from the British Empire was applied in 1922 to eliminate US competitors). See Edminster, *Control of Exports*, pp. 89 ff.

6 J. Viner, 'National Monopolies of Raw Materials', *Council on Foreign Affairs* 4 (1926), 585. Among the European powers still maintaining export restrictions during that time was Spain, which applied an export duty of 20 per cent on hides and skins similar to the 15 per cent *ad valorem* export duty maintained by England. Gorton, 'British Preferential Export Taxes', p. 58.

raw materials from the twentieth century.[7] The rationale associated with such measures was in most cases the protection of domestic industries and the promotion of higher value-added activities, together with – in the case of countries applying export taxes in particular – the generation of government revenues.[8]

After a temporary but intensive resurgence of export restrictions on primary commodities triggered by the two world conflicts,[9] the 1950s and 1960s were characterized by two distinct trends: on the one hand, developing countries were increasingly resorting to restrictions on the exportation of agricultural commodities in an attempt to raise government revenues and reduce volatility of markets which could significantly undermine their efforts towards economic development.[10] On the other hand, developed countries were starting to negotiate export-restraint arrangements[11] with developing countries expanding their textile

[7] Viner, 'National Monopolies of Raw Materials', p. 589 and E. Reubens, 'Commodity Trade, Export Taxes and Economic Development', *The Academy of Political Science* 71 (1956), 42.

[8] This rationale was particularly prevalent among Latin American countries. Ibid., pp. 45-8.

[9] Viner, 'National Monopolies', p. 585. In certain cases, countries have maintained export restrictions on raw materials for a number of decades. Switzerland, for instance, has continued to control exports of metal scrap to ensure adequate domestic supplies.

[10] These measures were used by developing countries that had undergone the process of decolonization. As countries relying on resource-led development, they counted on the exportation of agricultural commodities for a predominant share of their gross domestic product (GDP) and therefore aimed at stabilizing the price of such commodities in the world market. These types of issues continue to affect so-called monoeconomies and, more generally, least developed countries (LDCs) whose low level of economic diversification makes them heavily dependent on commodity exports. Radetzki, *A Handbook of Primary Commodities*, pp. 188–206; R. Piermartini, 'The Role of Export Taxes in the Field of Primary Commodities', World Trade Organization (WTO) Discussion Paper, Economic Research and Statistics Division, 2004, www.wto.org/english/res_e/booksp_e/discussion_papers4_e.pdf (accessed May 2014), pp. 1–2.

[11] The expression 'export-restraint arrangements' refers to a wide set of practices used to limit the volume of exports of a specific product to another country. These arrangements can be formal (i.e. legally binding agreements) or informal (i.e. statements or explanations of intent). They can be taken unilaterally or be concluded between the governments of the exporting country and the government of the importing country, between a government and non-governmental entities such as industry associations or groups of firms, or between industry groups. Depending on the type of actors involved and on the form of the arrangement, these measures are known as voluntary export restraints (VERs), voluntary restraint arrangements (VRAs) or orderly market arrangements (OMAs). J. H. Jackson, *The Jurisprudence of GATT & the WTO: Insights on Treaty Law and Economic Relations* (Cambridge University Press, 2000), pp. 69–72. Although such expressions suggest that the exporting country voluntarily enters into such arrangements, in practice it is often the case that it does so under the pressure that the importing country might otherwise adopt other types of restrictions having stronger negative impacts on its

sectors[12] as a way of protecting their own industry from increased import competition. In these cases, the restriction on exportation took the form of a quota on textile products but reflected the interests of importing trading partners rather than the national industrial strategies of exporting countries.[13] The United States was the first country to impose restrictions on Japan's exports of cotton textiles in 1955. Rapidly, however, many developing countries started using export-restraint arrangements – and, in particular, voluntary export restraints (VERs) – in order to cope with the surge of cheap textile products coming from industrializing Asian economies.[14]

It was, however, during the price spike of 1972–1975 that export restrictions were used systematically by exporting countries in an attempt to protect their own markets. The first wave of export restrictions developed in response to the commodity boom induced by the supply shocks that occurred in the cereal grain and vegetable oil international markets,[15] and exacerbated by the rise in oil prices that followed the embargo imposed by the Organization of the Petroleum Exporting Countries (OPEC) on the Western countries supporting Israel during the Yom

exports. Ibid., p. 72; P. Van den Bossche and W. Zdouc, *The Law and Policy of the World Trade Organization*, 3rd edn (Cambridge University Press, 2013), p. 491.

12 The textile sector has historically been a catalyst for industrialization in countries at early stages of development. A. J. H. Wohn, 'Towards GATT Integration: Circumventing Quantitative Restrictions on Textile and Apparel under the Multi-Fibre Agreement', *University of Pennsylvania Journal of International Economic Law* 22 (2001), 375, pp. 388–9.

13 J. M. Finger and A. Harrison, 'Import Protection for U.S. Textiles and Apparel: Viewed from the Domestic Perspective', in A. O. Krueger (ed.), *The Political Economy of Trade Protection* (University of Chicago Press, 1996), pp. 43–50.

14 Y. Yang and C. Zhong, 'China's Textile and Clothing Exports in a Changing World Economy', *The Developing Economies* 36 (1998), 3, pp. 3–23. This network of arrangements eventually led to the conclusion of the so-called Short-Term Cotton Agreement (1961), the Long-Term Arrangement (1962–1973) and, in 1974, the Multi-Fibre Agreement (MFA). P. C. Mavroidis, *Trade in Goods: The GATT and the Other Agreements Regulating Trade in Goods*, 2nd edn (Cambridge University Press, 2012), pp. 782–3. The MFA was a four-year agreement which was renewed four times over the decades with a progressively wider membership. The fourth Multi-Fibre Agreement was eventually replaced by the WTO Agreement on Textiles and Clothing (ATC) adopted at Marrakesh during the Uruguay Round in 1994. The ATC outlawed the introduction of new VERs and required that existing voluntary export restraints be phased out after a ten-year transition period (1995–2004). Ibid., pp. 783–7.

15 For an account of the causes of the 1970s commodity boom see R. N. Cooper and R. Z. Lawrence, 'The 1972–1975 Commodity Boom', *Brooking Papers on Economic Activity* 3 (1975), 671.

Kippur War.[16] Net food-exporting countries started to apply export restraints as price insulating instruments in reaction to high prices.[17] For the first time, world markets experienced the consequences of the domino effect of export restrictions,[18] namely vulnerability of supply and increased price volatility, while the poorest segments of the population in imposing countries often suffered negative income distribution effects.[19] The turbulence caused on the international agricultural markets did not, however, instigate any effective and timely reaction within the framework of the General Agreement on Tariffs and Trade (GATT).[20] The opportunity for improving multilateral disciplines on the export side within a

[16] WTO, 'World Trade Report 2013: Factors Shaping the Future of World Trade', www.wto.org/english/res_e/booksp_e/world_trade_report13_e.pdf (accessed November 2014), p. 171. OPEC came into existence in 1960 as a cartel of petroleum producers, and its original members were Venezuela, Iran, Iraq, Kuwait and Saudi Arabia. Gradually its membership has expanded to include Algeria, Angola, Ecuador, Libya, Nigeria, Qatar and the United Arab Emirates. Gabon and Indonesia also joined the Organization but eventually suspended their membership. OPEC, 'Member Countries', www.opec.org/opec_web/en/about_us/25.htm (accessed November 2014). OPEC has traditionally used its market share to influence international oil prices. While the 1973 embargo constitutes an extreme example of OPEC's abuse of its dominant position, its devastating effects on oil prices were caused by the increasing import dependence of the Western world on OPEC supplies during the 1970s, the low elasticity of oil demand and the lagged response of non-OPEC supply. Radetzki, *A Handbook of Primary Commodities*, p. 159.

[17] W. Martin and K. Anderson, 'Export Restrictions and Price Insulation During Commodity Price Booms', The World Bank Policy Research Working Paper No. 5645, 2011, www.imf.org/external/np/seminars/eng/2011/trade/pdf/session1-martin-paper.pdf (accessed May 2014); S. Mitra and T. Josling, 'Agricultural Export Restrictions: Welfare Implications and Trade Disciplines', International Food and Agricultural Trade Policy Council (IPC) Position Paper, Agricultural and Rural Development Policy Series, 2009, www.agritrade.org/documents/ExportRestrictions_final.pdf (accessed May 2014), pp. 3-8.

[18] The term 'domino effect' is used to describe the phenomenon by which the introduction of an export restriction by a country triggers the reaction of other countries sharing similar objectives so that the multiplication of export restrictions ultimately annuls their intended effects. In the case of agricultural commodities, in particular, export restrictions used as price insulating instruments have tangibly contributed to the rise of food prices. M. Ivanic, W. Martin and A. Mattoo, 'Welfare and Price Impacts of Price-Insulating Policies', Presented at the 14th Annual Conference on Global Economic Analysis (Venice, Italy, 2001), Global Trade Analysis Project (GTAP) Resource No. 3651, www.gtap.agecon.purdue.edu/resources/download/5580.pdf (accessed November 2014); P. Giordani, N. Rocha and M. Ruta, 'Food Prices and the Multiplier Effect of Export Policy', WTO Economic Research and Statistics Division, Staff Working Paper ERSD-2012-08, www.wto.org/english/res_e/reser_e/ersd201208_e.pdf (accessed November 2014).

[19] Piermartini, 'The Role of Export Taxes', pp. 15-8.

[20] General Agreement on Tariffs and Trade (1994), Marrakesh Agreement Establishing the World Trade Organization, Annex 1A, 1867 U.N.T.S. 187, entered into force 1 January 1995 [hereinafter GATT]. For an overview of the mixed reactions raised by the first wave

system traditionally centred on the goal of market access and, thus, on the removal of import barriers[21] was discussed during the Tokyo Round.[22] Concrete steps forward were hampered, however, by the continuous divergence between commodity net exporting and net importing countries.[23] Momentum was eventually lost following the normalization of international markets and the subsequent start of a two-decade span of commodities' price decline.[24]

During the 1980s and 1990s, the issue of export restrictions on raw materials was therefore not prominent on the international trade agenda. The main point of controversy concerning the use of measures restricting exports was linked to the surge of voluntary export restraints in manufacturing sectors other than textiles.[25] In particular, from the late 1970s through to the middle of the 1980s, VERs dominated Japan-US trade relations.[26] Several formal and informal export-restraint arrangements were concluded between the two countries to control the importation of a variety of manufactured goods of Nipponese origin into the US. Among the restricted products were automobiles, semiconductors, colour televisions and machine tools.[27] Many European states followed the example of the United States to restrain Japan's penetration into their markets.[28]

The wide diffusion of VERs reflected the intention of Western economies to protect those domestic manufacturing sectors most affected by the pressure of Japanese imports during a time of macroeconomic downturn. As soon as such sectors recovered, from the second half of the 1980s onwards, VERs were progressively phased out, and Western countries

of export restrictions on agricultural commodities and the consequently weak initiatives undertaken see Mitra and Josling, 'Agricultural Export Restrictions', pp. 14–15.

21 R. W. Staiger, 'Non-Tariff Measures and the WTO', WTO Economic Research and Statistics Division, Staff Working Paper ERSD-2012-01, www.wto.org/english/res_e/reser_e/ersd201201_e.pdf (accessed November 2014), p. 2.

22 See below, Chapter 7, Section 7.2. 23 GATT Doc. BISD 29/S/9, 29 November 1982.

24 Carter *et al.*, 'Commodity Booms and Busts', p. 93; Radetzki, *A Handbook of Primary Commodities*, p. 10.

25 Jackson, *The Jurisprudence of GATT & the WTO*, pp. 69–74.

26 W. Cline, *Trade Policy in the 1980s* (Washington, DC: Peterson Institute for International Economics, 1983).

27 M. Satake, 'Trade Conflict between Japan and the United States over Market Access: The Case of Automobiles and Automotives Parts', Pacific Economic Papers No. 310, Asian Pacific School of Economics and Governance (APSEG), 2000, https://crawford.anu.edu.au/pdf/pep/pep-310.pdf (accessed November 2014).

28 Cline, *Trade Policy in the 1980s*, pp. 529–30.

started focusing again on improving access to foreign markets as a way of increasing their own exports.[29]

The 'reversed' issue of access to primary supplies was therefore not a priority in the minds of the World Trade Organization (WTO) negotiators during the Uruguay Round (1986–1994). Two decades of low and declining commodity prices had led to a relatively moderate use of export restrictions. Any agreement on the topic, moreover, seemed hampered at the outset by the apparent insurmountable divide between net importing countries, interested in securing further trade liberalization, and net exporting countries, advocating the relevance of the principle of sovereignty over natural resources in this area.[30] Thus, while the use of VERs was explicitly outlawed in the new WTO Agreement on Safeguards,[31] the issue of export restrictions was not specifically dealt with, as countries were lacking the incentive to find a multilateral solution on the long-standing issue of access to primary commodities.

The unresolved questions at the basis of the stalemate in any discussions on export restrictions resurfaced abruptly at the beginning of the 2000s, when renewed use of export restraints soon reached an alarming dimension. The new wave again affected the agricultural sector, contributing to exacerbating the food crisis in 2007–2008 and the food price spike of 2010–2011.[32] Moreover, it also rapidly spread to various categories of commodities, including raw materials of mineral origin for use by other sectors of the economy.[33] This multi-directional expansion in the use of export

[29] Satake, 'Trade Conflict', pp. 6 ff.

[30] GATT Doc. MTN.GNG/NG2/W/40, 8 August 1989.

[31] Agreement on Safeguards, Multilateral Agreements on Trade in Goods, Annex 1A to the Agreement Establishing the World Trade Organization, 1869 U.N.T.S. 154, entered into force 1 January 1995. See Article 11.1(b).

[32] K. Anderson and S. Nelgen, 'Trade Barrier Volatility and Agricultural Price Stabilization', *World Development* 40 (2012), 36; Food and Agriculture Organization (FAO), International Fund for Agricultural Development (IFAD), International Monetary Fund (IMF), Organisation for Economic Co-operation and Development (OECD), United Nations Conference on Trade and Development (UNCTAD), World Food Programme (WFP), the World Bank, the WTO, International Food Policy Research Institute (IFPRI) and the United Nations High Level Task Force on the Global Food Security Crisis (UN HLTF), 'Price Volatility in Food and Agricultural Markets: Policy Responses', 2 June 2011, www.oecd.org/tad/agricultural-trade/48152638.pdf (accessed November 2014); A. L. Coulibaly, 'The Food Price Increase of 2010–2011: Causes and Impacts', Publication No. 2013-02-E, Ottawa, Canada, Library of Parliament, 2013, www.parl.gc.ca/Content/LOP/ResearchPublications/2013-02-e.pdf (accessed November 2014); P. Liapis, 'How Export Restrictive Measures Affect Trade in Agricultural Commodities', OECD Food, Agriculture and Fisheries Papers, No. 63 (OECD Publishing, 2013).

[33] See below, Section I.2.

restrictions reflected, on the one hand, the exceptional scope of the 2000s commodity boom[34] and the role played by hard commodities in the 2008 price spike[35] and, on the other hand, the sentiment of closure that countries started to develop in a phase of major macroeconomic downturn. By the end of the 2000s, export restrictions were the fastest growing component of the trade-restrictive measures introduced within the framework of the global financial and economic crisis.[36]

I.2 The panoply of export restrictions on primary commodities in recent times

The second wave of export restrictions on primary commodities represents a *unicum* in recent history due to its scope and length. It has extended to various categories of primary products, from foodstuffs and agricultural raw materials to hard commodities.[37] Directly linked to the 2000s commodity boom,[38] it has eventually taken its own course, becoming – at least for certain sectors – more persistent than price spikes. The unfolding of the financial and economic crisis at the end of the 2000s has moreover given further stimulus to the use of export restrictions, affecting in particular industrial raw materials in the context of a 'revival' of industrial policies.[39] Export restrictions indeed reached their peak

[34] Two main factors explain the commonality of price rises across commodity sectors experienced in the 2000s: first, the existence of common driving forces; second, the existence of linkages across markets. As to the first factor, analysts agree that the exceptional economic performances of developing Asia (and China in particular) played a key role in the 2000s commodity boom, boosting in particular energy and metals prices – for which demand skyrocketed to the highest level recorded in over twenty years. Radetzki, *A Handbook of Primary Commodities*, pp. 70–1. The second factor mainly refers to the 'contagion' effects that changes in energy (oil) prices produce in other commodity sectors and, particularly, in the agricultural sector. C. L. Gilbert, 'How to Understand High Food Prices', *Journal of Agricultural Economics* 61 (2010), 398, p. 401.

[35] There is general agreement on the fact that the demand shock affecting energy and metals from 2002-2003 onwards had a predominant role in boosting the entire range of commodities' prices. Carter *et al.*, 'Commodity Booms and Busts', p. 90.

[36] WTO Doc. WT/TPR/OV/14, p. 17.

[37] See WTO Docs. WT/TPR/OV/W/1–8 (various years).

[38] For a full account of the commodity boom see Radetzki, *A Handbook of Primary Commodities*, pp. 70-2; Gilbert, 'How to Understand'; Carter *et al.*, 'Commodity Booms and Busts'.

[39] The practice of introducing trade-restrictive measures during severe economic crises is not a new trend for either developed or developing countries. The use of export restraints has, however, raised particular concerns given their rapid diffusion and growing incidence. See WTO Doc. WT/TPR/OV/14, p. 17. For a comprehensive description of recent trends in the use of industrial policy instruments see, among others, J. Salzman and

between 2010 and 2011, registering an incidence of more than 150 per cent higher than the previous year.[40] While this increase still mainly concerned agricultural products,[41] minerals and metals gained prominence as one of the product categories most affected by the introduction of barriers applied to exportation.[42]

The number of new export restrictions applied to primary products slowly decreased in the subsequent years, going from thirty-two in the period between mid-October 2011 and mid-October 2012[43] to twenty-seven between mid-October 2012 and mid-November 2013.[44] In the period between mid-November 2013 and mid-May 2014, however, twenty-four new restrictive measures were already being introduced on the exportation, showing a reversed upward trend.[45] Although this number was still lower than the level recorded at the same point in time during the peak year 2010–2011,[46] export restrictions accounted for 27 per cent of the total number of trade-restrictive measures introduced in this timeframe by WTO Members,[47] the highest percentage registered

M. Wu, 'The Next Generation of Trade and Environment Conflicts: The Rise of Green Industrial Policy', *Northwestern University Law Review* 108 (2014), 401.

40 WTO Doc. WT/TPR/OV/14, p. 17. The number of new export restrictions had been steadily increasing since 2008, going from twenty new reported measures in the period between October 2008 and October 2009 to twenty-five between November 2009 and mid-October 2010, and finally sixty-four from mid-October 2010 to mid-October 2011 (of which thirty were introduced from mid-October 2010 to April 2011, and thirty-four from May 2011 to mid-October 2011). WTO Doc. WT/TPR/OV/W/5, p. 5, and WTO Doc. WT/TPR/OV/W/6, 28 June 2012, p. 1.

41 Food-related commodities were, in particular, severely affected by export restrictions as a reflection of the price spike of 2010–2011. See below, Section I.2.1.

42 WTO Doc. WT/TPR/OV/W/5, p. 16; WTO Doc. WT/TPR/OV/W/6, p. 19; B. Fliess, H. Gou and T. Mard 'Taking Stock of Measures Restricting the Export of Raw Materials: Analysis of OECD Inventory Data', OECD Trade Policy Working Paper No. 140 (OECD Publishing, 2012), p. 4. Because of the ambiguity of mineral export restraints – on the one hand, purporting to fulfil environmental goals such as resource depletion and, on the other hand, serving industrial purposes – the diffusion of these measures has generated further alarm. See below, Section I.2.2 and Section I.3.

43 WTO Doc. WT/TPR/OV/15, 29 November 2012, p. 2. Nineteen new export restrictions were applied in the period between mid-October 2011 and mid-May 2012. WTO Doc. WT/TPR/OV/W/6, p. 1.

44 WTO Doc. WT/TPR/OV/16, 31 January 2014, p. 23. Of the twenty-seven new measures reported in that year, sixteen were reported in the period between mid-October 2012 and April 2013. WTO Doc. WT/TPR/OV/W/7, 5 July 2013, p. 7.

45 WTO Doc. WT/TPR/OV/W/8, 27 June 2014, pp. 87–97.

46 In the first half of the 2010-2011 reporting period, WTO Members had introduced thirty new export restrictions. WTO Doc. WT/TPR/OV/W/5, p. 5.

47 WTO Doc. WT/TPR/OV/W/8, p. 23.

since the beginning of the crisis.[48] New measures were also introduced during the second half of the monitoring period, and by mid-October 2014 the number of export restrictions had grown to thirty-six.[49] In its last Report, the Directorate-General for Trade of the European Commission stressed the detrimental effects of a 'resurgence in the application of measures restricting *exports*'[50] for the global commodity markets, as practices that 'not only can affect or serve to regulate prices domestically but also worldwide'.[51] It accordingly considered 'the intensification of such a trend . . . particularly alarming as all countries are globally dependent on each other's natural resources'.[52]

Export restrictions have moreover been more resilient than other categories of trade-restrictive measures applied by WTO Members in the framework of the global crisis, accounting for a relatively marginal share of the overall trade-facilitating measures adopted during each reporting period. This share has furthermore slightly decreased over time, going from 18 per cent in the period between 2011 and 2012 to 6 per cent between 2012 and 2013, and to 5 per cent from mid-November 2013 to mid-October 2014.[53] Trade-facilitating measures applied on the export side (e.g. reduction of export duties or elimination of export bans or quotas) occur for the most part in the agricultural sector, with only a negligible proportion intervening in the mineral sector.[54]

These data show the pervasiveness of the export restrictions on raw materials introduced in the last decade as well as the persistence of such measures. Furthermore, they anticipate the emergence of a substantial upward trend in the use of restrictive measures applied on the exportation of minerals and metals. As explained below, this new element has

[48] Between 2010 and 2011, the number of export restrictions accounted for less than 20 per cent of the total number of trade-restrictive measures introduced by WTO Members. WTO Doc. WT/TPR/OV/14, p. 17.

[49] WTO Doc. WT/TPR/OV/17, 24 November 2014, p. 20.

[50] European Commission 'Eleventh Report on Potentially Trade-Restrictive Measures Identified in the Context of the Financial and Economic Crisis', 2014, http://horizon.hozint.com/2014/11/dg-trade-11th-report-on-potentially-trade-restrictive-measures/ (accessed December 2014), p. 13. Based on the European Commission's calculations, the use of export restrictions increased by 39 per cent between 1 June 2013 and 30 June 2014 compared with previous monitoring periods.

[51] Ibid., p. 4. [52] Ibid.

[53] WTO Doc. WT/TPR/OV/16, p. 18 and WTO Doc. WTO Doc. WT/TPR/OV/17, p. 19.

[54] WTO Doc. WT/TPR/OV/W/8, pp. 57–67.

proved to be one of the most controversial features of this second wave of export restrictions.

I.2.1 Agricultural export restrictions and food security concerns

The use of export restrictions in the agricultural sector is not a new phenomenon. Low-income commodity-dependent developing countries (CDDCs) have quite regularly restricted the exportation of food-related agricultural commodities and, to a lesser extent, agricultural raw materials of both plant and animal origin[55] in recent decades in order to pursue price stabilization, control inflation, counteract tariff escalation and/or raise government revenues.[56] The first wave of export restrictions triggered by the commodity boom of 1972–1975 also mainly affected agricultural products. Governments of the imposing countries used them as instruments to ensure an adequate domestic supply at insulated prices, and thus maintain food security.[57] However, export restrictions seldom proved effective at protecting domestic consumers from external shocks in the international food markets, while they unambiguously exacerbated price volatility.[58]

Agricultural commodities have again been widely affected by export restrictions since the second half of the 2000s. The bulk of newly introduced export restrictions have been applied to food products and, in particular, to cereal grains (e.g. rice, wheat and maize), oilseeds (e.g. soybeans and sunflower seeds), and vegetable oils (e.g. cottonseed oil, linseed oil, sunflower oil, palm oil and coconut oil).[59] Agricultural raw

[55] For a full account of definitional issues see below, Chapter 1, Section 1.2.

[56] Piermartini, 'The Role of Export Taxes', pp. 7–15.

[57] According to the definition of food security given by the Committee on World Food Security of the FAO, 'food security exists when all people, at all times, have physical, social and economic access to sufficient, safe and nutritious food to meet their dietary needs and food preferences for an active and healthy life. The four pillars of food security are availability, access, utilization and stability'. FAO, 'Report of the Thirty-fifth Section of the Committee on World Food Security' (Rome, 14-15 and 17 October 2009), www.fao.org/fileadmin/templates/cfs/Docs0910/ReformDoc/CFS_2009_2_Rev_2_E_K7197.pdf (accessed November 2014), p. 1.

[58] Anderson and Nelgen, 'Trade Barrier Volatility', p. 37.

[59] Liapis, 'How Export Restrictive Measures', p. 2; WTO Docs. WT/TPR/OV/W/14–17 (various years); J. Kim, 'Recent Trends in Export Restrictions on Raw Materials', in OECD, *The Economic Impact of Export Restrictions on Raw Materials* (OECD Publishing, 2010) pp. 13–59; B. Fliess, C. Arriola and P. Liapis, 'Recent Developments in the Use of Export Restrictions in Raw Materials Trade', in OECD, *Export Restrictions in Raw Materials Trade: Facts, Fallacies and Better Practices* (OECD Publishing, 2014), pp. 1–45. The restrictions on fishery products are generally considered in isolation as they are regularly

materials of plant and animal origin[60] have also been subject to export restrictions although not as systematically,[61] with some notable exceptions such as cotton.[62]

Similar to the case of 1972–1975, the second wave of export restrictions in the food sector developed in reaction to a complex mix of cyclical and structural factors[63]: (i) a situation of tight world food markets following two decades of very low agricultural prices and investment decline[64];

applied by a certain number of developing countries in order to preserve exhaustible stocks (and impose a seasonal break). OECD, 'Analysis of Non-Tariff Measures: The Case of Export Duties', TD/TC/WP(2002)54/FINAL, 31 January 2003, http://search.oecd.org/officialdocuments/displaydocumentpdf/?doclanguage=en&cote=td/tc/wp(2002)54/ final (accessed November 2014), pp. 13–5; OECD 'Analysis of Non-Tariff Measures: The Case of Export Restrictions', TD/TC/WP(2003)7/FINAL, 4 April 2003, http://search.oecd.org/officialdocuments/displaydocumentpdf/?doclanguage=en&cote=TD/TC/WP(2003)7/FINAL (accessed November 2014), pp. 16–20.

60 See below, Chapter 1, Section 1.2.

61 The number of export restrictions on forestry products (e.g. logs, timber) did not follow a trend comparable to that of food-related export restrictions, and mainly reflected the prior existence of such measures in a group of countries that continued to maintain them during the commodity boom. OECD, 'Analysis of Non-Tariff Measures: The Case of Export Duties', pp. 13–5; OECD, 'Analysis of Non-Tariff Measures: The Case of Export Restrictions', pp. 16–20. Interestingly, among these countries are also the United States, Australia and Canada (where, however, these measures are usually imposed by selected provinces and not at the federal level). The measures mainly include non-automatic licensing regimes, although in some cases export bans have also been applied. Fliess *et al.*, 'Recent Developments', p. 22. Leather, hides and skins have also been subject to export restrictions quite regularly by a certain number of WTO Members including Argentina and Brazil. OECD, 'Analysis of Non-Tariff Measures: The Case of Export Duties', pp. 13–4; OECD, 'Analysis of Non-Tariff Measures: The Case of Export Restrictions', pp. 15–7. These measures generally purport to encourage downstream processing and are still maintained, and sometimes reinforced. European Commission, 'Eleventh Report', pp. 97–106.

62 India has repeatedly resorted to a number of short-term export restraints (mainly export taxes) on cotton and cotton yarn. See, e.g. National Cotton Council, 'Influence of India's Export Restrictions on the World Cotton Market', 2010, www.cotton.org/issues/2010/upload/NCC-Report_India-and-World-Cotton-Market_October-26-2010_FINAL-3.pdf (accessed November 2014), p. 6.

63 For the distinction between cyclical and structural causes of food crises see B. Karapinar, 'Introduction: Food Crises and the WTO', in B. Karapinar and C. Häberli, (eds.), *Food Crises and the WTO* (Cambridge University Press, 2010), pp. 6–7. See also J. von Braun and E. Díaz-Bonilla, 'Globalization of Agriculture and Food: Causes, Consequences and Policy Implications', in J. von Braun and E. Díaz-Bonilla (eds.), *Globalization of Food and Agriculture and the Poor* (Cambridge University Press, 2006).

64 R. Sharma, 'Food Export Restrictions: Review of the 2007–2010 Experience and Considerations for Disciplining Restrictive Measures', FAO Commodity and Trade Policy Research Working Paper No. 32, 2011, www.fao.org/fileadmin/templates/est/PUBLICATIONS/Comm_Working_Papers/EST-WP32.pdf (accessed November 2014), p. 6.

(ii) increasing food consumption due to population growth and economic improvements in developing countries[65]; (iii) a more general rise in energy and metal prices, exposed to exogenous demand shocks in the international markets due to the rapid expansion of developing Asia[66] and, (iv) exchange rate movements.[67] The concurrence of these factors determined a situation of inelasticity of supply, whereby any exogenous shock would expose prices to great variations.[68] Food prices thus experienced high volatility in the 2000s.

Food commodities in international markets started to rise in 2003, but dramatically increased in 2007 following Russia's announcement of suspension of grain exports due to unfavourable weather conditions compromising the harvests.[69] When world food prices spiked in 2008, doubling the 2005 levels,[70] export restrictions on agricultural commodities also peaked.[71] By mid-2008, out of seventy-seven countries surveyed by the Food and Agriculture Organization (FAO) (sixty of which were low-income countries), one quarter had imposed export restrictions on food commodities. The majority of them were located in East and South Asia (where 40 per cent of the countries under scrutiny applied export restrictions) and in Europe and Central Asia (35 per cent of the covered countries).[72] These

[65] Mitra and Josling, 'Agricultural Export Restrictions', p. 3.

[66] Agricultural commodities' prices moved concurrently with energy and metal prices during the 2000s commodity boom. Carter *et al.*, 'Commodity Booms and Busts', p. 91. The commonality of price rises was not an immediate result of China's exceptional growth rates, which reflected the expansion of the construction sector and only marginally affected the demand for imported agricultural products (see below, Section I.2.2). Rather, it resulted from the linkage between the energy sector and the food sector, which contributed to the escalation of food prices through the transmission of higher transport costs, freight rates and fertilizer prices into agricultural production costs. Gilbert, 'How to Understand', pp. 411–2. Moreover, international food markets were put under pressure by increased biofuel demand. R. Howse and T. Josling, 'Agricultural Export Restrictions and International Trade Law: A Way Forward', IPC Position Paper, 2012, http://www.agritrade.org/Publications/documents/Howse_Josling_Export_Restriction_final.pdf (accessed November 2014), p. 4.

[67] The price of agricultural commodities is denominated in US dollars. Changes in the value of the dollar can therefore affect the prices of such commodities. Gilbert, 'How to Understand', p. 405.

[68] Ibid., pp. 401–4. [69] Anderson and Nelgen, 'Trade Barrier Volatility', p. 36.

[70] Gilbert, 'How to Understand', p. 399.

[71] Giordani *et al.*, 'Food Prices', pp. 18–25; Howse and Josling, 'Agricultural Export Restrictions', pp. 5–7; Liapis, 'How Export Restrictive Measures', p. 3.

[72] In all other regions considered (namely, Africa, Latin America and the Caribbean, and the Middle East and North Africa), approximately 20 per cent of the countries included in the survey applied export restrictions. In some cases, however, a single country implemented several measures (e.g. Argentina). FAO, 'Soaring Food Prices: Facts, Perspectives, Impacts

countries mostly resorted to bans and taxes, quotas (including export licensing systems aimed at administering such quotas) and, to a limited extent, minimum export prices.[73] In the intention of the imposing governments, these measures would shield domestic consumers from price increases and surges of exports in a context of high commodity prices, thereby ensuring food security.[74] Yet, the spiral of export restrictions eventually led to an increase in price volatility, contributing to a worsening of the food crisis.[75]

The use of export restrictions substantially declined in 2009 with the onset of recession,[76] but then regained prominence as prices of food commodities rose to a new high in early 2011.[77] In May 2011, the FAO estimated that 33 countries out of 105 surveyed (31 per cent) had resorted to at least one export restriction on foodstuffs since 2007, with a total of 87 measures.[78] Among them were large exporters such as Argentina, China, India, Egypt, Pakistan, the Russian Federation,

and Actions Required', 2008, HLC/08/INF/1, www.fao.org/fileadmin/user_upload/foodclimate/HLCdocs/HLC08-inf-1-E.pdf (accessed November 2014).

[73] Liapis, 'How Export Restrictive Measures', pp. 4–9.

[74] Mitra and Josling, 'Agricultural Export Restrictions', p. 3. The vast majority of foodstuffs affected by export restrictions consisted of products providing most of the calorie needs of developing country populations, whose total budget is predominantly spent on less processed commodities. Liapis, 'How Export Restrictive Measures', p. 4; FAO *et al.*, 'Price Volatility', p. 6.

[75] Giordani *et al.*, 'Food Prices', pp. 10-7; Martin and Anderson, 'Export Restrictions and Price Insulation', pp. 9–13; Mitra and Josling, 'Agricultural Export Restrictions', pp. 11–2; FAO *et al.*, 'Price Volatility'; D. D. Headey, 'Rethinking the Global Food Crisis: The Role of Trade Shocks', IFPRI Discussion Paper No. 958/2010, www.ifpri.org/sites/default/files/publications/ifpridp00958.pdf (accessed November 2014); A. Bouet and D. Laborde Debucquet, 'The Economics of Export Taxes in the Context of Food Security', in OECD, *The Economic Impact of Export Restrictions on Raw Materials* (OECD Publishing, 2010), pp. 59–79. Some authors argued that, in spite of their beggar-thy-neighbour effects, export restrictions can be considered a second-best policy to protect poor food consumers during times of high food prices. The first-best option would be a welfare-maximizing insurance policy between food producers and food consumers based on the principle of income redistribution from producers to consumers when prices are high and vice versa. Q.-T. Do, A. Levchenko and M. Ravallion, 'Trade Insulation as Social Protection', The World Bank Policy Research Working Paper No. 6448, 2013, http://elibrary.worldbank.org/doi/book/10.1596/1813-9450-6448 (accessed November 2014). For an account of the impact of high food prices on poverty see W. Martin and M. Ivanic, 'The Food Price Crisis, Poverty and Agricultural Trade Policy', in B. Karapinar and C. Häberli (eds.), *Food Crises and the WTO* (Cambridge University Press, 2010).

[76] Liapis, 'How Export Restrictive Measures', pp. 3–9.

[77] Coulibaly, 'The Food Price Increase of 2010–2011', pp. 2–3.

[78] Sharma, 'Food Export Restrictions', p. 8.

Ukraine and Vietnam.[79] These measures were in most cases of a temporary nature, but they were frequently extended for longer than initially declared or reactivated after being lifted, depending on the situation of the international food markets.[80] Similarly to what happened during the food crisis of 2007–2008, they significantly exacerbated the price spike of 2010–2011.[81]

The two food crises of 2007–2008 and 2010–2011 placed the issue of food security squarely on the agenda of numerous multilateral agencies and intergovernmental fora, including the World Trade Organization.[82] The discourse on food security has been broad and complex, including a wide array of topics linked to the need to (i) increase investment in agriculture, especially in developing countries; (ii) improve the resilience of poor farmers to weather and price shocks; (iii) increase access to food and nutrition support for vulnerable populations and, (iv) improve the performance of international food markets.[83] Within this framework, the escalation of export policy responses in connection with the food crises and the substantial body of evidence attesting to the linkages between the former and the latter concurred to give prominence to the issue of export restrictions on agricultural commodities.

The connection between export restrictions and price volatility was first acknowledged in the Comprehensive Framework for Action elaborated by the High Level Task Force on the Global Food Security Crisis (HLTF) established by the United Nations (UN) Secretary-General in 2008, representing more than twenty UN agencies, several International Financial Institutions (IFIs) and the WTO.[84] Accordingly, the Task Force recommended that the use of export restrictions be minimized in order

[79] Ibid., pp. 29–31; see also Liapis, 'How Export Restrictive Measures', pp. 4–9.

[80] Sharma, 'Food Export Restrictions', pp. 29–32.

[81] Giordani *et al.*, 'Food Prices', pp. 10–7; Martin and Anderson, 'Export Restrictions and Price Insulation', pp. 9–13.

[82] A synthetic yet comprehensive overview of the various multilateral initiatives and reactions of intergovernmental institutions to food insecurity is given in C. Häberli, 'God, the WTO – and Hunger', in K. Nadakavukaren Schefer (ed.), *Poverty and the International Economic Legal System: Duties to the World's Poor* (Cambridge University Press, 2013), pp. 79–106. See also Howse and Josling, 'Agricultural Export Restrictions', pp. 7–10.

[83] Howse and Josling, 'Agricultural Export Restrictions', pp. 7–10.

[84] UN HLTF (2008), *Comprehensive Framework for Action* (United Nations: 2008), www.un.org/en/issues/food/taskforce/Documentation/CFA%20Web.pdf (accessed November 2014). An updated version of the Comprehensive Framework for Action was released in August 2010: http://un-foodsecurity.org/sites/default/files/UCFA_English.pdf (accessed November 2014).

to reduce the risks of price instability and pave the way for 'progress towards multilateral trade reforms and free trade in the agriculture sector'.[85] The issue of how to strengthen existing WTO disciplines on agricultural export restrictions seemed then to regain momentum in the WTO as a phase of renewed engagement in the topic arose within the context of the Doha agricultural negotiations in the aftermath of the food crisis.[86]

At the heart of the discussions was the need to widen and strengthen the scope of Article 12 of the WTO Agreement on Agriculture,[87] namely the only provision in the Agreement specifically covering the issue of export restrictions on agricultural commodities. Article 12 AoA elaborates on existing GATT disciplines on export restrictions contained in Article XI GATT. Under Article XI:1, export duties or taxes are allowed, while export prohibitions (i.e. bans) and other forms of quantitative restrictions (quotas, minimum export prices, etc.) are prohibited. Article XI:2(a), however, makes an exception for those restrictions which are 'temporarily applied to prevent or relieve critical shortages of foodstuffs or other products essential to the exporting contracting party'.[88] Article 12:1 AoA clarifies that when a Member institutes a new export prohibition or restriction on foodstuffs in accordance with

[85] UN HLTF, *Comprehensive Framework for Action*, p. 12.

[86] The topic of export restrictions had figured prominently in Phase 1 of the Agricultural Negotiations initiated in 2000 under Article 20 of the WTO Agreement on Agriculture and then carried over in the Doha Round in November 2001. WTO, 'Agriculture Negotiations: Phase 1 (2000–2001)', www.wto.org/english/tratop_e/agric_e/negoti_ph1_e.htm (accessed November 2014). A large number of negotiating proposals on how to improve relevant WTO disciplines were submitted by a number of Members including the United States, Japan, Switzerland and the Cairn Group (a group of 19 agricultural exporting countries, composed of Argentina, Australia, Bolivia, Brazil, Canada, Chile, Colombia, Costa Rica, Guatemala, Indonesia, Malaysia, New Zealand, Pakistan, Paraguay, Peru, the Philippines, South Africa, Thailand and Uruguay). For a detailed account, see Mitra and Josling, 'Agricultural Export Restrictions', pp. 16–18. By the time the Doha Mandate on Agriculture was formulated, however, the topic of export restrictions had disappeared from the agenda. WTO Doc. WT/MIN(01)/DEC/1, 20 November 2001, paras. 13 and 14. The Doha Negotiating Group on Agriculture only sporadically addressed the issue, which figured only marginally in the so-called July package of 2004. Under the heading of 'Other Issues', Members generally agreed that '[d]isciplines on export prohibitions and restrictions in Article 12.1 of the Agreement on Agriculture will be strengthened', with any further detail being left for future discussion. See WTO Doc. WT/L/579, 2 August 2004.

[87] Agreement on Agriculture, Multilateral Agreements on Trade in Goods, Annex 1A to the Agreement Establishing the World Trade Organization, 1867 U.N.T.S. 410, entered into force 1 January 1995 [hereinafter AoA].

[88] For a more detailed analysis of Article XI GATT see below, Chapter 5.

Article XI:2(a) GATT, the Member shall (i) give due consideration to the effects of such prohibition or restriction on importing Members' food security; (ii) give prior notice in writing, as far in advance as practicable, to the Committee on Agriculture indicating the nature and the duration of such measure and, (iii) upon request, consult with any other Member having a substantial interest as an importer with respect to any matter related to the measure in question, and provide such a Member with any necessary information. As per Article 12:2 AoA, moreover, food-importing developing countries are exempted from such consultation and notification requirements 'unless the measure is taken by a developing country which is a net food exporter of the specific foodstuff concerned'.[89]

Against this backdrop, some WTO Members (in particular, Japan and Switzerland) circulated in 2008 an informal paper detailing possible avenues to strengthen existing AoA disciplines on export restrictions.[90] The proposal aimed at setting more stringent requirements for WTO Members seeking to impose a new export prohibition or restriction under Article 12:1 AoA, but did not address the issue of whether new binding disciplines should be introduced to regulate the use of export taxes on agricultural products. In particular, the proposed rules required that '"any new export prohibition or restriction be limited to the extent strictly necessary" for the imposing country, in light of production, stocks and domestic consumption'.[91] Stricter consultation and notification requirements were also envisaged, with WTO Members required to (i) give due consideration to the effects of export restrictions on importers' food security, also taking into account what trade patterns would have emerged in the absence of such measures; (ii) show the impact of the measures on food aid for net food-importing developing countries (NFIDCs); (iii) notify the WTO Committee on Agriculture before the institution of export restrictions, giving account of their nature, lifetime and rationale and, (iv) consult with importers on any matter related to the proposed measure, suspend its implementation during consultations, and accept the binding arbitration of a 'standing committee of experts' in case of failure of consultations within a reasonable timeframe.[92]

[89] Although there is hardly any application record of this provision, no real WTO case law has yet emerged on the implementation of Article 12 AoA. Mitra and Josling, 'Agricultural Export Restrictions', p. 15.

[90] The 2008 informal paper was not publicly circulated. It is, however, discussed in Mitra and Josling, 'Agricultural Export Restrictions', p. 18.

[91] Ibid. [92] Ibid.

The spirit of this proposal was kept in the recommendations on export restrictions elaborated by the FAO and nine other institutions including the WTO in the 2011 Report on Price Volatility of Food and Agricultural Markets, compiled at the request of the G20.[93] In particular, the report envisaged a '"fast track" consultation process ... to discuss whether the [export restriction] can be avoided and how' and regular consultations so that 'the measure, once in place, is removed at the earliest possible moment'.[94] Perhaps as a reflection of the participation of the WTO Secretariat, the report also recommended the development of an 'operational definition of a critical food shortage that might justify consideration of an export restrictive measure' and suggested that 'an export ban be defined as a time-limited measure of last resort, allowed only when other measures, including triggering domestic safety net measures for the poorest, have been exhausted, and taking into account, in particular, the food security needs of least developed countries and net food-importing developing countries'.[95] This latter recommendation was intended to remedy the absence of any definition of 'critical shortage' and 'temporarily' in Article XI:2(a) GATT.[96]

All these efforts notwithstanding, the latest draft negotiating text under consideration in the Doha Negotiating Committee on Agriculture – the so-called revised draft modalities – does not incorporate all the recommended improvements.[97] Overall, it includes more flexible notification and consultation requirements. WTO Members are required to notify to the Committee on Agriculture existing export prohibitions or restrictions under Article XI:2(a) within 90 days of the coming into force of the new provisions,[98]

[93] FAO *et al.*, 'Price Volatility'. [94] Ibid., p. 26. [95] Ibid.

[96] Article XI:2(a) GATT was then interpreted by the Appellate Body in 2012 in *China–Raw Materials* (*China–Raw Materials*, AB Reports, paras. 308–44). The Appellate Body gave a quite narrow interpretation of this exception, indicating that, should a dispute over export restrictions arise, a high threshold would need to be adopted in order to assess whether it responds to a critical shortage concern. For a detailed account see below, Section 5.3.1. Although dealing with export restrictions applied to non-agricultural raw materials, the AB ruling arguably contains important indications as to the room of manoeuvre left to WTO Members under Article XI:2(a) GATT to use export restrictions when food security is at risk. For an excellent analysis in this respect see Howse and Josling, 'Agricultural Export Restrictions', pp. 13–19.

[97] See WTO Doc. TN/AG/W/4/Rev.4, 6 December 2008. The proposed rules were not based on Japan and Switzerland's informal paper, but rather on a proposal from the G20 group of developing countries. Mitra and Josling, 'Agricultural Export Restrictions', p. 18.

[98] See WTO Doc. TN/AG/W/4/Rev.4, para. 172.

and give notice of the reasons for introducing and maintaining any new measures.[99] The consultation procedure as envisaged under Article XI:2(a) GATT is not significantly improved other than the requirement imposed on Members to 'report the progress made in the consultations to the Committee on Agriculture'.[100] The Committee is also required to provide an 'annual notification update and surveillance of these obligations'.[101] The suspension of the measure at issue during the consultation phase is, however, not foreseen, nor are WTO Members obliged to abide by an arbitration decision in case consultations fail. A temporal limit of one year is anyway set for existing export prohibitions and restrictions under Article XI:2(a) GATT,[102] while new export restrictions could last up to 18 months with the agreement of the affected importing Members.[103] The issue of export taxes on agricultural products is once again left unanswered.

The December 2008 revised draft modalities remain the basis for future negotiations on the issue of export restrictions on agricultural commodities. However, as the Doha Round has stalled, attempts to raise the level of ambition of the negotiations on the topic have languished. In the wake of the second food crisis in 2011, renewed engagement in the issue of export restrictions mainly focused on humanitarian exemptions for food purchased for non-commercial purposes. Already in 2009, the G8 Summit had adopted the 'Aquila Food Security Initiative', where governments 'call[ed] upon all countries to remove food export restrictions or extraordinary taxes, especially for food purchased for humanitarian purposes, and to consult and notify in advance before imposing any new restrictions'.[104] This plea was then reiterated at various gatherings. At the 2009 World Summit on Food Security organized in Rome in November 2009, all FAO members agreed to 'remove food export restrictions for food purchased for non-commercial humanitarian purposes, and to consult and notify in advance before imposing any such new restrictions'.[105] Although significantly narrower than the one sought in L'Aquila, this commitment was

[99] Ibid., para. 173. Cross-notifications are also allowed under para. 177, in accordance with Article 18:7 AoA.

[100] WTO Doc. TN/AG/W/4/Rev.4, para. 175. LDCs and NFIDCs are, however, exempted from this requirement. Ibid., para. 180.

[101] Ibid., para. 176. [102] Ibid., para. 178. [103] Ibid., para. 179.

[104] G8, 'L'Aquila Joint Statement on Global Food Security: L'Aquila Food Security Initiative' (L'Aquila, 10 July 2009), www.ifad.org/events/g8/statement.pdf (accessed November 2014), para. 6.

[105] FAO, 'Draft Declaration of the World Summit on Food Security' (Rome, 16–18 November 2009), www.fao.org/fileadmin/templates/wsfs/Summit/Docs/Declaration/WSFS09_Draft_Declaration.pdf (accessed November 2014), para. 21.

finally incorporated into the Action Plan on Food Price Volatility adopted by the G20 Agriculture Ministers in 2011; the Ministers, in particular, agreed to remove these measures and 'not to impose them in the future', while seeking 'support within the United Nations agencies' and 'recommend[ing] consideration of the adoption of a specific resolution by the WTO for the Ministerial Conference in December 2011'.[106] However, no decision to exempt humanitarian food purchases from export restrictions or extraordinary taxes was adopted on that occasion. The topic was only briefly mentioned in the Chairman's Concluding Statement,[107] reflecting some divergence of interests among WTO Members.[108] The matter of food security was generally left for further action to be taken by the Ninth Ministerial Conference, but there seemed to be no consensus as to whether the issue of export restrictions was to be part of such action.[109]

The so-called Bali Package adopted during the Ninth Ministerial Conference in December 2013 in the attempt to revive the Doha Round did address at various levels the issue of food security. Yet, the topic of export restrictions virtually disappeared,[110] in spite of the request to seek 'a resolution of this issue in the WTO' put forward in the Report of the G20 Vice Ministers of Agriculture of 2012.[111] The decision not to include export restrictions in the list of 'Bali issues' reflects, on the one hand, the change of approach promoted under the leadership of the new WTO Director-General, Roberto Azevêdo, aimed at creating the conditions necessary to get to an expedited close of the Doha Round albeit at the expense of a less

[106] G20, 'Action Plan on Food Price Volatility and Agriculture' adopted by the G20 Agriculture Ministers (Paris, 22-23 July 2011), www.g8.utoronto.ca/g20/agriculture/index.html (accessed October 2014), para. 40.

[107] WTO Doc. WT/MIN(11)/11, 17 December 2011.

[108] The Chairman noted that '[m]any Ministers urged WTO Members to commit to remove, and not to impose in the future, food export restrictions or extraordinary taxes for food purchased for non-commercial humanitarian purposes by the World Food Programme. Other Ministers stressed the importance of addressing the root causes of food insecurity and underlined the importance of allowing Members to use their rights under WTO Agreements . . . Several Ministers noted that the issue of food security was multifaceted and needed to be looked at in its entirety, including the impact of export restrictions on international prices'. Ibid., p. 5.

[109] Ibid.

[110] WTO, 'Ninth Ministerial Conference: Ministerial Declaration and Decisions' (Bali, 3–6 December 2013), http://wto.org/english/thewto_e/minist_e/mc9_e/bali_texts_combined_e.pdf (accessed November 2014).

[111] G20, 'Report of the Agriculture Vice Ministers/Deputies Meeting' (Mexico City, 18 May 2012), www.g20.utoronto.ca/agriculture/index.html (accessed November 2014), para. vi.

ambitious deal.[112] On the other hand, it echoes a relative loss in importance of the topic in the more general discourse carried on by the relevant international institutions engaged in the food security discourse. [113] Since the 2011 price spike, the number of newly introduced export restrictions on agricultural commodities has in fact not again reached critical levels,[114] while many measures have been dismantled in the last few years, reflecting the relative stabilization of agricultural commodities' prices.[115]

Such circumstances notwithstanding, growing evidence suggests that food export restrictions are likely to remain a challenge in the future due to the structural irreversible changes that are affecting international food markets.[116] Possibly the 'permanent' solution sought for public stockholding for food security purposes in developing countries could include

[112] The 'Bali issues' are four agriculture issues that were selected out of the more comprehensive agenda pursued in the Doha Round agricultural negotiations in the attempt to make some tangible progress after several years of stalemate. See WTO Doc. WT/MIN (11)/W/2, 1 December 2011.

[113] Food security is still high on the agenda of a number of multilateral agencies, which continue to assess the state of food insecurity in the world and to stress the need for action to eradicate hunger and malnutrition. Among the proposed strategies, however, the regulation of export restrictions is no longer included. See, e.g. FAO, WFP and IFAD, 'The State of Food Insecurity in the World', 2014, www.fao.org/3/a-i4030e.pdf (accessed November 2014). The topic of export restrictions also stopped being explicitly addressed by G20 Summits. At the 2013 Summit, which took place in St. Petersburg, G20 governments simply 'reaffirm[ed the] determination to implement all previous G20 commitments and existing initiatives including that stated in the Action Plan on Food Price Volatility and Agriculture which the G20 endorsed in 2011'. See G20, 'Leaders' Declaration' (St. Petersburg, 6 September 2013), www.g20.utoronto.ca/2013/2013-0906-declaration.html (accessed November 2014), para. 82. Neither was the topic of export restrictions on food commodities addressed at the last G20 Summit in Brisbane, Australia in November 2014. See G20, 'Leaders' Communiqué' and, in particular, the 'Food Security and Nutrition Framework' (Brisbane, 16 November 2014), www.g20.utoronto.ca/2014/2014-1116-communique.html (accessed November 2014).

[114] Liapis, 'How Export Restrictive Measures', pp. 3–9.

[115] WTO Doc. WT/TPR/OV/W/8, pp. 57–67. Overall, prices of food commodities remain higher than they were in 2005. However, their levels have been slightly but steadily decreasing in the last few years. For the latest trends see FAO Food Price Index, www.fao.org/worldfoodsituation/foodpricesindex/en/ (accessed November 2014).

[116] As noted by Bellmann and Wilke, 'population and income growth keep exceeding productivity and area growth' so that, in the medium- to long-run, world food markets are to grow tighter and prices to increase. See C. Bellmann and M. Wilke, 'Trade Policies for Resource Security: Rethinking Export Restrictions', in R. Meléndez-Ortiz, C. Bellmann, and M. Rodriguez Mendoza (eds.), *The Future and the WTO: Confronting the Challenges* (Geneva: ICTSD, 2012), p. 198.

something about export restrictions.[117] However, the need for substantial new disciplines on agricultural export restrictions is here to stay.

I.2.2 Export restrictions on energy commodities between new and old concerns

Hard commodities led the commodity boom of the 2000s.[118] Oil prices almost doubled in real terms between 2005 and 2008 and metals reached a record high with a 125 per cent increase in the same period.[119] Analysts and scholars agree that this upward move originated from common factors. On the demand side, world markets experienced a severe shock due to the global macroeconomic expansion of the mid-2000s[120] and, in particular, the exceptional growth rates of China and other emerging Asian economies[121] driven by the expansion of the construction sector.[122] On the supply side, a two-decade span of low prices made investment in the extractive sector decline, producing a situation of tight spare capacity.[123] Such constraints on the supply front amplified the pressure on hard commodities prices caused by the demand shock. The typical low elasticity of supply in the extractive sector also contributed to further price escalation.[124]

Within this general framework, scholars have also identified energy-specific factors underlying the price spikes of the mid-2000s. Radetzki, in

[117] WTO, 'The Bali Decision on Stockholding for Food Security in Developing Countries', https://www.wto.org/english/tratop_e/agric_e/factsheet_agng_e.htm (accessed November 2014).

[118] Carter *et al.*, 'Commodity Booms and Busts', p. 90.

[119] Gilbert, 'How to Understand', p. 399.

[120] Energy demand is strongly correlated to increases in GDP. Energy Charter Treaty (ECT), 'Putting a Price on Energy: International Pricing Mechanisms for Oil and Gas', 2007, www.encharter.org/fileadmin/user_upload/document/Oil_and_Gas_Pricing_ 2007_ENG.pdf (accessed November 2014), p. 21.

[121] Radetzki, *A Handbook of Primary Commodities*, pp. 70–71. Inasmuch as they are at earlier stages of development, emerging economies use primary supplies much more intensively than mature economies. Ibid., pp. 7–12.

[122] The construction sector relies on the massive use of energy and metals. Ibid., p. 71. During the course of the 2000s, demand for energy and metals rose to its highest level in over twenty years. M. Farooki and R. Kaplinsky, *The Impact of China on Global Commodities Prices: The Global Reshaping of the Resource Sector* (Routledge Studies in Global Competition, 2012), pp. 70–103.

[123] ECT, 'Putting a Price on Energy', p. 21; G. Peeling, P. Stothart and B. Toms 'Increasing Demand for and Restricted Supply of Raw Materials', in OECD, *The Economic Impact of Export Restrictions on Raw Materials* (OECD Publishing, 2010), p. 157.

[124] In the extractive sector, the response of supply to demand expansion is typically lagged. Ibid., pp. 117–39. For a full account see below, Chapter 1, Section 1.3.1.

particular, observes that the oil price increases occurred 'initially in consequence of a rising assertiveness by the [OPEC], combined by disciplined production cuts, soon amplified by political supply problems in the Middle East (the 2003 war in Iraq)'.[125] This statement identifies two main factors influencing trends in world energy prices: first, the exposure of world global supply to disruptions caused by situations of geopolitical tension; second, the ability of OPEC to exert upward pressure on oil prices through supply management policies.[126] These factors have been at the heart of the peculiar dynamics affecting international energy markets and, not surprisingly, energy prices have usually spiked much more frequently than just during commodities booms.[127]

Sector-specific developments have also substantially influenced the peculiar parabola of energy prices in recent years. Following the price bust of 2008–2009 determined by the eruption of the financial and economic crisis, oil prices started rising again from 2010 and continued hovering around 100 USD/barrel between 2011 and 2014 due to the combination of civil war in Libya, political turmoil in Iraq and the effects of sanctions imposed on Iran.[128] By late 2014, however, the scenario had changed sharply. The so-called shale revolution gradually transformed the United States, traditionally a top world energy importer, into the largest global producer of oil and natural gas.[129] This major breakthrough, associated with the effects of the economic slowdown in China and Germany, determined a situation of excess supply in the world markets.[130] Between

[125] Radetzki, *A Handbook of Primary Commodities*, p. 160.

[126] Still today, OPEC Members produce approximately 40 per cent of the global oil supply and export around 60 per cent of internationally traded oil. EIA, 'What Drives Crude Oil Prices?', 2014, www.eia.gov/finance/markets/supply-opec.cfm (accessed November 2014). OPEC has traditionally used its market share to influence international oil prices and, up to the mid-1980s, it directly controlled oil prices through a system of official selling prices and coordinated supply disruptions. ECT, 'Putting a Price on Energy', p. 55. Now that oil is globally traded as a commodity, OPEC sets production targets in order to adjust oil production in its member countries based on current and projected supply and demand in the world market. EIA, 'What Drives Crude Oil Prices?'.

[127] ECT, 'Putting a Price on Energy', pp. 52–62.

[128] B. Plumer, 'Why Oil Prices Keep Falling – and Throwing the World Into Turmoil', *Vox*, 23 January 2015, www.vox.com/2014/12/16/7401705/oil-prices-falling (accessed January 2015).

[129] A. Sieminski, 'Implications of the US Shale Revolution', Presentation given at the Canada Energy Summit (Chicago, 17 October 2014), www.eia.gov/pressroom/presentations/sieminski_10172014.pdf (accessed January 2015).

[130] Plumer, 'Why Oil Prices Keep Falling'; N. Butler, 'After the Oil Price Fall, Is Natural Gas Next?' *Financial Times*, 4 January 2015.

December 2014 and January 2015, oil prices reached a 5-year low of 49 USD, halving within a 6-month period,[131] while natural gas prices also plunged to their lowest levels since 2012, and are projected to continue decreasing considerably.[132]

Against this backdrop, the discourse on export restrictions on energy commodities presents distinct characteristics, which only partially reflect the contingent dynamics of the 2000s commodity boom, more generally resulting from the long-standing tension between energy security and fuel export dependence on the part of import-dependent countries and net exporting countries, respectively.[133] In this regard, export restrictions in the energy sector have mainly consisted of export taxes applied by energy exporters, generally aimed at capturing the gains linked to high energy prices and/or protecting domestic consumers from price increases or supply shortages, and export quantitative restrictions employed by relatively well-endowed net importing countries attempting to reduce import reliance. While such measures mostly pre-date the 2000s energy price spikes, in many instances they have either been tightened following price increases or gained prominence in the international trade agenda due to a number of contingent factors more or less linked to the commodity boom.

A number of energy net exporting countries have increased export taxes on petroleum, natural gas and/or related products in the last few years, including top energy exporters such as the Russian Federation. The Russian government has regularly upwardly revised the export duties applied on energy products since the beginning of the 2000s.[134] In 2010, the government increased the rate of export duty on oil to reflect the rise in world prices and, in 2011, it introduced a new method for calculating the export duties to be applied on petroleum products whereby heavy and light petroleum products would be taxed at 46.7 per cent and 67 per cent of the rate for crude oil, respectively.[135] In 2013, the rate of export duty imposed on petrol was also increased at around 415 USD/tonne, while the export duty for natural gas was set at a flat rate of 30 per cent.[136]

Other large energy exporters have introduced and/or updated the export duties imposed on energy products. Kazakhstan reintroduced an

131 A. Raval, 'Oil Prices Slide To Five-Year Low', *Financial Times*, 8 December 2014.

132 Butler, 'After the Oil Price Fall'.

133 Saudi Arabia and the Russian Federation, the top energy exporters, rely on energy export revenues for a large part of their GDP (around 60 per cent of the budget revenues and 90 per cent of fiscal revenues, respectively). C. Giles, 'Winners and Losers of Oil Price Plunge', *Financial Times*, 15 December 2014.

134 European Commission, 'Eleventh Report', p. 103. 135 Ibid., p. 103. 136 Ibid.

export duty on crude oil in 2010, and doubled it in 2011 to a 40 USD/tonne rate. In the same year, it also elevated the export duty on light and heavy oil to 76.03 and 114.05 USD/tonne, respectively.[137] In 2012, furthermore, the Kazakh government elevated the export duties on petroleum products by 15 per cent.[138] Argentina has also recently increased the export duty rates applied on energy products. Since 2012, it has introduced prohibitive duty rates on exports of natural gas (100 per cent), propane gas (45 per cent) and LPG (minimum rate of 45 per cent, varying in accordance with international prices).[139] It has also taxed exports of crude oil based on a formula that takes into account world price trends and imposed a cap at 45 per cent.[140] Other net exporting WTO Members have applied non-prohibitive duty rates on exports of oil products, for instance, Nigeria, Mexico and Malaysia. Nigeria has imposed an administrative levy of 3 USD/tonne and a 0.5 per cent levy on all exports of raw materials since 1992.[141] Mexico and Malaysia have taxed exports of various petroleum products at a rate of 25 per cent[142] and 10 per cent, respectively.[143]

These countries remain free to impose export duties as per their WTO obligations. All but the Russian Federation are in fact exclusively bound by GATT provisions and Article XI:1 GATT expressly allows Members to use export duties on any product to the extent that they are not in excess of any scheduled bound levels negotiated for specific products.[144] The Russian Federation has negotiated export duty concessions for more than 700 tariff lines in its GATT schedule in the context of its WTO accession.[145] In the case of oil and gas products, however, it did not commit to bind export duty rates at a fixed rate. Rather, it incorporated into its GATT Schedule specific formulas for determining the maximum applicable rate based on, respectively, the world price of crude oil and the

[137] Ibid., p. 102.

[138] Ibid. Kazakhstan also banned exports of various gas products from May 2010 to July 2012.

[139] WTO Doc. WT/TPR/S/277, 13 February 2013, pp. 96, 145. [140] Ibid., p. 149.

[141] WTO Doc. WT/TPR/S/247, 24 May 2011, p. 33.

[142] WTO Doc. WT/TPR/S/279, 27 February 2013, p. 84.

[143] A. Chelliah, 'Tax and the Upstream O&G Sector', *Malaysian Business*, 16 August 2011, www.taxand.com.my/pdf/publications/Tax%20and%20the%20Upstream%20OnG%20Sector%20-Malaysian%20BusinessAugust.pdf (accessed January 2015). WTO Doc. WT/TPR/S/292, 27 January 2014, pp. 58–9.

[144] For a full account see below, Chapter 4, Section 4.3.

[145] See Chapter 4, Section 4.4.1.8.

average LPG price at DAF Brest.[146] Kazakhstan is not yet a WTO Member.[147]

Beside export taxes, export quantitative restrictions are also applied to energy commodities. The most notable examples are implemented by large world energy consumers such as the United States and China in order to reduce import dependence. In the case of China, exports of coal, crude oil and processed oil are subject to a quota that is administered by the Chinese Ministry of Commerce through a non-automatic export licensing scheme.[148] Exports of such commodities are also subject to state trading.[149] The measures allegedly aim to ensure adequate domestic supply and exhaustible natural resources conservation.[150] They reflect the fact that, despite still being a top energy producer, China has recent transformed from net exporter into the world's largest energy consumer *and importer.*[151]

The United States has implemented a comprehensive export regime on crude oil and natural gas since the 1970s. Interestingly, the crude oil regime was construed as a reaction to the 1972–1973 OPEC oil embargo in an attempt to relieve the US economy from excessive reliance on Arab oil.[152] Under this scheme, crude oil exports are prohibited unless they are granted explicit presidential authorization under one of the specific statutory control regimes provided in the Export Administration Regulations on Short Supply Controls of the US Bureau of Industry and Security (BIS).[153] To this end, the US President will make and publish an explicit finding that such exports, among others, will not diminish the total quantity or quality of petroleum available to the

[146] See footnote # and footnote ## in Part V of the Russian Federation's Schedule, WTO Doc. WT/MIN(11)/2/Add.1, 17 November 2011.

[147] WTO, 'Accessions', www.wto.org/english/thewto_e/acc_e/acc_e.htm (accessed January 2015).

[148] WTO Doc., WT/TPR/S/300, 27 May 2014, pp. 81–2. See also below, Chapter 2, Section 2.5.2.

[149] WTO Doc., WT/TPR/S/300, p. 104. [150] Ibid., p. 103.

[151] Since 2010, China has been the top world energy consumer, accounting for 17.5 per cent of global energy consumption. US Energy Information Administration, *Countries Profiles: China*, 2014, www.eia.gov/countries/analysisbriefs/China/china.pdf (accessed December 2014). China's growing dependence on fossil fuel imports originates from the exceptional macroeconomic expansion. For instance, China's oil consumption growth accounted for one-third of global oil demand growth in 2013. Ibid.

[152] I. Espa and K. Holzer, 'Negotiating an Energy Deal Under TTIP: Drivers and Impediments to US Shale Exports to Europe?' *Denver Journal of International Law and Policy* 43 (2015), 358, pp. 366–7.

[153] Ibid, pp. 367–8.

United States and are in the national interest.[154] Natural gas exports are subject to a stringent non-automatic licensing regime administered by the US Department of Energy (DoE). According to the scheme, applications to export natural gas are approved by the DoE so long as they are in the public interest.[155] Yet, a list of eligibility requirements in this respect has not been formalized. The DoE has on occasions posited that criteria as varied as adequacy of domestic supply, economic considerations, environmental impacts, and energy security should govern the evaluation.[156]

The shale revolution has brought the issue of removal of US export barriers on crude oil and natural gas back onto the radar. The underlying narrative is that, although booming production could make the United States self-sufficient, falling domestic prices induced by local oversupply situations may eventually hamper the profitability of fracking and horizontal drilling in the medium to long term.[157] At the same time, the geopolitical tension between the Russian Federation and Ukraine has prompted the EU to seek alternative reliable suppliers and, to this end, it has started to explore opportunities to facilitate US shale imports in the context of the Transatlantic Trade and Investment Partnership (TTIP) negotiations.[158] The hypothesis of dismantling existing export barriers on crude oil and natural gas, however, remains controversial in the US at the time of writing.[159]

Finally, due to the recent shrinkage of world oil prices, the issue of production quotas on the part of OPEC has also regained prominence in the international trade debate. The use of supply management policies is a long-standing practice in the oil sector, regularly implemented by OPEC based on current and projected levels of world demand and supply as a way of influencing world oil prices.[160] However, whether production quotas constitute export restrictions for the purposes of WTO regulation

[154] A report containing such findings and explaining the reasons for justifying crude oil exports must then be submitted to Congress, which has sixty days to veto the presidential permission. Ibid. Under the US Export Administration Act, the president also has the authority to restrict exports of refined petroleum products but has refrained from using it since 1981.

[155] Exports to FTA partners are deemed to be in the public interest. See M. Ratner, P. W. Parfomak, L. Luther, and I. F. Fergusson, 'U.S. Natural Gas Exports: New Opportunities, Uncertain Outcomes', US Congressional Research Service (CRS) Report for Congress, 8 April 2013, www.fas.org/sgp/crs/misc/R42074.pdf (accessed December 2014), p. 14.

[156] Ibid., p. 15. [157] Plumer, 'Why Oil Prices Keep Falling'.

[158] Espa and Holzer, 'Negotiating an Energy Deal Under TTIP', pp. 365–6.

[159] For a full account of the multifaceted debate in the US see ibid, pp. 372–8.

[160] EIA, 'What Drives Crude Oil Prices?'

is still a matter of much debate. While such measures have never been challenged before the WTO dispute settlement bodies, scholars still disagree as to their admissibility under Article XI:1 GATT.[161] At the time of writing, Saudi Arabia and other OPEC Members have not yet agreed on production cuts.[162] Yet, should oil prices continue to plunge, petroleum production quotas may eventually be implemented,[163] and the question resurface in the WTO.

I.2.3 Export restrictions on raw materials of mineral origin and the revival of industrial policies

As discussed above, common supply and demand dynamics in the world markets triggered an upward move in the price of hard commodities in the 2000s. The parabola of mineral and metal prices, however, followed a path of its own. Minerals and metals prices grew faster than energy prices,[164] and often peaked earlier rather than simply moving in line with them.[165] Mineral commodities reached historic price levels between 2007 and mid-2009;[166] after a period of decline following the outbreak of

[161] T. Cottier, G. Malumfashi, S. Matteotti-Berkutova, O. Nartova, J. de Sépibus and S. Z. Bigdeli, 'Energy in WTO Law and Policy', in T. Cottier and P. Delimatsis (eds.), *The Prospects of International Trade Regulation: From Fragmentation to Coherence* (Cambridge University Press, 2011), pp. 233–7; M. Cossy, 'Energy Trade and WTO Rules: Reflections on Sovereignty Over Natural Resources, Export Restrictions and Freedom of Transit', *European Yearbook of International Economic Law* 3 (2012), 281; R. Bhala, 'Saudi Arabia, the WTO, and American Trade Law and Policy', *The International Lawyer* 34 (2004), 741; Mavroidis, *Trade in Goods*, pp. 62–5, noting *inter alia* that '. . . the Working Party Report on [Saudi Arabia's] accession reflects a rather benign discussion of this issue, where it was made clear that petroleum production quotas were not, at least in the view of the acceding country, legally challengeable before a WTO Panel'.

[162] Plumer, 'Why Oil Prices Keep Falling'. Saudi Arabia has objected to production cuts, presumably in an attempt to preserve its market shares in the EU and Asian energy markets. The decision also seems directed at recapturing market share in the US, where shale projects will stop being economically profitable with prices below 80 USD per barrel. A. Critchlow, 'US Oil Production Surge to Break Saudi Arabia's Grip on World Energy', *The Telegraph*, 9 November 2014. The dramatic effects of low prices on the Russian economy may also be part of the explanation in light of the current geopolitical situation. Giles, 'Winners and Losers'.

[163] Analysts have estimated that its foreign exchange reserves notwithstanding, Saudi Arabia would run into a fiscal deficit corresponding to 14 per cent of its GDP if oil prices keep hovering around 60 USD in 2015. Ibid.

[164] Radetzki, *A Handbook of Primary Commodities*, p. 71.

[165] Carter *et al.*, 'Commodity Booms and Busts', p. 90.

[166] The World Bank, 'Commodity Markets Outlook: Global Economic Prospects', July 2014, www.worldbank.org/content/dam/Worldbank/GEP/GEPcommodities/commodity_markets_outlook_2014_july.pdf (accessed December 2014), p. 9.

the financial crisis, prices started to rise again, reaching pre-crisis levels in 2011.[167] In the subsequent 3 years, they showed an overall slow decline, although remaining significantly higher than the levels preceding the mineral boom and the 2009 price bust.[168] From the third quarter of 2014, however, an upward trend was again registered, reflecting expectations of tightened supply conditions.[169]

Triggered by these upward price moves, the use of export restrictions in the minerals and metals sector has been on the rise. Yet, mineral export restraints have not solely reflected the evolution of mineral prices.[170] The number of export restrictive measures applied in the sector increased significantly at the time of deepest economic downturn, that is, the period of most evident price decline.[171] Between 2009 and 2010, export restrictions in the minerals and metals sector were 'pervasive'[172]; the total number of countries restricting exports of raw materials of mineral origin rose by five,[173] and only a negligible proportion of these raw materials was not subject to any form of export restriction by any country.[174] In 2010, 231 tariff lines corresponding to materials in their raw state (Harmonized System 6-digit level) and 629 corresponding to semi-processed products were subject to restrictions on their exportation,[175] and in each case by at least two of the five largest producers of each product.[176]

167 Farooki and Kaplinsky, *The Impact of China*, p. 118.

168 The World Bank, 'Commodity Markets Outlook', July 2014, p. 9.

169 This increase has mainly concerned base metals, while the price of iron ore has been steadily decreasing. The World Bank, 'Commodity Markets Outlook: Global Economic Prospects', October 2014, www.worldbank.org/content/dam/Worldbank/GEP/GEPcommodities/commodity_markets_ outlook_2014_october.pdf (accessed December 2014), p. 8.

170 For a full account of the rationales of mineral export restraints see below, Chapter 3, Section 3.3.

171 From 2009 onwards, minerals and metals started being reported as the product category most affected by export restrictions, second to agricultural products. Kim, 'Recent Trends in Export Restrictions', pp. 15–17; WTO Doc. WT/TPR/OV/W/5, p. 5 and WTO Doc. WT/TPR/OV/W/6, p. 1.

172 Fliess *et al.*, 'Taking Stock of Measures', p. 4. 173 Ibid.

174 K. C. Fung and J. Korinek, 'Economics of Export Restrictions as Applied to Industrial Raw Materials', OECD Trade Policy Papers, No. 155 (OECD Publishing, 2014), pp. 7–13.

175 These data are based on the OECD Inventory on Raw Materials covering the export restrictions applied to 75 non-energy minerals and metals by a total of 100 countries. Although the sample of countries is limited, it always includes the five largest suppliers in the world, for each of the products considered. Fliess *et al.*, 'Taking Stock of Measures', p. 12.

176 Fung and Korinek, 'Economics of Export Restrictions', p. 8.

The vast majority of these export restrictions were not recorded as temporary measures.[177] Since then, only a small portion of mineral export restrictions have been phased out (as at 31 December 2014),[178] while the minerals and metals sector remains one of the product categories most affected by the introduction of new measures by WTO Members.[179] Among the countries most frequently using mineral export restraints are resource-rich developing countries and emerging economies such as China, the Russian Federation, Argentina, India, Indonesia, Kazakhstan, South Africa, and Vietnam.[180] These countries often purport to be adopting such measures to fulfil non-economic public policy goals directly related to the negative externalities associated with the exploitation of raw materials of mineral origin, such as resource conservation and environmental protection.[181] Yet, for the most part, these measures are adopted within the context of comprehensive industrial policy programmes aimed at boosting industrial base growth and economic diversification.[182] Mineral export restraints have thus proved to be extremely contentious, and ambivalent at the least.[183]

The use of export restrictions on industrial raw materials of mineral origin has raised great concerns in the part of the Western world historically dependent on the importation of minerals and metals, whose use is essential for those manufacturing sectors already hit by the effects of the global economic crisis.[184] Two major WTO disputes arose from these tensions and were concluded in the last few years.[185] Meanwhile, the

177 Fliess *et al.*, 'Taking Stock of Measures', p. 18.

178 Export restrictions in the minerals and metals sector are being removed at a far slower pace than those in other sectors (e.g. agriculture). WTO Doc. WT/TPR/OV/W/8, pp. 57–67. China, however, has recently removed a number of export restraints applied on minerals and metals in pursuance of two WTO AB Reports (*China–Raw Materials* and *China–Rare Earths*). See below, Part II.

179 European Commission, 'Tenth Report on Potentially Trade-Restrictive Measures Identified in the Context of the Financial and Economic Crisis', 2013, http://trade.ec.europa.eu/doclib/docs/2013/september/tradoc_151703.pdf (accessed November 2014); Fliess *et al.*, 'Recent Developments', pp. 5–10.

180 Fliess *et al.*, 'Taking Stock of Measures', p. 13; European Commission, 'Tenth Report'.

181 Fliess *et al.*, 'Taking Stock of Measures', pp. 7–18.

182 European Commission, 'Ninth Report of Potentially Trade-Restrictive Measures Identified in the Context of the Financial and Economic Crisis', 2012, http://trade.ec.europa.eu/doclib/docs/2012/june/tradoc_149526.pdf (accessed November 2014), pp. 9–11.

183 Fliess *et al.*, 'Taking Stock of Measures', pp. 4–5.

184 For an excellent overview of the historical reasons at the basis of the rich world's dependence on the importation of raw materials see Radetzki, *A Handbook of Primary Commodities*, pp. 37–40.

185 For a full account of *China–Raw Materials* and *China–Rare Earths* see below, Part II.

issue of export restrictions on mineral resources is as alive as ever in the international trade debate as these measures continue to proliferate.[186]

All these elements show how mineral export restraints emerged from the mid-2000s to the early 2010s as a distinct new element of the present wave of export restrictions compared with the past cycles and the other sub-waves of the 2000s, in terms of triggers, trends, impacts and rationales. Key factors explaining the phenomenon include the mineral boom, the revival of industrial policies and the global economic crisis. As illustrated below, the impact of these factors has been higher in the case of 'critical' minerals and metals.

I.3 'Critical' minerals and metals and the debate over the use of export restrictions

The intensification in the use of export restrictions on raw materials of mineral origin during a time of macroeconomic downturn has ignited tensions between industrialized countries historically reliant on the massive importation of minerals and metals (Europe and the United States in particular, but also countries such as Japan) and resource-rich countries resorting to mineral export restraints. Such tensions came to a head as the former started claiming that a proliferation of export restrictions was affecting 'critical' minerals and metals.[187] The emergence of a 'criticality' component in the debate over mineral export restraints gave rise to a general sense of urgency and alarm, and substantially contributed to a strengthening of the call to reform the current WTO rules in this domain.[188]

The discourse on export restrictions assumes distinctive contours when restricting the analysis to the segment of 'critical' minerals and metals, reflecting not only the peculiarities of the mineral sector

[186] European Commission, 'Eleventh Report', p. 4; OECD, *Export Restrictions in Raw Materials Trade: Facts, Fallacies and Better Practices* (OECD Publishing, 2014).

[187] Closing Statement of the United States at the Second Substantive Meeting of the Panel with the Parties, *China–Measures Related to the Exportation of Rare Earths, Tungsten and Molybdenum* (DS431), 19 June 2013, p. 1. The European Union and the United States have in particular elaborated a methodology for determining the level of mineral criticality and have accordingly compiled lists of critical raw materials. Interestingly, these lists overlap for the most part. See below, Chapter 1, Section 1.3.3.

[188] See, e.g. B. Karapinar, 'Export Restrictions and the WTO Law: How to Reform the "Regulatory Deficiency"', *Journal of World Trade* 45 (2011), 1139; B. Gu, 'Mineral Export Restraints and Sustainable Development – Are Rare Earths Testing the WTO's Loopholes?', *Journal of International Economic Law* 14 (2012), 765.

considered as a whole, but more specifically a sense of vulnerability on the part of net importing countries regarding the effects of export restrictions. This is because minerals and metals considered to be critical are, on the one hand, geographically concentrated, and net exporting countries are often large countries that enjoy a dominant position in the international market;[189] on the other hand, they are only substitutable to a limited extent, especially in the short term.[190] Combined with the typically lagged response of supply in the mining sector,[191] these factors mean that access to critical mineral resources depends upon unrestricted trade, while major exporting countries can *de facto* exercise control over prices and quantities made available on world markets through the regulation of export flows. The imposition of export restrictions on critical minerals and metals could thus not only condition world market prices but in principle also impair availability of supply. More stringent disciplines have thus been advocated by net importing countries to secure fair access to critical minerals and metals.

At the same time, resource-rich countries have claimed the right to maintain sufficient policy space to use export restrictions for achieving legitimate public policy goals and, in particular, environment-related objectives, directly calling into question the role that international law principles such as the principle of sustainable development should play in the WTO system as well as the degree of openness to environmental values that should be admitted by the WTO Agreement.[192] The main problem here lies in the inherent 'duality' of export restrictions in the mineral sector. Standard economic theory predicts that they may, under certain conditions, contribute to mitigating the negative environmental externalities associated with mining activities by slowing the pace of extraction and, consequently, the rate of depletion of finite mineral resources.[193] However, by lowering the domestic price of the restricted materials, export restrictions may also serve industrial policy goals inasmuch as they 'reserve' low-price supplies for domestic manufacturing

[189] Large countries are conventionally considered price setters for their ability to influence world prices. Export restrictions can be a means for large countries to exercise this control. Piermartini, 'The Role of Export Taxes', p. 3.

[190] Fliess *et al.*, 'Recent Developments', p. 4.

[191] The mining sector is characterized by high capital intensity and long-term payback perspectives, so new capacity expansion needs a time lag of five or even seven to eight years to become operational. See below, Chapter 1, Section 1.3.1.

[192] For more on these aspects see below, Chapter 6.

[193] For the economic implications of export restrictions see below, Chapter 3, Section 3.2.

sectors, to the detriment of foreign competitors.[194] In the case of critical minerals and metals, the incentive to use export restrictions for industrial purposes may be high, being essential materials for high value-added manufacturing such as hi-tech sectors and environmentally friendly technologies.[195]

Export restrictions fulfilling industrial policy goals have been strenuously opposed by affected countries as beggar-thy-neighbour instruments distorting trade and competition.[196] Yet, developing countries have strongly defended such instruments as legitimate developmental tools. This line of argument has been supported by emerging economies and low-income CDDCs alike. Its underlying premise is that export restrictions do not just aim for industrial promotion as a goal in and of itself, but are a means to move from a model of resource-led development – potentially leading to a resource curse scenario or the 'Dutch disease'[197] – to a model of economic development based on export diversification and industrial base growth.[198] This development link has especially been invoked to defend export restrictions applied to critical minerals and metals whose exploitation would trigger the expansion of key manufacturing sectors and rapidly expanding industries, thus facilitating a faster and more solid transition towards a path of strong economic development.

The insistence on a development link also explains why the discourse on mineral export restrictions has, finally, reinvigorated the debate over the rights and duties enshrined in the principle of sovereignty over natural resources and the role that such a principle should play in the WTO legal framework. Emerging economies have in particular resorted to this principle to defend export restrictions imposed on critical minerals and metals whose alleged environmental purpose was being questioned. Accordingly, they have contended they have the sovereign right to restrict the exportation of national mineral resources to achieve sustainable economic development as recognized under international law.[199]

In light of all the foregoing, focusing on the narrow segment of critical minerals and metals allows the most quintessential elements of the debate over (mineral) export restrictions, as they have emerged within the

194 See, e.g. Fung and Korinek, 'Economics of Export Restrictions', pp. 15-20; AB Reports, *China-Rare Earths*, para. 5.156.

195 Peeling *et al.*, 'Increasing Demand', p. 158.

196 See below, Chapter 7, Section 7.4.

197 See below Chapter 3, Section 3.3.1.

198 For an analysis of the economic literature documenting a negative correlation between high growth rates and resource-led development see ibid.

199 See below, Chapter 6, Section 6.3.2 and Chapter 7, Section 7.3.1.

multilateral trading system, to be addressed, while still permitting the broader systemic issues raised by the arguments in favour of export restrictions as sustainable economic development tools to be tackled.

I.4 Goals, method and structure of this study

The resurgence of export restrictions on primary commodities was not anticipated in all its breadth by the multilateral trading system. By the time the phenomenon started to be closely monitored, the first food crisis had already occurred and, at the dawn of the greatest financial and economic crisis since the 1930s, export restrictions were already about to peak.[200] Under the pressure of unprecedented challenges shaking the solidity of the multilateral trading system in an era of 'crisis of multilateralism',[201] export restrictions became the archetypal symbol of the paucity of existing WTO rules in a changed world where access to supplies proved as vital as market access but much more uncertain. Accordingly, many have started denouncing the proliferation of export restrictions and connecting it to a fundamental problem of under-regulation in the WTO system.[202] Gradually, a conventional wisdom has arisen that relevant WTO disciplines do not provide for an effective, sufficient and credible legal framework capable of preventing abuses in the use of export restrictive measures.[203]

The aim of this study is to test this conventional wisdom. In particular, this study will investigate the nature and intensity of the relationship between the proliferation of export restrictions and the alleged deficiency of WTO disciplines on the export side arising in the narrow segment of critical minerals and metals. Because of the strategic importance of such materials, the greatest amount of tension over the use of export

[200] See Chapter 2, Section 2.3.

[201] During the last years of his mandate, the former Director General of the World Trade Organization, Pascal Lamy, expressed concerns about the future of multilateralism on several occasions, including the Ninth Annual Global Conference on Commodities Finance with the speech on the topic of 'Global Commodities Trade: Perspective of the WTO', www.wto.org/english/news_e/sppl_e/sppl236_e.htm (accessed November 2014), and the inaugural speech given at the World Trade Forum 2012, 'Multilateralism is Struggling', www.wto.org/english/news_e/sppl_e/sppl244_e.htm (accessed November 2014).

[202] General references to the paucity of WTO rules on the export side and the proliferation of export restrictions are mentioned in many WTO textbooks. See, e.g. Van den Bossche and Zdouc, *The Law and Policy*, p. 470.

[203] The expression 'conventional wisdom' with reference to a presumed insufficiency of existing WTO rules on export restrictions is borrowed from Howse and Josling, 'Agricultural Export Restrictions', p. 10.

restrictions has in fact concerned this specific category of products. However, a certain degree of rhetoric has influenced the debate, so that the 'criticality' argument has at times become more of a label to urge for the reinforcement of existing WTO rules rather than a specific criterion upon which to assess the actual implications of current disciplines on the use of mineral export restraints. As the discourse was triggered by general alarm over access to critical mineral resources, the substantial weakness of current WTO rules on export restrictions to effectively counteract the proliferation of such measures and their effects has yet to be substantiated.[204] Moreover, the central premise on the proliferation of mineral export restraints has been predicated on an incomplete and often biased understanding of the phenomenon.

This study aims to test the adequacy of existing WTO disciplines to specifically address export restrictions on critical minerals and metals. A comprehensive overview of the export restrictions applied to such materials is first provided, with the aim of identifying any distinctive features in the proliferation of such measures. The legal analysis of existing WTO rules – scope, gaps and inconsistencies – draws on this empirical exercise. Various solutions to bring more coherence to existing WTO rules will also be suggested based on the main findings of this two-step analysis.

The issues covered by this study can be broken down into more specific questions, including the following: are export restrictions on critical minerals and metals proliferating, and, if so, do they exhibit any peculiar features? Do they escape existing WTO disciplines, and, if so, are the regulatory gaps greater than in other sectors? What are the implications of these gaps for trade in critical minerals and metals? What solutions, if any, could allow for a greater reach of existing WTO disciplines? Would incremental changes and/or additions to already existing disciplines suffice or are newly drafted disciplines on export restrictions to be preferred? If so, should they be sectoral?

In order to respond to these questions, the study starts by mapping export restrictions on critical minerals and metals (Part I). Chapter 1 addresses definitional issues. It identifies relevant criteria for determining the criticality of minerals and metals within the framework of the major transformations occurring in the international minerals and metals markets, followed by an illustration of the relevant trends and emerging

[204] Howse and Josling have provided an excellent analysis of the meaning and implications of existing WTO rules for agricultural export restrictions, going beyond the general paradigm which would normally consider them 'meaningless'. Ibid., pp. 10–21.

patterns in the distribution, production, utilization and trade of critical minerals and metals. Chapter 2 gives an overview of the different types of export restrictions instituted and/or maintained on critical minerals and metals since the mid-2000s. For each category of export restrictive measures, an account is given of the macro-trends emerging in the use of such measures. This general picture is then complemented with a detailed analysis of the export restrictions specifically affecting critical minerals and metals. The qualification of the proliferation of export restrictions in this sector is finally completed with an analysis of the economic effects of these measures and a review of the public policy goals attached to them (Chapter 3). The aim of this exercise is to understand the magnitude of the phenomenon of export restrictions on critical minerals and metals in light of the coverage of the different types of measures adopted by each of the actors involved, their economic implications, and their rationale.

This is followed by an examination of the existing WTO disciplines on export restrictions relevant to critical minerals and metals (Part II). For categorization purposes, the analysis will distinguish between the rules applicable to export duties or tariffs arising out of the GATT and a number of post-1994 WTO accession protocols (Chapter 4), and the GATT provisions applicable to different types of quantitative restrictions on exports (Chapter 5).[205] Chapter 6 will examine the set of relevant exceptions available in the GATT, elaborating on the issue of the policy space left to WTO Members to resort to export restrictions and the implications of the current configuration of existing WTO disciplines in light of the most recent case law. The analyses inevitably link to the main findings of Part I, and lead to some preliminary conclusions on the adequacy of existing WTO rules for regulating the use of export restrictions on critical minerals and metals as a segment of the category of non-agricultural products. The key finding is that, while the proliferation of export restrictions on minerals and metals cannot be directly attributed to a regulatory deficiency in the WTO legal system *per se*, its peculiar features have been shaped by the lack of consistency in WTO obligations on the export side. The most evident shortcoming in this regard is the absence of a system-wide discipline on export duties. The fragmentation of WTO rules on export taxes carries implications of systemic relevance for the WTO system, ranging from the existence of substantial asymmetries in the scope and coverage of the

[205] This categorization reflects the distinction resulting from Article XI:1 GATT. See below, Chapter 4.

obligations assumed by different original and newly acceded WTO Members to the varying degrees of (in)flexibility enjoyed by such Members in the use of export duties. In some notable cases, including the case of China, the rigidity of commitments intervenes irrespective of whether export duties address legitimate public policy objectives, and could lead to incentives towards the use of more trade-restrictive types of export restrictions. Another element of inconsistency is the asymmetrical treatment of import and export quantitative restrictions in the GATT. Relevant economic development and emergency exceptions applicable to Article XI:1-inconsistent import restrictions are unavailable on the export side, thus discouraging engagement in the negotiation of export tariff bindings to be incorporated in GATT schedules – yet, the only form of protection explicitly admitted in the GATT.

Based on these findings, the study will finally attempt to identify regulatory prospects for reforming WTO rules on export restrictions (Part III). Chapter 7 will give an account of the interplay between the conflicting interests that continue to oppose net importing WTO Members and resource-endowed developing country Members over the use of export restrictions on industrial (mineral) raw materials. An understanding of the issues at stake will serve as a foundation to critically analyse the proposals tabled during the course of the Doha Round in the Non-agricultural Market Access (NAMA) Negotiating Group (Chapter 8). Finally, the conceptual framework will inform the discussion on the ideal scope and targets of a reform of WTO rules on the export side. Chapter 9 will thus attempt to identify the ideal constitutive elements and the potentially available implementing options for designing more coherent and equitable WTO disciplines on export restrictions.

PART I

Minerals and metals critical to the world economy and the recent proliferation of export restrictions

1

Evolving world trade patterns in critical minerals and metals

1.1 Introduction

This chapter sets the stage for the empirical investigation conducted in Part I of this study, elaborating on the notion of 'critical' minerals and metals. It first gives a definition of mineral raw materials and identifies the mineral products included in the definition (Section 1.2). It then discusses the relevant criteria for determining the 'criticality' of minerals and metals (Section 1.3) and accordingly gives an account of the recent trends in the distribution, production and utilization of such raw materials (Section 1.4). This is followed by an analysis of emerging patterns of international trade in critical minerals and metals in light of the more general dynamics affecting the international minerals and metals markets (Section 1.5). The main findings indicate that critical minerals and metals present distinctive features that expose them to specific supply risk factors. In particular, due to heavy geographical concentration, these materials may be subject to trade distortions potentially disrupting the availability of supply and contributing to the acceleration of the ongoing transformations affecting the international minerals and metals markets.[1]

1.2 Defining mineral raw materials

1.2.1 A definition of raw materials

When addressing the issue of export restrictions, the existing literature seems to use expressions such as 'raw materials', 'natural resources', 'primary products' and 'primary commodities' interchangeably.[2] The

[1] US National Research Council of the National Academies, *Minerals, Critical Minerals and the US Economy* (The National Academies Press, 2008), www.nma.org/pdf/101606_nrc_study.pdf (accessed December 2014), p. 22.

[2] The WTO Secretariat recognizes in particular that 'most people have an intuitive idea of what natural resources are, but "common sense" definitions cannot be relied upon since

main reason for this is that export restrictions are widely conceived and understood as measures generally intervening at the first stages of the value chain, while not affecting higher value-added products whose competitiveness would otherwise be hampered.[3] Yet, exploring the meaning of these notions remains fundamental in order to understand some basic features of the materials targeted in this study.

The Glossary of Statistical Terms elaborated by the Organisation for Economic Cooperation and Development (OECD) contains a definition of 'natural resources' which overlaps with the concept of 'raw materials'. According to this definition, natural resources are 'natural assets (raw materials) occurring in nature that can be used for economic production or consumption'.[4] When investigating the context relevant to the definition, the Glossary specifies that 'natural resources' should be intended as:

> the naturally occurring assets that provide use benefits through the provision of raw materials and energy used in economic activity (or that may provide such benefits one day) and that are subject primarily to quantitative depletion through human use. They are subdivided into four categories: mineral and energy resources, soil resources, water resources and biological resources.

It follows that 'raw materials' are finite assets directly obtainable from the mineral, energy, soil, water and biological resources naturally available on earth and that are, or may be, economically valuable. They therefore cannot exist without natural resources.

they eventually run into problems when dealing with ambiguous cases'. World Trade Organization (WTO), 'World Trade Report 2010: Trade in Natural Resources', www.wto.org/english/res_e/booksp_e/anrep_e/world_trade_report10_e.pdf (accessed December 2014), p. 42.

3 Mitra and Josling explain the reasons why export restrictions are normally applied on unprocessed products: 'The rationale for restricting exports ... is straightforward: by exporting ... products in their raw form, a country foregoes the processing margin, which then accrues instead to the importing countries. By withholding raw products from the international market, a country can augment its production and exports of processed products and increase income generated from exports'. S. Mitra and T. Josling, 'Agricultural Export Restrictions: Welfare Implications and Trade Disciplines', International Food and Agricultural Trade Policy Council (IPC) Position Paper, Agricultural and Rural Development Policy Series, 2009, p. 4. This picture is of course reversed in the case of export-restraint arrangements: these measures have affected higher value-added products but precisely to serve the interests of the importing countries rather than those of the exporting country imposing the 'voluntary' restriction. See above, Introduction, Section I.1.

4 OECD, *Glossary of Statistical Terms* (OECD Publishing, 2008), p. 352.

Interestingly, the same criteria are also identified by the World Trade Organization (WTO) Secretariat in the 2010 World Trade Report on Trade in Natural Resources, which espouses the following definition of 'natural resources':

> stocks of materials that exist in the natural environment that are both scarce and economically useful in production and consumption, either in their raw state or after a minimal amount of processing.[5]

Once again, the criterion of economic usefulness is pivotal. Naturally occurring assets do not qualify *per se* as natural resources unless they are needed in economic activity. Scarcity is another fundamental feature that contributes to attaching economic value to natural assets. The World Trade Report also distinguishes between immobile natural resources and internationally tradable natural resources:

> [i]t is important to distinguish between natural resources as factors of production and natural resources as goods that can be traded internationally. For instance, minerals, oil, and various other materials can be extracted and enter into trade. However, other resources may form the economic basis for various sectors of the domestic economy, and therefore are only involved in trade in an indirect way. For example . . ., agricultural land, which is the archetypal 'fixed, immobile' natural resource, can be exported through agricultural commodities grown on that land. Hence, at a fundamental level, natural resources are often a reason for trade rather than tradable goods in their own right.[6]

Matched with the previous definitions, this distinction indirectly clarifies that the concept of raw materials incorporates the notion of natural resources inasmuch as scarce economically valuable assets available in the natural environment can be traded directly or after a minimal amount of processing.

Finally, 'raw materials' are used as inputs in the manufacturing sector.[7] In this respect, 'primary products' and 'commodities' do not necessarily belong to the category of raw materials, since the former generally encompass all homogenous products[8] of the primary sector, which conventionally comprises agriculture (including hunting, forestry and fishing), mining and utilities.[9] Food-related agricultural commodities are thus

[5] WTO, 'World Trade Report 2010', p. 46. [6] Ibid.

[7] M. Radetzki, *A Handbook of Primary Commodities in the Global Economy* (Cambridge University Press, 2010), p. 24.

[8] As homogenous products, commodities can be subject to centralized trading with a unified price. WTO, 'World Trade Report 2010', p. 59.

[9] Radetzki, *A Handbook of Primary Commodities*, pp. 22–3.

conventionally not considered as raw materials, but it is common to distinguish between agricultural raw materials – either of plant (e.g. rubber and cotton) or animal origin (e.g. leather, hides and skins) – and industrial raw materials, which are derived from extracted rather than cultivated resources[10] and mainly include fuels and mining products.[11] This feature further distinguishes the concept of raw materials from that of natural resources, as the former includes those marketable products that have undergone additional processing at the beginning of the value chain. A typical example is the process of metal extraction from metallic ores.[12] Industrial raw materials as here defined, moreover, 'exist in finite quantities, so [that] every unit consumed today reduces the amount available for future consumption'.[13] In other words, they are non-renewable by definition.[14]

In conclusion, raw materials are products obtained from naturally occurring assets that can be traded in their unprocessed or semi-processed forms and are essential to industrial production processes. The economic value of raw materials arises from their economic usefulness in industrial production, their scarcity and, in the case of industrial raw materials, their exhaustibility.

1.2.2 Raw materials of mineral origin

Based on the above definitions, mineral raw materials are industrial raw materials obtained from mineral resources, which are traded internationally in their unprocessed form or after a minimal amount of processing because of their economic value. This definition targets the so-called 'mine output'. Mine output conventionally consists of ores, which are 'a concentration of

[10] Cultivated resources are goods whose production requires human activity, whereas extracted resources exist in the natural environment and can be traded in their natural state or after a minimal amount of processing. WTO, 'World Trade Report 2010', p. 46.

[11] Radetzki, *A Handbook of Primary Commodities*, pp. 22–3; K.C. Fung and J. Korinek, 'Economics of Export Restrictions as Applied to Industrial Raw Materials', OECD Trade Policy Papers, No. 155 (OECD Publishing, 2014), p. 6. More ambivalent is the category of forestry products, which can either be taken from naturally existing stocks or be cultivated. For the purposes of WTO trade statistics, however, forestry products are conventionally classified under the 'Agriculture' heading. WTO, 'World Trade Report 2010', p. 46.

[12] See below, Section 1.2.2.

[13] WTO, 'World Trade Report 2010', p. 47.

[14] The OECD defines non-renewable resources as those 'exhaustible natural resources such as *mineral resources* that cannot be regenerated after exploitation'. OECD, *Glossary of Statistical Terms*, p. 352 (emphasis added). Similarly, the WTO Secretariat posits that they 'do not grow or otherwise renew themselves over time'. WTO, 'World Trade Report 2010', p. 47.

mineralization located either proximate to the surface or underground [which are] liberated from surrounding rock and brought to the surface via energy-intensive and capital-intensive mechanical means',[15] and of 'the output [obtained] from processing at or near the mines'.[16] In the case of metallic ores, for instance, this alludes to the upgrading of ores to concentrates at or near the extraction site.[17] Ores are in fact almost never traded internationally before undergoing at least some processing at or near the mines in order to increase their price per unit weight before being shipped.[18] Exports of mine output (e.g. concentrates obtained from metallic ores) therefore reflect domestic primary mine production.[19]

For the reasons outlined earlier, however, the expression mineral raw materials includes, in its most comprehensive sense, marketable products obtained after the first processing stages[20] and used in the production of metals, alloys and metal compounds. Metals are also conventionally classified as industrial raw materials.[21] Exports of higher value-added

[15] G.A. Davies, 'Trade in Mineral Resources', WTO Economic Research and Statistics Division, Staff Working Paper ERSD-2010-01, www.wto.org/english/res_e/reser_e/ersd201001_e.pdf (accessed December 2014), p. 2. The author observes that '[m]ining is therefore very similar to manufacturing. The main difference is that with manufacturing raw materials are brought to fixed locations where they are combined with labor and capital, while in mining labor and capital are brought to the raw materials and combined at the location of the rock'.

[16] See World Mining Congress, 'World Mining Data Report 2014', www.wmc.org.pl/sites/default/files/WMD2014.pdf (accessed December 2014), p. 2.

[17] Ibid.; Davies, 'Trade in Mineral Resources', p. 2. [18] Ibid.

[19] Mine output data, however, do not necessarily reflect the distribution of mineral concentrations in the Earth's crust, as the geological configuration of mineral resources significantly influences their exploitability. In this respect, official classifications distinguish between (i) 'world resources', defined as 'the concentration of naturally occurring solid, liquid, or gaseous material in or on the Earth's crust in such form and amount that economic extraction of a commodity from the concentration is currently or potentially feasible' (ii) 'identified resources', which are those 'resources whose location, grade, quality and quantity are known or estimated from specific geological evidence' (iii) 'reserve base', which is 'that part of an identified resource that meets specified minimum physical and chemical criteria related to current mining and production practices, including those for grade, quality, thickness, and depth' and, (iv) 'reserves', defined as 'that part of the reserve base which could be economically extracted or produced at the time of determination'. US Geological Survey (USGS), *Mineral Commodity Summaries 2014* (Washington, DC: US Government Printing Office, 2014), 'Appendix C, Part A: Resource/Reserve Classification for Minerals', pp. 191–2.

[20] World Mining Congress, 'World Mining Data Report 2014', p. 2; WTO, 'World Trade Report 2010', p. 46.

[21] See World Mining Congress, 'World Mining Data Report 2014', p. 1; B. Fliess, H. Gou and T. Mard, 'Taking Stock of Measures Restricting the Export of Raw Materials: Analysis of OECD Inventory Data', OECD Trade Policy Working Papers, No. 140 (OECD

raw materials of mineral origin may thus not necessarily occur in the country responsible for primary mine production; yet, sending such materials may be more cost-effective for primary producers at least in those cases where the export destination is a distant location, inasmuch as it avoids the erosion of profit margins by transportation costs.[22] From this perspective, big mine producers may be tempted to restrict the exportation of ores and concentrates to stimulate value-added processing domestically. The same holds true at further stages in the value chain, especially where mineral raw materials are used as inputs in key manufacturing sectors.

1.2.3 A classification of minerals and metals

Mineral raw materials are conventionally classified under various categories: (i) iron and ferro-alloy metals (i.e. iron, chromium, cobalt, manganese, molybdenum, nickel, niobium, tantalum, titanium, tungsten and vanadium); (ii) non-ferrous metals (aluminium, antimony, arsenic, bauxite, bismuth, cadmium, copper, gallium, germanium, lead, lithium, mercury, rhenium, rare earth elements,[23] selenium, tellurium, tin and zinc); (iii) precious metals (gold, platinum-group metals

Publishing, 2012), p. 4; B. Fliess, C. Arriola and P. Liapis, 'Recent Developments in the Use of Export Restrictions in Raw Materials Trade', in OECD, *Export Restrictions in Raw Materials Trade: Facts, Fallacies and Better Practices* (OECD Publishing, 2014), pp. 1–45.

22 Davies, 'Trade in Mineral Resources', p. 3. It should be noted, however, that the sharp abatement of transportation costs has triggered a dramatic expansion of international trade in minerals and metals and importantly contributed to the globalization of primary commodity markets. Radetzki, *A Handbook of Primary Commodities*, pp. 12–15. Significantly, world mineral production grew at an average rate of 2.7 per cent from 1950 to 2003, while in the same period the volume of world exports grew by 4.1 per cent annually and the value of mineral and metal exports rose from $23 billion to $671 billion. P. Maxwell, 'Trade in Minerals', in P. Maxwell (eds.), *Australian Mineral Economics: A Survey of Important Issues* (Melbourne: The Australian Institute of Mining and Metallurgy, 2006), pp. 27–34.

23 Rare earth elements are a group of seventeen elements in the periodic table, namely fifteen elements belonging to the family of lanthanides (lanthanum, praseodymium, neodymium, promethium, samarium, europium, gadolinium, terbium, dysprosium, holmium, erbium, thulium, ytterbium and lutetium) plus scandium and yttrium. Rare earth elements are commonly distinguished between light rare earth elements (LREEs) and heavy rare earth elements (HREEs). This category includes lanthanum, praseodymium, neodymium, promethium and samarium. The group of HREEs comprises europium, gadolinium, terbium, dysprosium, holmium, erbium, thulium, ytterbium, lutetium, and yttrium. Scandium is considered as constituting a category of its own due to its different production and utilization patterns. European Commission, 'Report on Critical Raw Materials for the EU: Report of the Ad-hoc Working Group on Defining Critical Raw

(PGMs)[24] and silver); and (iv) industrial minerals[25] (asbestos, barite, bentonite, boron minerals, diamond, diatomite, feldspar, fluorspar, graphite, gypsum and anhydrite, kaolin, magnesite, perlite, phosphates, potash, salt, sulphur, talc, vermiculite and zircon).[26]

While this classification mostly reflects the economic functions of the various mineral products, two official trade classifications have been used to categorize, among others, exports and imports of minerals and metals: the Standard International Trade Classification (SITC) and the Harmonized System Nomenclature (HS). The SITC was developed by the United Nations (UN) Statistics Division and approved in its fourth revision by the UN Statistical Commission at its thirty-seventh session in March 2006.[27] Originally created for statistical reasons, it classifies goods based on the stage of production and its aim is 'to provide the commodity aggregates required for purposes of economic analysis and to facilitate the international comparison of trade-by-commodity data'.[28] Studies of trade in mineral commodities usually refer to the following SITC aggregates: Section 27 (crude industrial minerals), Section 28 (metalliferous ores and scrap), Section 67 (iron and steel) and Section 68 (non-ferrous metals).[29]

Materials', May 2014, http://ec.europa.eu/enterprise/policies/raw-materials/files/docs/crm-report-on-critical-raw-materials_en.pdf (accessed December 2014), p. 19.

[24] The expression 'platinum group metals' (PGMs) indicates a group of six precious metals with similar chemical and physical properties: platinum, palladium, rhodium, ruthenium, iridium and osmium. European Commission, 'Report on Critical Raw Materials for the EU: Critical Raw Materials Profiles', May 2014, http://ec.europa.eu/enterprise/policies/raw-materials/files/docs/crm-critical-material-profiles_en.pdf (accessed December 2014), p. 123.

[25] The category of industrial minerals is defined as 'mineral[s], which may be used in an industrial process directly due to [their] chemical/physical properties'. European Commission, 'Report on Critical Raw Materials for the EU', p. 17. This feature distinguishes industrial minerals from other categories of mineral raw materials, which are derived from metallic ores.

[26] See World Mining Congress, 'World Mining Data Report 2014', p. 1. Another category is that of mineral fuels, which includes steam coal (including anthracite and sub-bituminous coal), coking coal, lignite, natural gas, crude petroleum, oil sands, oil shales and uranium. Mineral fuels, however, are not mineral resources in the sense of this study, but energy resources. See above, Section 1.2.1.

[27] UN, 'Standard International Trade Classification, Revision 4', http://unstats.un.org/unsd/trade/sitcrev4.htm (accessed December 2014).

[28] OECD, *Glossary of Statistical Terms*, p. 507.

[29] See, e.g. Radetzki, *A Handbook of Primary Commodities*, p. 24. According to Davies, Sections 66 (non-metallic mineral manufactures), 96 (coin, other than gold coin), and 97 (non-monetary gold, excluding ores and concentrates) should also be included in minerals and metals aggregates. See Davies, 'Trade in Mineral Resources', p. 3.

The SITC categories are reorganized in the HS in order to provide further detail on commodity trade flows. The Harmonized System is an international nomenclature governed by the Convention on the Harmonized Commodity Description and Coding System, elaborated under the auspices of the World Customs Organization (WCO) and regularly reviewed.[30] Based on the nature of the commodities, the HS provides very precise information through a breakdown of individual product categories into approximately 5,000 commodity groups identified by a six-digit code.[31] Mineral raw materials appear in various subheadings of Section V on 'Mineral Products', Chapters 25 (salt, sulphur, earths and stone; plastering materials, lime and cement), and 26 (ores, slag and ash); Section VI on 'Products of the Chemical or Allied Industries', Chapter 28 (inorganic chemicals; organic or inorganic compounds of precious metals, of rare earth metals, of radioactive elements or of isotopes);[32] and various chapters (from 72 to 81) under Section XV on 'Base Metals and Articles of Base Metal'.[33]

Both the SITC and the HS include waste and scrap metals in the relevant subheadings regrouping mineral raw materials. These are mineral raw materials of secondary origin, which result from metal production or recycling. Similarly to primary materials, they are used as inputs in the manufacturing sector and are thus economically valuable.[34]

1.3 Identifying 'critical' minerals and metals

1.3.1 Demand and supply determinants for mineral raw materials

Minerals and metals account for a relatively minor share of the global goods trade.[35] They are generally used in small quantities in the

[30] There are currently 138 Contracting Parties to this Convention, and more than 200 administrations worldwide make use of the HS, including the WTO. HS 6-digit categories are often further detailed through the use of 8-digit or 10-digit level subheadings by national governments and/or institutions using this classification. WCO, 'HS Nomenclature 2012 Edition', www.wcoomd.org/en/topics/nomenclature/instrument-and-tools/hs_nomenclature_2012/hs_nomenclature_table_2012.aspx (accessed December 2014).

[31] OECD, *Glossary of Statistical Terms*, p. 241; WCO, 'HS Nomenclature 2012 Edition'.

[32] This chapter includes rare earth elements (subheading 28.05), different by-products of the metal extraction process (e.g. metal oxides) and various industrial minerals. WCO, 'HS Nomenclature 2012 Edition'.

[33] Ibid.

[34] The economic importance of secondary supply increases as metals become scarcer. International trade in secondary raw materials is indeed on the rise. Davies, 'Trade in Mineral Resources', p. 3.

[35] Radetzki, *A Handbook of Primary Commodities*, p. 27.

production of manufactured goods, and their value is declining compared with the value of technically sophisticated products.[36] Yet, they remain essential to the manufacturing sector.[37] Demand for mineral raw materials is comparatively higher in emerging economies, which are at a much more commodity-intensive stage of development than highly industrialized economies, already at a mature stage of economic growth.[38]

Various demand and supply factors may hamper the availability of mineral raw materials. Some are inherent to the typical functioning of the mining industry and are thus common to all mineral products. These include exhaustibility and low short-term supply and demand elasticity. Exhaustibility derives from the fact that mineral resources are fixed endowments that cannot be regenerated after exploitation. As with any non-renewable resource, the exploitation of mining products inevitably brings their point of depletion closer.[39] Sudden and persistent shifts in demand may therefore expose mineral resources to unsustainable rates of extraction and eventually lead to their exhaustion.[40]

Physical availability of mineral raw materials may also be challenged as a result of a lack of investment in the mining sector, which occurs quite frequently at times of low mineral prices due to the high risks entailed in capital-intensive and long-term payback extraction activities. Under

[36] As Radetzki notes, 'the path of economic development can be seen as a process of dematerialization [so that] the higher the value, the less primary material inputs will be needed per dollar value represented by the items. The essence of economic development is to move the center of the economy's gravity down the list, towards goods with ever higher value per kilogram.' Ibid., p. 10.

[37] J. Korinek and J. Kim, 'Export Restrictions on Strategic Raw Materials and their Impact on Trade and Global Supply', in OECD, *The Economic Impact of Export Restrictions on Raw Materials* (OECD Publishing, 2010), p. 104.

[38] Radetzki, *A Handbook of Primary Commodities*, pp. 37–43.

[39] New discoveries and technological improvements have significantly upgraded the prospects of availability of finite mineral resources. European Commission, 'Critical Raw Materials for the EU: Report of the Ad-hoc Working Group on Defining Critical Raw Materials', 30 July 2010, http://ec.europa.eu/enterprise/policies/raw-materials/files/docs/report-b_en.pdf (accessed August 2014), pp. 14–21. A comprehensive account of how societies have reacted to resource scarcity throughout history is given by E. B. Barbier, *Scarcity and Frontiers: How Economies Have Developed Through Natural Resource Exploitation* (Cambridge University Press, 2011). The author notes that societies have kept growing, notwithstanding resource scarcity, thanks to the so-called expansion of the frontier, where the frontier is defined as 'a geographic region adjacent to the unsettled portions of the continent in which a low man-land ratio and unusually abundant, unexploited natural resources provide an exceptional opportunity for social and economic betterment to small-propertied individuals'. Ibid., p. 10.

[40] WTO, 'World Trade Report 2010', p. 47.

such circumstances, any shift in demand occurring at times of macroeconomic expansion may put pressure on existing levels of spare capacity,[41] and lead to booming mineral prices.[42] This is because in the short to medium term it might be difficult to replace materials experiencing high price volatility without compromising the quality or the performances of a good. Even if mining companies decide to increase investment, furthermore, actual capacity expansion requires at least five years and sometimes even decades to become operational[43] due to decision and implementation lags.[44] In other words, demand and, in particular, supply react quite slowly to changes in mineral prices.[45]

The effects of such lagged responses became evident in the mid-2000s, when the general demand shock triggered by the high growth performance of developing Asia generated a substantial increase in the prices of minerals and metals, and accelerated a process of deep transformation of traditional trade patterns in the international markets.[46] In particular, dominant players historically reliant on the importation of mineral raw materials for their top-ranking industries started to suffer from increased competition over access to mineral supplies exposed to particularly severe demand shocks.[47] As a response to the sudden exposure to supply disruptions, some industrialized economies have developed reference

[41] G. Peeling, P. Stothart and B. Toms, 'Increasing Demand for and Restricted Supply of Raw Materials', in OECD, *The Economic Impact of Export Restrictions on Raw Materials* (OECD Publishing, 2010), pp. 156–7.

[42] US National Research Council of the National Academies, *Minerals, Critical Minerals*, p. 136.

[43] Korinek and Kim, 'Export Restrictions on Strategic Raw Materials', p. 104.

[44] Once a mining company decides to invest in the expansion of production capacity as a result of persistent price increases, it is confronted with a growing number of regulatory constraints that in most cases delay the implementation of the project. More precisely, 'regulatory requirements facing new projects in both developed and developing countries have generally become more burdensome as public values such as environmental protection and the need for post mining reclamation have captured certain externalities and turned them into development costs while also driving environmental assessment processes to be more inclusive of social issues'. Peeling *et al.*, 'Increasing Demand', p. 159.

[45] The Ad-hoc Working Group on Defining Critical Raw Materials recalls that, even in conditions of booming prices, 'mining companies normally only invest what they require for their short-term needs to prove reserves and thus to justify commercial investment decisions over a period of, say, 20 years. They don't necessarily aim at proving the full ore body.' European Commission, 'Critical Raw Materials for the EU', p. 16.

[46] See Introduction, Section I.2.2.

[47] M. Farooki and R. Kaplinsky, *The Impact of China on Global Commodities Prices: The Global Reshaping of the Resource Sector* (Routledge Studies in Global Competition, 2012), pp. 117–39.

frameworks to identify *ex ante* specific demand and supply factors likely to put the availability of supply at particular risk.

1.3.2 *Specific factors of mineral criticality*

In the past few years, industrialized players have repeatedly stressed the need to identify critical minerals and metals in view of elaborating medium- to long-term strategies aimed at mitigating their heavy dependence upon the importation of strategically important mineral raw materials.[48] These efforts have led to the adoption of a number of expert reports elaborating standard methodologies for determining mineral criticality. The precursor in this respect was the United States. In 2008, a committee of experts formed under the auspices of the US National Research Council of the National Academies released a report which developed a 'criticality matrix' and identified a group of critical raw materials for the US economy.[49] In a similar fashion, the European Commission created an Ad-hoc Working Group on defining critical raw materials in the context of the so-called Raw Materials Initiative.[50] The Ad-hoc Working Group had the mandate to devise a methodology for determining the criticality of raw materials and accordingly establish a list of critical raw materials for the European Union. To this end, it released a first report in 2010[51] and then updated the list of materials in a second report which appeared in 2014 and was then approved by the European Commission.[52] Reflecting the diffusion of concerns over potential shortages of supply among industrialized countries,[53] the OECD also identified a selected number of minerals and

48 See Peeling *et al.*, 'Increasing Demand', p. 169.

49 US National Research Council of the National Academies, *Minerals, Critical Minerals.*

50 The Raw Materials Initiative, launched in 2008, is an integrated policy based on three pillars: (i) reducing the EU's consumption of primary raw materials; (ii) increasing the availability of raw materials sourced within the EU; (iii) promoting the sustainable supply of raw materials from third countries. See Communication from the Commission to the European Council and Parliament, 'The Raw Materials Initiative – Meeting our Critical Needs for Growth and Jobs in Europe', COM (2008) 699, http://eur-lex.europa.eu/LexUriServ/LexUriServ.do?uri=COM:2008:0699:FIN:en:PDF (accessed June 2013). Follow-ups include the Communications 'Tackling the Challenges in Commodity Markets and on Raw Materials', COM (2011) 0025, and 'On the Implementation of the Raw Materials Initiative', COM (2013) 0442.

51 European Commission, 'Critical Raw Materials for the EU'.

52 European Commission, 'Report on Critical Raw Materials for the EU'.

53 These concerns led to a number of other initiatives and strategies in countries such as Japan, Australia and Canada. See European Commission, 'Report on Critical Raw Materials for the EU', pp. 11–12.

metals which were defined as 'strategic' for their 'particular interest' on the basis of various demand and supply factors.[54] Significantly, all these initiatives consider a combination of similar demand and supply risk factors leading to mineral criticality.

On the demand side, the key guiding criterion is the economic importance or importance in use, albeit measured using different indicators. The Ad-hoc Working Group of the European Commission, for instance, adopted a sophisticated methodology which consisted in calculating the contribution of each mineral raw material to seventeen different manufacturing mega-sectors of the EU economy.[55] The US Committee elaborated a similar notion of economic importance, whereby the economic value of the US consumption of each identified mineral for a particular end use was matched with an estimation of the importance of growth in emerging uses that could affect the existing production capacity in the short term.[56] Moreover, in the US methodology, the economic importance factor incorporates an additional indicator, the substitutability rate, which is defined as 'the percentage of US consumption in existing uses for which substitution is difficult or impossible'.[57] In both methodologies, however, the higher the demand for raw materials, the higher the economic importance component grows, indicating that any supply disruption would have a greater impact on the economy.[58] Among the most powerful forces influencing the economic importance factor is technological change, as the diffusion of new technologies and/or new applications for known technologies can trigger an expansion in the demand

[54] Korinek and Kim, 'Export Restrictions on Strategic Raw Materials', p. 104. For the purposes of this study, the expressions 'strategic' and 'critical' will be used interchangeably.

[55] The Ad-hoc Working Group explains that 'a mega-sector represents the value-added chain which will be affected by a shortage of the material upstream'. European Commission, 'Critical Raw Materials for the EU', p. 54. Hence, it clarifies that '[c]onceptually the mega-sectors are thus defined in order to aggregate all sectors or sub-sectors belonging to the same value chain. As raw materials go into different value chains of heterogeneous economic importance to the EU economy we can evaluate economic importance based on the raw materials' contribution to different mega-sectors.' Ibid., p. 55.

[56] US National Research Council of the National Academies, *Minerals, Critical Minerals*, p. 77.

[57] In the words of the US Committee: 'some minerals are more essential than others in the sense that they have few if any substitutes capable of providing similar functionality at similar costs. The greater the difficulty, expense, or time it takes for material substitution to occur, the more critical a mineral is to a specific application or product.' Ibid., p. 134.

[58] Ibid., p. 132; European Commission, 'Critical Raw Materials for the EU', p. 42.

for certain raw materials.[59] This factor is reflected in the OECD list of strategic raw materials, which includes those minerals and metals used as inputs into high-technology sectors, including emerging environmentally friendly technologies and strategically important industrial sectors.[60]

On the supply side, mineral criticality is assessed against the so-called supply risk factor. According to the US Committee, the supply risk factor is determined by a combination of four different indicators: geological availability, technological availability, social and environmental availability and political availability.[61] The first two components depend upon the level and location of mine exploration and production, while the latter two determinants reflect, on the one hand, the environmental spillovers of the mining sector and the impact of mines on local communities and, on the other hand, the influence of national political decisions on the availability of mineral supply.[62] The EU Ad-hoc Working Group calculated the supply risk factor based on the following indicators: the concentration of primary production in countries with poor governance, substitutability prospects and end-of-life recycling rates.[63] The first determinant, which combines the geological and the political availability components identified by the US Committee, has a predominant bearing on the final calculation of the level of supply risk. The latter two indicators are considered as risk-reducing filters.[64] Both studies, furthermore, adopted a short term and medium term perspective, thus considering the level of production concentration rather than the distribution of geological reserve bases.[65] This holds true for the OECD analysis of strategic raw materials. The Organization selected a number of minerals and metals using the geographical concentration of supply, along with the low substitutability rate in the short term, as additional guiding criteria on the supply side.[66]

[59] Ibid., p. 39; US National Research Council of the National Academies, *Minerals, Critical Minerals*, p. 78.

[60] Korinek and Kim, 'Export Restrictions on Strategic Raw Materials', p. 106.

[61] US National Research Council of the National Academies, *Minerals, Critical Minerals*, pp. 135–7.

[62] Ibid., p. 137.

[63] European Commission, 'Report on Critical Raw Materials for the EU', pp. 21–3.

[64] Ibid., p. 22. The end-of-life recycling rate indicator was introduced by the EU Ad-hoc Working Group as part of the supply risk assessment in the 2014 Report, in substitution for the Environmental Performance Index (EPI) used in the 2010 Report. The expert group identified several inconsistencies between 'the environmental performance in the mining sector of relevant producing countries and the EPI indicator values'. Ibid., p. 24.

[65] Ibid., p. 21; US National Research Council of the National Academies, *Minerals, Critical Minerals*, p. 136.

[66] Korinek and Kim, 'Export Restrictions on Strategic Raw Materials', p. 106.

In conclusion, critical minerals and metals share a number of defining characteristics which, in the short- to medium-run, render them more exposed to risk of supply shortages and increase the likelihood that any supply disruptions (caused by either a demand or supply shock) will severely impact on the economy. These include essential economic importance to industry, geographical concentration of primary production, and low substitutability prospects and recycling rates.

1.3.3 *A list of critical minerals and metals*

The criticality methodology developed by the expert groups instigated by the United States and the European Union led to the identification of a restricted number of critical raw materials of mineral origin. In the case of the United States, the committee appointed by the US National Research Council selected eleven critical minerals or mineral groups: copper, gallium, indium, lithium, manganese, niobium, PGMs, rare earth elements, tantalum, titanium and vanadium.[67] The committee did not intend the list to be exhaustive, but specified that those minerals 'illustrate[d] the range of circumstances that the matrix methodology accommodates and considers', and therefore 'would likely be included in a more comprehensive assessment of all potentially critical minerals'.[68] Such comprehensive assessment, however, had not been conducted as at December 2014. Rather, the matrix developed by the US committee has continued to be selectively applied and, in particular, it has been used by the US Department of Energy (DoE) to assess the level of criticality of those materials deemed necessary for clean energy technologies (e.g. wind turbines, electric vehicles, photovoltaic cells and fluorescent lighting).[69] According to the DoE report, five rare earth elements (dysprosium, neodymium, terbium, europium and yttrium) were assessed as most critical in 2010 in the short (i.e. 0–5 years) to medium

[67] US National Research Council of the National Academies, *Minerals, Critical Minerals*, p. 132.

[68] Ibid. In other words, the committee considered those minerals which exhibited particularly high values of the indicated factors of criticality: '[f]or example, the committee considered minerals used in large quantities throughout the economy in traditional applications and others used in limited quantities in a small number of (often emerging) applications, minerals produced largely as byproducts, and other minerals for which recycling of scrap is an important source of supply'.

[69] US Department of Energy, 'Critical Materials Strategy', 2010, http://energy.gov/sites/prod/files/edg/news/documents/criticalmaterialsstrategy.pdf (accessed December 2014).

term (i.e. 5–15 years);[70] depending on the timeframe (either short term or medium term), cerium and lanthanum (two other rare earth elements), as well as indium, lithium and tellurium, were considered to be critical or near-critical.[71] This was substantially confirmed in 2011, when the DoE released an updated version of its report which maintained that dysprosium, neodymium, terbium, europium and yttrium would continue to be critical in both the short and medium term.[72]

The Ad-hoc Working Group of the European Commission identified a group of fourteen critical minerals and metals: antimony, beryllium, cobalt, fluorspar, gallium, germanium, graphite, indium, magnesium, niobium, platinum group metals, rare earth elements, tantalum and tungsten.[73] They revised their list in 2014 to include six new raw materials: borates, chromium, coking coal, magnesite, phosphate rock and silicon metal (silicium).[74] The category of rare earth elements was moreover subdivided into heavy rare earth elements (HREE) and light rare earth elements (LREE) to better reflect existing differences in supply and demand factors.[75] Tantalum was the only product excluded from the 2014 list, compared with 2010, as a result of the diversification of primary production sources.[76]

Significantly, the two lists not only contain overlapping elements, but those are all included on the list of twenty minerals and metals covered by the OECD study on strategic raw materials,[77] together with the majority of materials identified by the Unites States or the European Union only.[78] Moreover, this list adds a number of other materials (in particular, manganese, molybdenum and nickel), reflecting the more diverse and composite needs of industrialized economies considered as a whole.[79]

70 Ibid., p. 8. 71 Ibid.

72 US Department of Energy, 'Critical Materials Strategy', 2011, http://energy.gov/sites/prod/files/DOE_CMS2011_FINAL_Full.pdf (accessed December 2014), p. 4. However, the DoE estimated that in the medium term only lithium and tellurium will remain near-critical, while indium, cerium and lanthanum will only be so in the short term.

73 European Commission, 'Critical Raw Materials for the EU', p. 6.

74 European Commission, 'Report on Critical Raw Materials for the EU', pp. 15–6.

75 Ibid., p. 17. The analysis of the criticality of scandium was conducted separately from the two above-mentioned categories and, as a result, scandium does not appear among the critical raw materials for the European Union. Ibid., p. 24.

76 Ibid., p. 25.

77 The only exception is niobium (columbium), which is, however, partially incorporated through the examination of tantalum. The latter, in fact, often occurs with niobium. European Commission, 'Report on Critical Raw Materials for the EU', p. 188.

78 The only element missing is tellurium, which was identified by the US DoE as near-critical. US Department of Energy, 'Critical Materials Strategy', 2011, p. 4.

79 The OECD list contains, for instance, all the materials identified by the Japanese 'Strategy for Ensuring Stable Supplies of Rare Metals' elaborated by the Ministry of Economy,

Finally, the critical mineral raw materials are predominantly metallic minerals. Most of them fall into the category of iron and ferro-alloy metals (chromium, cobalt, manganese, molybdenum, nickel, niobium, rhenium, silicon metal, tantalum, titanium, tungsten and vanadium); a significant number belong to the category of non-ferrous metals (antimony, beryllium, copper, gallium, germanium, indium, lithium, rare earth elements (HREEs and LREEs) and tellurium); and a few of them are precious metals (PGMs and silver). The EU Ad-hoc Working Group moreover identified a number of raw materials belonging to the category of industrial minerals: borates, fluorspar, natural graphite, magnesite, magnesium and phosphate rock. These thirty mineral raw materials constitute the nucleus of critical minerals and metals covered for the purposes of this study.[80]

1.4 Distinctive features of critical minerals and metals

The selected critical minerals and metals present distinctive features of particular interest for this study. First, these materials are essential to industry and, in particular, in key sectors such as construction, metallurgy, electronics and telecommunication, equipment manufacturing, transportation (including automotive, aeronautics, train and shipbuilding), chemicals, plastic and glass.[81] These sectors cover a predominant share of the total value added for the manufacturing sector.[82] Importantly, the economic importance of critical minerals and metals is influenced by their irreplaceable role in emerging technologies. These

Trade and Industry in 2009 (i.e. cobalt, nickel, manganese, molybdenum, rare earth elements, tungsten and vanadium). US Department of Energy, 'Critical Materials Strategy', 2010, pp. 61–3.

80 The only material excluded from the scope of this study is coking coal, which was considered critical in the 2014 Report of the Ad-hoc Working Group of the European Commission. European Commission, 'Report on Critical Raw Materials for the EU', pp. 15–6. Coking coal is in fact classified as a mineral fuel, thus belonging to the category of energy resources under the approach espoused in this study. See above, Section 1.2.1.

81 An overview of the main applications of critical minerals and metals is given by Korinek and Kim, 'Export Restrictions on Strategic Raw Materials', p. 106. More detailed profiles covering the main end-use markets of each material were prepared by the European Commission. See European Commission, 'Critical Raw Materials Profiles', and European Commission, 'Report on Critical Raw Materials for the EU: Non-Critical Raw Materials Profiles', May 2014, http://ec.europa.eu/enterprise/policies/raw-materials/critical/index_en.htm (accessed December 2014).

82 For instance, the seventeen mega-sectors identified by the EU Ad-hoc Working Group cover approximately 90 per cent of the total value added for the EU's manufacturing sector. European Commission, 'Critical Raw Materials for the EU', p. 58.

include new high-tech applications in information and communication technology, automotive and aerospace engineering, electrical equipment manufacturing, optical technologies and in the metallurgical and chemical industries.[83] For instance, rare earth elements, indium and tantalum are used in portable computers, plasma screens, handheld electronic devices and in military applications; gallium is used in integrated circuits in wireless communication, smartphones, defence and aerospace applications, as well as in the production of light-emitting diodes (LEDs) and laser diodes; copper and silver are used in radio-frequency identification technology, and germanium is essential for fibre optics, infrared optics and solar cells used in space-based applications.[84] Various critical materials are also used to make alloys and superalloys for a variety of metallurgical applications, especially instrumental for the nuclear, defence and aircraft industries (e.g. manganese, molybdenum, beryllium, niobium, rhenium, tantalum and various PGMs), while many others are employed in medical devices such as orthopaedic implants and clinical diagnostics (e.g. beryllium, chromium, molybdenum, silver, tantalum, titanium and vanadium).[85]

Moreover, many critical mineral raw materials are essential for the manufacture of clean energy technology components, such as permanent magnets used in wind turbines and electric vehicles or the phosphors needed for fluorescent lighting (HREEs and LREEs), the thin film of photovoltaic cells (indium, gallium and tellurium) and the batteries of electric and hybrid vehicles (cobalt, lithium and LREEs).[86] The DoE has estimated that approximately 20 per cent of global consumption of critical minerals goes into the production of such components.[87] Critical minerals and metals are also more generally employed in a variety of environmentally friendly technologies aimed at reducing pollution and greenhouse gas emissions. Relevant examples include chromium, manganese, titanium and palladium used for seawater desalination; copper, which is essential for the production of efficient electric motors; nickel, which is needed for the development of carbon

[83] European Commission, 'Critical Raw Materials for the EU', pp. 39–46.

[84] European Commission, 'Critical Raw Materials Profiles' and 'Non-Critical Raw Materials Profiles'; USGS, *Mineral Commodity Summaries 2014,* (Washington, DC: US Government Printing Office, 2014).

[85] USGS, *Mineral Commodity Summaries 2014.*

[86] US Department of Energy, 'Critical Materials Strategy', 2010, p. 14.

[87] Moreover, this percentage is destined to grow as the deployment of clean energy technologies continues to expand. Ibid., p. 6.

capture and storage technologies, solar power tower systems, and second generation biofuels; and platinum group metals, which are employed in autocatalyst converters and fuel cells.[88]

Within such a framework, the primary production of critical minerals and metals is heavily geographically concentrated. Out of thirty critical minerals and metals, in six cases (i.e. copper, manganese, nickel, silver, tantalum and titanium), the top three world mine producers account for nearly two-thirds of the total primary production, and in any case they are responsible for almost half of the world's mine output.[89] In a few cases, production is so concentrated that over half of the world's production, and sometimes the quasi-totality of primary production, occurs in a single country: for instance, 92 per cent of the primary production of niobium is based in Brazil; South Africa accounts for more than 60 per cent of the total production of PGMs; the United States produces 90 per cent of total primary beryllium; and the Democratic Republic of Congo is responsible for 56 per cent of the global mine production of cobalt.[90] In these cases, therefore, the top two or three producers account for almost the entire world mine production.[91]

However, China is the country that most frequently enjoys a dominant position, if not a *de facto* monopolistic power, in the world's mine production of critical minerals and metals. Significantly, many of the materials whose production is concentrated in China are essential inputs in the manufacture of cutting-edge technologies and clean energy components. Typical examples include rare earth elements, both light and heavy, gallium, germanium, indium and lithium.[92] China is nearly always among the top primary producers of critical minerals and metals, with notable exceptions being chromium, niobium, platinum group metals

[88] USGS, *Mineral Commodity Summaries 2014;* European Commission, 'Critical Raw Materials for the EU', p. 43.

[89] World Mining Congress, 'World Mining Data Report 2014'; European Commission, 'Report on Critical Raw Materials for the EU', p. 29.

[90] Other cases of significantly concentrated production include chromium in South Africa (43 per cent) and molybdenum (42 per cent) in China. European Commission, 'Report on Critical Raw Materials for the EU', p. 30.

[91] For instance, in the case of beryllium, 99 per cent of the total production takes places in the United States (90 per cent) and China (9 per cent). The same is true for niobium, with Brazil, Canada and the Russian Federation accounting for more than 99 per cent of world mine production. PGMs are predominantly mined in South Africa (61 per cent) and in the Russian Federation (27 per cent). Ibid.

[92] World Mining Congress, 'World Mining Data Report 2014'; European Commission, 'Report on Critical Raw Materials for the EU', p. 29.

and tantalum. China's dominance in mine production is therefore a main driver of criticality (see Table 1.1 below).

While China is the country contributing most to the worldwide supply of critical mineral raw materials, it is noteworthy that the residual primary production is currently concentrated in a relatively small number of regular producers. These include mainly developing countries and emerging economies located in Asia (the Russian Federation, India, Turkey, Indonesia and Kazakhstan), Latin America (Brazil and Chile) and Africa (South Africa and Democratic Republic of the Congo).[93] Countries such as the Philippines, Vietnam, Ukraine, Argentina, Mexico and Peru are among the top producers of a few critical raw materials whose mine output is either relatively less concentrated or dominated by the top one or two primary producers: for instance, the Philippines is the top world mine producer of nickel (16 per cent); Argentina is the third highest producer of borates and lithium (respectively, 13 and 9 per cent); Vietnam is the fifth highest producer of titanium (7 per cent); Ukraine is the third highest producer of gallium (14 per cent) and is one of the only six countries producing germanium; Peru is the third highest mine producer of silver and copper (14 per cent and 8 per cent, respectively); Mexico is the top producer of silver (21 per cent) and the second highest producer of fluorspar (20.5 per cent). The latter two countries are also among the top five producers of molybdenum, accounting for approximately 7 and 5 per cent, respectively.[94] Among the advanced economies, the United States has the highest rate of contribution, along with Australia and Canada.[95] In many cases, not only is primary production geographically concentrated, but a few companies control worldwide production.[96]

The current geographical concentration of primary production does not always reflect the global distribution of the reserve base and, therefore, possible future trends in world production. The countries with the most abundant world reserves are generally among the top mine producers, although their individual contributions may not reflect the exact

93 World Mining Congress, 'World Mining Data Report 2014'; European Commission, 'Report on Critical Raw Materials for the EU', p. 29.
94 World Mining Congress, 'World Mining Data Report 2014'.
95 European Commission, 'Report on Critical Raw Materials for the EU', p. 29.
96 US National Research Council of the National Academies, *Minerals, Critical Minerals*, p. 94.

Table 1.1 *China's dominance in mine production of critical minerals and metals*

Product	China's ranking	Percentage of China's contribution to world primary production (%)	Criticality
Antimony	1	87	EU, OECD
Beryllium	2	9	EU
Borates	5	3	EU
Chromium	13	0.7	EU, OECD
Cobalt	2	5	EU, US, OECD
Copper	2	10	US, OECD
Fluorspar	1	56	EU
Gallium	1	69	EU, US, OECD
Germanium	1	59	EU, OECD
Indium	1	58	EU, US, OECD
Lithium	4	9	US, OECD
Magnesite	1	69	EU
Magnesium	1	86	EU
Manganese	2	20	US, OECD
Molybdenum	1	42	OECD
Natural graphite	1	69	EU
Nickel	8	5	OECD
Niobium	NA	NA	EU, US
Phosphate rock	1	38	EU
Platinum group metals	NA	NA	EU, US, OECD
Rare earths (heavy)	1	99	EU, US, OECD
Rare earths (light)	1	87	EU, US, OECD
Rhenium	6	5	OECD
Silicon metal	1	56	EU, OECD
Silver	2	14.5	OECD
Tantalum	NA	NA	EU, US, OECD
Titanium	4	11	US, OECD
Tellurium	1	18	US
Tungsten	1	85	EU, OECD
Vanadium	1	49	US, OECD

Source: World Mining Congress, 'World Mining Data Report' (2014) and European Commission, 'Report of the Ad-hoc Working Group' (2014).

same proportions.[97] There are two main exceptions to this: first, it often occurs that, when China is among the world's top mine producers, the distribution of reserves is much more widely dispersed (for example, with antimony, rare earth elements and tungsten).[98] In certain cases (e.g. copper), moreover, China is a top mine producer but is not among the countries with the most abundant deposits.[99] This seems to suggest that China is currently overexploiting its resources in order to feed the primary needs of its growing industry. Second, the United States and other industrialized countries such as Australia and Canada are much better endowed with critical mineral resources than the current geographical concentration of production may suggest. For instance, the United States holds one of the largest reserves of molybdenum, rhenium, rare earth elements, silver and titanium, whereas Australia is rich in tantalum, cobalt and nickel, and Canada in rhenium and tungsten.[100] While this situation may reflect economic considerations related to the cost-effectiveness of expanding production capacity, the increasing regulatory costs associated with compliance with environmental regulations have often acted as an additional deterrent for companies from investing in advanced economies.[101]

In some cases, data on current reserves are not available, as some critical materials are so-called by-products or coupled minerals derived from ores of 'carrier' metals. Examples include germanium and indium, which are found in zinc ores; gallium, which is extracted in addition to bauxite; and tellurium, derived from copper and lead ores.[102] Critical metals may also occur as coupled elements without a carrier metal, so

[97] For instance, reserves of chromium are more abundant in Kazakhstan than in South Africa. Korinek and Kim, 'Export Restrictions on Strategic Raw Materials', p. 108; European Commission, 'Critical Raw Materials Profiles' and 'Non-Critical Raw Materials Profiles'.

[98] European Commission, 'Critical Raw Materials Profiles'.

[99] Korinek and Kim, 'Export Restrictions on Strategic Raw Materials', p. 108.

[100] Ibid. In Europe, some countries have significant deposits of minerals such as lithium (Portugal), silver (Poland) and titanium (Germany).

[101] See, e.g. European Commission, 'Annex V to the Report on Critical Raw Materials for the EU on Defining Critical Raw Materials', 30 July 2010, https://ec.europa.eu/eip/raw-materials/en/system/files/ged/52%20annex-v-b_en.pdf (accessed December 2014), p. 200; B. Gu, 'Mineral Export Restraints and Sustainable Development – Are Rare Earths Testing the WTO's Loopholes?', *Journal of International Economic Law* 14 (2011), 765, p. 774.

[102] European Commission, 'Critical Raw Materials for the EU', p. 16. Sometimes, critical mineral raw materials are produced as a by-product from other by-products. This is the case with rhenium, which is a by-product of molybdenum, which is itself obtained from copper.

they are generally mined and processed together (e.g. PGMs, rare earth elements and tantalum-niobium).[103] Normally occurring in low concentrations, the extraction of these metals often depends on the demand for carrier metals,[104] although in some cases it may be economical to extract them as target metals on their own (e.g. cobalt, molybdenum, silver, PGMs and tantalum).[105]

Furthermore, substitutability is a major obstacle for most critical minerals and metals. In some cases, substitution is not a viable solution due to the unique properties of critical materials, at least with respect to their most important and/or strategic applications (e.g. beryllium, chromium, cobalt, gallium, lithium, magnesite, manganese, phosphates, rhenium and titanium).[106] Exceptions include antimony and fluorspar.[107] More commonly, substitutes may exist for some applications, but they either result in performance losses (e.g. beryllium, copper, germanium, molybdenum and rare earth elements) or higher prices (e.g. molybdenum, niobium and tungsten). Available substitutes are often other critical raw materials (e.g. PGMs substitutes are other PGMs and this is true for rare earth elements as well; manganese, molybdenum, niobium and titanium are substitutes for vanadium; tantalum and niobium can replace molybdenum; and cobalt, tungsten and PGMs can be used instead of rhenium).[108]

Finally, the average recycling rate of critical minerals and metals is also generally low, partly because they are often employed in dissipative applications (e.g. antimony, borates and tellurium), but most often because of the small size of the components and/or the low metal content in such components (e.g. beryllium, cobalt and tellurium).[109] Notable

[103] Ibid.

[104] According to the EU Ad-hoc Working Group, '[t]he economic driver for mining here is clearly the major metal. However, by-products metals can generate additional revenue, if they can be extracted economically; in some cases, however, they are regarded as impurities that drive up production costs'. Ibid.

[105] Ibid.

[106] European Commission, 'Critical Raw Materials Profiles' and 'Non-Critical Raw Materials Profiles'.

[107] European Commission, 'Critical Raw Materials Profiles', pp. 9, 55.

[108] European Commission, 'Critical Raw Materials Profiles'.

[109] It should be noted, however, that information on recycling varies considerably by region, product, application, and available technology, which makes it very difficult to identify final values for global average recycling rates. United Nations Environment Programme (UNEP), 'Recycling Rates of Metals: A Status Report', 2011, www.unep.org/resourcepanel/Portals/24102/PDFs/Metals_Recycling_Rates_110412-1.pdf (accessed December 2014), p. 18.

exceptions are copper, nickel, PGMs, silver and tungsten.[110] Pre-consumer and post-consumer scrap or waste phases, moreover, vary considerably (e.g. gallium, germanium and indium have high rates of pre-consumer waste recycling but almost null recovery from end-of-life components).[111] Technological innovations, supply concerns and/or increases have in some cases encouraged great improvements in recycling (e.g. for indium, lithium, PGMs and rare earth elements).[112]

1.5 Main patterns of international trade in critical minerals and metals

Due to their distinctive features, critical minerals and metals have been exposed to similar trade patterns. A first common trend is that world demand for most critical materials is expected to grow for the rest of the decade and exceed existing levels of world production.[113] Current estimations predict that the highest rates of demand growth will be experienced by niobium, gallium and HREEs (more than 8 per cent per year), as well as cobalt, LREEs, indium and magnesium metal (between 4.5 and 8 per cent per year). Significantly, primary production of these materials is heavily geographically concentrated.[114] The vast majority of the remaining critical mineral raw materials are forecast to experience steadier rates of demand growth.[115] Technical innovations and the ensuing emergence of new technologies and applications have a 'clearly visible dominance' on future demand patterns of critical minerals and metals.[116]

Against this backdrop, the price of most critical minerals and metals in the international markets has increased, often sharply, in the last few years. In most cases, prices experienced a significant decline between 2009 and 2010 due to the macroeconomic downturn, but started to rise again after 2011.[117] The expansion of world demand and the parallel rise in world prices has generated renewed investment interest and engagement in the

110 European Commission, 'Critical Raw Materials Profiles' and 'Non-Critical Raw Materials Profiles'.

111 Ibid. 112 Ibid.

113 European Commission, 'Critical Raw Materials for the EU', p. 42; European Commission, 'Report on Critical Raw Materials for the EU', p. 33.

114 See above, Section 1.4.

115 European Commission, 'Report on Critical Raw Materials for the EU', p. 34.

116 European Commission, 'Critical Raw Materials for the EU', p. 42.

117 European Commission, 'Critical Raw Materials Profiles' and 'Non-Critical Raw Materials Profiles'.

part of mining companies.[118] Under these circumstances, the physical availability of the vast majority of critical minerals and metals is not in danger, although, in a few cases, situations of large (e.g. antimony) or small (e.g. indium, PGMs and HREEs) deficits are predicted up to 2020.[119] Large surpluses are expected for materials such as gallium, magnesium, natural graphite, phosphates and LREEs.[120] In these cases, the expansion of primary production is mostly attributable to the intensification of extraction by dominant mine producers, which in most cases includes China.[121]

China is not only a dominant primary producer of most critical minerals and metals, but has also increased its imports of a number of such materials.[122] Significantly, in addition to those situations of relative scarcity (e.g. chromium and cobalt), this trend is observable in several cases where China already enjoys a major share in world primary production (e.g. gallium, fluorspar, germanium, indium, magnesite, magnesium and natural graphite). Such an increase of Chinese imports suggests that the massive exploitation of China's production capacity is still not sufficient to meet its internal demand for critical minerals and metals, and that China needs to resort to the international markets to this end. This is confirmed by the rapid expansion of China's metals industry, as well as of those manufacturing sectors intensive in critical minerals and metals (e.g. the construction sector).[123]

Accordingly, industrialized countries traditionally reliant on the importation of primary mineral supplies are increasingly facing competition from emerging economies such as China, on two levels: first, they are competing internationally for access to critical minerals and metals; second, their industrial fabric is struggling to remain competitive as newly industrializing players increase their export share of higher value-added products in the international markets. The European Union is extremely vulnerable, as only 3 per cent of the global primary production of critical raw materials is indigenous to the EU.[124] In fact, the EU relies almost completely on imports for over half of the minerals that are critical for its economy (i.e. antimony, beryllium, borates, cobalt,

[118] European Commission, 'Critical Raw Materials Profiles'.

[119] European Commission, 'Report on Critical Raw Materials for the EU', p. 35.

[120] Ibid. [121] See above, Table 1.1.

[122] European Commission, 'Critical Raw Materials Profiles' and 'Non-Critical Raw Materials Profiles'.

[123] Radetzki, *A Handbook of Primary Commodities*, p. 71.

[124] The data refer to the total supply calculated for the twenty critical raw materials for the European Union. China's contribution amounts to 49 per cent. European Commission, 'Report on Critical Raw Materials for the EU', pp. 30–1.

magnesium, natural graphite, niobium, PGMs, phosphate rock, HREEs, LREEs, molybdenum, rhenium, tantalum, titanium and vanadium), including materials at risk of deficit in the short to medium term. In all other cases, however, the EU remains a net importer, with the internal supply of primary raw materials covering, at most, up to 40 per cent of its needs (e.g. in the case of chromium, copper and silver). The EU has, moreover, a long-standing history of import dependence in those sectors where it has traditionally had a top-ranking industry (e.g. the cobalt and tungsten industries).[125] The United States contributes more significantly to the total supply of critical minerals and metals.[126] In only a few cases, including indium, HREEs, manganese and tantalum, the US relies almost entirely on imported supplies, while in some cases it imports over half of its primary supply needs (antimony, cobalt, gallium, germanium, rhenium, platinum, tellurium, and vanadium).[127] However, due to the expansion of world demand and the parallel rise in the world price, the US is expected to intensify mining activities and exploit its abundant deposits of many critical minerals.[128]

In conclusion, the main patterns of international trade in critical minerals and metals not only reflect but exemplify the major geoeconomic changes structurally affecting the traditional configuration of international trade in mineral and metals. In the case of critical minerals and metals, the upward trends in world prices and demand derive from both their essential economic importance and the emergence of new global players whose rapid industrialization is altering the traditional allocation of world primary supply. These players hold a predominant share of the current production capacity of such materials and are increasingly challenging the traditional domain of industrialized countries by becoming competitive in those industrial sectors which make intensive use of critical minerals and metals.[129]

1.6 Conclusions

Critical minerals and metals have been exposed to the more general dynamics currently affecting the international markets of minerals and

[125] European Commission, 'Annex V', pp. 52–3, 130–1, 209–10.

[126] For the twenty critical raw materials for the European Union, the contribution of the United States to world primary supply is 9 per cent. European Commission, 'Report on Critical Raw Materials for the EU', p. 28.

[127] USGS, *Mineral Commodity Summaries 2014.*

[128] See above, Section 1.4.

[129] See above, Introduction, Section I.2.3.

metals at the root of the 2000s mineral boom. However, they present distinctive features ultimately putting their availability at particular risk, at least for certain players. In particular, they are heavily geographically concentrated, with the bulk of production capacity located in those countries which are mainly responsible for the structural transformations occurring in international trade patterns.

Accordingly, two key interrelated developments could be observed, relevant to this study: on the one hand, unrestricted trade becomes vital for import-dependent countries to secure access to critical minerals and metals essential to their industrial fabric; on the other hand, as a few emerging economies gain control over prices and quantities of critical minerals and metals made available on world markets, they may be tempted to resort to export barriers in an attempt to increase their export share of higher value-added products, and thus reap the benefits arising from mining exploitation[130] (and compensate for the negative environmental externalities thereby produced).[131] The effects of any such behaviour would be amplified in the case of critical minerals and metals, and any supply disruptions would thus potentially contribute to the acceleration of the ongoing transformations shaking the international markets.[132]

The combination of these factors partially explains the escalation of tension between newly industrializing resource-rich economies and the Western world over the use of export restrictions on minerals and metals among the wide range of restrictive measures adopted within the framework of the economic and financial crisis.

[130] US National Research Council of the National Academies, *Minerals, Critical Minerals*, pp. 67, 136.

[131] J. Salzman and M. Wu, 'The Next Generation of Trade and Environment Conflicts: The Rise of Green Industrial Policy', *Northwestern University Law Review* 108 (2014), 401, pp. 426–30.

[132] US National Research Council of the National Academies, *Minerals, Critical Minerals*, p. 22.

2

The landscape of export restrictions on critical minerals and metals

2.1 Introduction

This chapter gives an overview of the different types of export restrictions instituted and/or maintained on critical minerals and metals since the mid-2000s. Section 2.2 gives a definition of export restrictions covered in this study, modelled on the classification used in the World Trade Organization (WTO), followed by an overview of the various surveillance mechanisms launched and/or reinforced in the last few years to identify coverage, distribution and patterns of use of the most common categories of restrictive measures applied on exportation (Section 2.3). Different types of export restrictions are then examined with regard to the basic distinction between export duties or taxes (Section 2.4) and quantitative forms of export restrictions (Section 2.5). For each, a description of the emerging macro-trends is complemented with a detailed account of the use of export restrictions specifically affecting critical minerals and metals. The aim of this exercise is to understand the magnitude of the proliferation of export restrictions in light of the evolving patterns of international trade in critical minerals and metals identified in Chapter 1 (Section VII). This chapter thus provides the basis for the analysis of the economic effects and rationales of export restrictions in the sector (Chapter 3).

2.2 A definition of export restrictions

Export restrictions are so-called at-the-border measures imposed on national goods at the point of exportation.[1] Similar to trade barriers

[1] See WTO, 'World Trade Report 2012: Trade and Public Policies: A Closer Look at Non-Tariff Measures in the 21st Century', www.wto.org/english/res_e/booksp_e/anrep_e/world_trade_report12_e.pdf (accessed December 2014), p. 51. In this respect, they differ from 'behind-the-border' measures, which are applied domestically on national and foreign (imported) goods.

imposed on importation, they are normally classified into different categories: (i) price or tax measures, which operate by increasing the prices of exports; (ii) quantity measures, which work by limiting the quantity of a good to be exported; and, (iii) administrative measures, which are imposed in respect of certain conditions for the exportation of a certain product.[2] All types of export restrictions ultimately limit the volume of exports of the targeted products, either explicitly (e.g. in the case of an export quota setting a numerical ceiling) or implicitly (e.g. in the case of export taxes and administrative measures that make exports more burdensome).[3]

From a strict economic standpoint, it is always possible to 'convert' quantitative forms of export restrictions into their equivalent export taxes, that is, to determine the level of export tax yielding the same level of exports entailed by a direct numerical ceiling.[4] Moreover, there is in principle a 'ban-equivalent' for every type of export restriction. In other words, notwithstanding the formal qualification of an export restriction, an excessively high level of export taxes or low level of export quotas, as well as the imposition of very tight conditions upon exportation, could produce the same economic effects as an export ban.[5]

For the purposes of WTO regulation, export restrictions are explicitly addressed in Article XI:1 of the General Agreement on Tariffs and Trade (GATT), labelled 'General Elimination of Quantitative Restrictions'. Article XI:1 outlaws any '*prohibitions* or *restrictions* other than *duties, taxes or other charges*, whether made effective through *quotas*, ... *export licences* or *other measures*, ... instituted or maintained by any contracting party on the ... exportation' (emphasis added).[6] The wording of Article XI:1 GATT establishes a dichotomy between 'duties, taxes or other charges' and 'prohibitions or restrictions'. The former are price measures and fall outside the scope of the general elimination obligation imposed as per Article XI:1 GATT, similar to the reciprocal measures on the import side

[2] Ibid.

[3] R.W. Staiger, 'Non-Tariff Measures and the WTO', WTO Economic Research and Statistics Division, Staff Working Paper ERSD-2012-01, www.wto.org/english/res_e/reser_e/ersd201201_e.pdf (accessed December 2014), p. 2.

[4] S. Mitra and T. Josling, 'Agricultural Export Restrictions: Welfare Implications and Trade Disciplines', International Food and Agricultural Trade Policy Council (IPC) Position Paper, Agricultural and Rural Development Policy Series, 2009, p. 24. For a full account see below, Chapter 3.

[5] R. Piermartini, 'The Role of Export Taxes in the Field of Primary Commodities', WTO Discussion Paper, Economic Research and Statistics Division, 2004, www.wto.org/english/res_e/booksp_e/discussion_papers4_e.pdf (accessed May 2014), p. 8.

[6] For a thorough analysis of Article XI:1 GATT see below, Chapter 5, Section 5.2.1.

(i.e. import tariffs).[7] The latter are the 'quantitative restrictions' (QRs) subject to the elimination obligation under Article XI:1 GATT and are defined in a way so as to encompass any type of measure having a limiting effect on the volume of exports, be it a total prohibition or, more generally, a restriction imposed through an export quota, an export licence or any other measure. Any quantity measures and/or administrative measures imposed on exportation may thus qualify as 'quantitative restrictions' for the purposes of Article XI:1 GATT depending on their effects on the volume of exports.[8]

In line with the comprehensive scope of Article XI:1 GATT, an all-encompassing definition of 'export restriction' was adopted by the WTO Panel in *US–Export Restraints*: 'a border measure that takes the form of a government law or regulation which expressly limits the quantity of exports or places explicit conditions on the circumstances under which exports are permitted, or takes the form of a government-imposed fee or tax on exports of the products calculated to limit the quantity of export'.[9]

2.3 The explosion of attention towards export restrictions

Since the end of the 2000s, a number of multilateral institutions have created and/or reinforced various surveillance mechanisms over the introduction of new potentially restrictive measures in response to the global financial and economic crisis. In 2009, a first Joint Report was released by the Secretariats of the WTO, the Organisation for Economic Co-operation and Development (OECD) and United Nations Conference on Trade and Development (UNCTAD) with the aim of monitoring trade and investment measures adopted in the context of the crisis.[10] The report was instigated at the request of the G20 to verify the implementation of the commitments undertaken in Washington in 2008, according to which G20 governments agreed to 'refrain from raising new barriers to investment or to trade in goods and services, imposing new export restrictions, or implementing World Trade Organization

[7] For a full account of the WTO rules on export 'tariffs' see below, Chapter 4.

[8] See below, Chapter 5. [9] Panel Report, paras. 8.16–7.

[10] The Joint Report is currently issued twice a year and includes information on any newly established trade-restrictive and trade-facilitating measures adopted by G20 countries. The first report was distributed in January 2009 and the most recent one on 5 November 2014. At the outbreak of the global crisis, it was released four times per year. WTO, 'Trade Monitoring', www.wto.org/english/news_e/archive_e/trdev_arc_e.htm (accessed December 2014).

inconsistent measures to stimulate exports'[11] and continues to be prepared twice a year. In the same spirit, the Directorate-General for Trade of the European Commission has also compiled a report on potentially trade-restrictive measures, as of 2008,[12] which includes information on any trade-restrictive measures as well as any trade-facilitating measures introduced in the context of the crisis. The WTO has also closely surveyed the use of new restrictions through periodic trade monitoring reports issued by the Director General to the Trade Policy Review Body (TPRB),[13] which give an overview of the trade-related developments implemented by WTO Members and observers that have an impact on the multilateral trading system.[14]

These reports have followed, with growing concern, the proliferation of export restrictions, which constituted the fastest growing component of trade restrictive measures between 2010 and 2011[15] and have continued to rise in the last few years.[16] In its last report, the Directorate-General for Trade of the European Commission denounced the 'particularly alarming . . . resurgence in the application of measures restricting *exports*',[17] with the total number of newly introduced export restrictions increasing by 39 per cent between 1 June 2013 and 30 June 2014.[18] Along

[11] G20, 'Declaration of the Summit on Financial Markets and the World Economy' (Washington DC, 15 November 2008), www.g20.utoronto.ca/2008/2008declaration1115.html (accessed December 2014), para. 13.

[12] For the last report see European Commission, 'Eleventh Report on Potentially Trade-Restrictive Measures Identified in the Context of the Financial and Economic Crisis', 2014, http://horizon.hozint.com/2014/11/dg-trade-11th-report-on-potentially-trade-restrictive-measures/ (accessed December 2014).

[13] The last report was distributed on 24 November 2014 (see WTO Doc. WT/TPR/OV/17). The preparation of these reports is envisaged under Paragraph G of the Trade Policy Review Mechanism mandate and aims 'to assist the TPRB to undertake an annual overview of developments in the international trading environment which are having an impact on the multilateral trading system'. Trade Policy Review Mechanism, Annex 3 to the Agreement Establishing the World Trade Organization, 1869 U.N.T.S. 480, entered into force 1 January 1995. Since 2009, these reports have been preceded by the release of a preparatory contribution to the report by the Director-General. The last one was issued on 27 June 2014 (see WTO Docs. WT/TPR/OV/W/1–8).

[14] Since the eruption of the crisis, the report has been released several times a year and is currently issued twice a year. WTO, 'Trade Monitoring'. The WTO Secretariat has also created a database covering information on the trade-restrictive measures contained in the Trade Monitoring Reports. WTO, 'Trade Monitoring Database', http://tmdb.wto.org (accessed December 2014).

[15] WTO Doc. WT/TPR/OV/14, 21 November 2011, p. 17.

[16] WTO Doc. WT/TPR/OV/17, pp. 20–1.

[17] European Commission, 'Eleventh Report', p. 13.

[18] Ibid.

the same lines, the Director General of the WTO reported that the number of export restrictions introduced between 2013 and 2014 accounted for 27 per cent of total restrictions applied during the monitoring period, the highest ever incidence since the outbreak of the crisis.[19] Moreover, these reports have consistently criticized the substantial lack of transparency surrounding the use of export restrictions.[20] Until recently, the only verified source of information concerning the use of export restrictions was in fact the Trade Policy Reviews (TPRs) conducted by the TPRB,[21] which include a specific section on 'Measures directly affecting exports' within Part III, devoted to 'Trade policies and practices by measure'. However, information on the use of export restrictions in TPRs has generally proved inconsistent, reflecting the varying degree of detail provided by governments on the issue.

Against this backdrop, the OECD has in the last few years intensified its monitoring efforts through a number of studies aimed at identifying patterns of use, coverage and impacts of different types of export restrictions on raw materials.[22] In particular, the OECD Trade Committee has recently created the first factual inventory of export restrictions applied to industrial raw materials, in order to fill 'a void at the international level, where systematic and comparable data are unavailable as far as export regulation is concerned'.[23] Initially compiled through the 'migration' of data put together by the Directorate-General for Trade of the European Commission in the framework of the Raw Materials Initiative,[24] the OECD Inventory now covers the export restrictions applied to more than eighty

19 WTO Doc. WT/TPR/OV/W/8, 27 June 2014, p. 23.

20 See, e.g. WTO Doc. WT/TPR/OV/W/5, 9 June 2011, pp. 16–17.

21 The TPRB undertakes periodic reviews of each WTO Member in order to closely monitor the trade policies implemented at national levels. Each review follows a standard template. The frequency varies according to each country's share of world trade. See WTO, 'Trade Policy Review', www.wto.org/english/tratop_e/tpr_e/tpr_e.htm (accessed December 2014).

22 OECD, 'Analysis of Non-Tariff Measures: The Case of Export Duties', TD/TC/WP(2002)54/FINAL, 31 January 2003, http://search.oecd.org/officialdocuments/displaydocumentpdf/?doclanguage=en&cote=td/tc/wp(2002)54/final (accessed November 2014); OECD, 'Analysis of Non-Tariff Measures: The Case of Export Restrictions', TD/TC/WP(2003)7/FINAL, 4 April 2003, http://search.oecd.org/officialdocuments/displaydocumentpdf/?doclanguage=en&cote=TD/TC/WP(2003)7/FINAL (accessed November 2014); OECD, *The Economic Impact of Export Restrictions on Raw Materials* (OECD Publishing, 2010).

23 B. Fliess, H. Gou and T. Mard, 'Taking Stock of Measures Restricting the Export of Raw Materials: Analysis of OECD Inventory Data', OECD Trade Policy Working Papers, No. 140 (OECD Publishing, 2012), p. 19.

24 European Commission, 'EU Trade Policy for Raw Materials: Second Activity Report', 30 May 2012, http://trade.ec.europa.eu/doclib/docs/2012/may/tradoc_149515.pdf (accessed December 2014), p. 14.

industrial raw materials in their primary, semi-refined and waste and scrap forms at the HS6 level of product classification[25] by eighty-four countries[26] for the period 2009–2012.[27] The database generally includes data for the five leading producers of each of the raw materials considered and therefore covers a large portion of world production and related global trade.[28] Based on the OECD Inventory, sixty countries applied at least one export-restrictive measure on almost all industrial raw materials covered in the database over the period 2009–2012.[29]

Notwithstanding the considerable progress in the monitoring of export restrictions, the OECD has, significantly, continued to warn against the '[l]ack of transparency in the use of export restrictions, evidenced by the relative paucity of information published on governmental websites and the fact that not all measures are notified in the WTO'.[30] In its latest report, in particular, the Organization defined such lack of transparency as 'disconcerting', and admitted that 'there remain many gaps in our knowledge about the use of these instruments' even after the launch of the Inventory.[31]

2.4 Recent trends in the use of export taxes

2.4.1 A definition of export taxes or duties

For the purposes of classification, the category of 'export taxes' or 'export duties' is generally distinguished from the category of export quantitative restrictions. As discussed above, this distinction is drawn from the

[25] In addition to mineral raw materials as defined in this study, the OECD database also covers wood products, which are classified as industrial raw materials. B. Fliess, C. Arriola and P. Liapis, 'Recent Developments in the Use of Export Restrictions in Raw Materials Trade', in OECD, *Export Restrictions in Raw Materials Trade: Facts, Fallacies and Better Practices* (OECD Publishing, 2014), p. 22. Along the same lines, the OECD has also created an inventory for agricultural commodities. P. Liapis, 'How Export Restrictive Measures Affect Trade in Agricultural Commodities', OECD Food, Agriculture and Fisheries Papers, No. 63 (OECD Publishing, 2013).

[26] The European Union is counted as a single region. Fliess *et al.*, 'Recent Developments', p. 6.

[27] While eighty-four countries are covered by the database, current data for the whole period 2009–2012 are available for seventy-two countries. Ibid.

[28] According to the OECD, the Inventory 'covers 80% of world production volume of minerals, metals and wood in their primary state and a large share of related global trade (67% of 2012 total value of exports of primary raw materials, 45% of total exports of primary and semi-processed materials combined, and over 90% of exports of metals waste and scrap)'. Ibid.

[29] Ibid., p. 7. [30] Ibid., p. 31. [31] Ibid., p. 33.

language of Article XI:1 GATT, which refers to 'duties, taxes or other charges'. Various GATT provisions also contain similar expressions without offering any specific guidance on whether they are used with the exact same meaning. For instance, Article I GATT on the most-favoured nation treatment makes reference to 'customs duties and charges of any kind imposed on or in connection with . . . exportation' (and importation); Article VI on customs valuation refers to 'duties or other charges . . . on exportation' (and importation); and Article VIII on fees and formalities rendered in connection with (importation and) exportation excludes 'export duties' from its scope of application.[32] None of the GATT provisions, however, define these expressions.

The GATT Analytical Index and TPRs use the expressions 'export taxes' and 'export duties' interchangeably, although the former expression generally prevails.[33] Based on the definition espoused by the OECD, they both refer to a 'tax collected on goods or commodities when they leave a customs territory'.[34] This is in line with the general definition of export duties given by Van den Bossche: 'an export duty, be it a customs duty or another duty or charge on exports, is a financial charge or tax on exported products, due because of their exportation'.[35] As measures levied at the customs level, export duties are also often labelled export 'tariffs' similar to those on the import side.[36] Other expressions generally defining the same concept include export 'levies' and export 'charges'.[37] However, a different meaning is attached to export-related 'fees and charges', which include the range of fees and formalities imposed in connection with exportation (as well as importation). These measures are specifically subject to Article VIII GATT, according to which they shall correspond to the approximate cost of the service rendered.

[32] OECD, 'Analysis of Non-Tariff Measures: The Case of Export Duties', p. 6.

[33] Ibid., pp. 6–7. See also WT/TPR/OV/W/5, 9 June 2011, p. 17 and Fliess *et al.*, 'Recent Developments', p. 40.

[34] Fliess *et al.*, 'Recent Developments', p. 40. Accordingly, 'customs duties' are defined as 'taxes on the importation and exportation of commodities, merchandise and other goods'. See OECD 'Analysis of Non-Tariff Measures: The Case of Export Duties', p. 7.

[35] P. Van den Bossche and W. Zdouc, *The Law and Policy of the World Trade Organizations*, 3rd edn (Cambridge University Press, 2013), p. 470.

[36] The symmetry between import and export taxes was first formally articulated by Lerner. See A. P. Lerner, 'The Symmetry Between Import and Export Taxes', *Economica* 3 (1936), 306. Notwithstanding the analogy between import and export tariffs in terms of price and trade effects (see below, Chapter 3), export duties are not subject to the same legal regime under WTO law. For a full account see below, Chapter 4.

[37] See WT/TPR/OV/W/5, 9 June 2011, p. 17 and Fliess *et al.*, 'Recent Developments', p. 40.

Export duties may consist of statutory or applied rates; the latter tend to be lower than the former. They can be *ad valorem* (i.e. expressed in the form of a percentage of the value of the product), specific (i.e. a fixed amount to pay on a per unit basis) or compound (i.e. when the two components are combined).[38] Export taxes can also be progressive or differential, meaning that the tax rate increases the higher the value of the product.[39] Finally, export duties may be imposed on specific products or horizontally applied to all exports.[40]

2.4.2 The proliferation of export duties or taxes

Export duties or taxes are among the most frequently used types of export restrictions and their incidence has grown sharply in the last few years. The OECD undertook a first comprehensive survey of the export taxes applied to raw materials in 2009, based on information collected through the TPRs prepared by the WTO Secretariat.[41] It concluded that sixty-five WTO Members introduced and/or maintained export duties during the years 2003–2009 out of the 128 countries reviewed by the TPRB in that period (that is, almost 51 per cent).[42] The data showed an ambiguous upward trend, as in the period 1995–2002 the number of WTO Members imposing export duties only accounted for 39 per cent of the Members under scrutiny.[43] As shown in Figure 2.1, the number of WTO Members resorting to export duties has continued to grow during the years 2009–2014. Based on the information available in the TPRs carried out up to December 2014, seventy-nine WTO Members applied export taxes on raw materials, accounting for approximately 53 per cent of the total number of countries under review.[44]

38 Van den Bossche and Zdouc, *The Law and Policy*, p. 422.

39 In these cases, export taxes are the reverse of what tariff escalation is on the import side. Mitra and Josling, 'Agricultural Export Restrictions', p. 16.

40 Among the countries that have used horizontal export duties are Argentina, Bangladesh, Chad, Gambia, and Niger. J. Kim, 'Recent Trends in Export Restrictions on Raw Materials', in OECD, *The Economic Impact of Export Restrictions on Raw Materials* (OECD Publishing, 2010), pp. 36–45.

41 Kim, 'Recent Trends in Export Restrictions', pp. 15–16.

42 Ibid., p. 16.

43 OECD, 'Analysis of Non-Tariff Barriers: The Case of Export Duties', p. 13.

44 For the period 2010–2014, the analysis was conducted following the same methodology used by the OECD in its studies of 2003 and 2010. During the years 2003–2014, some WTO Members were reviewed multiple times, but were counted as one. The latest information was considered. The EU was counted as 28. See Kim, 'Recent Trends in Export Restrictions', pp. 15–16. The new countries included in the survey are Cambodia, Cote d'Ivoire, Croatia (as a European Union Member State), Democratic Republic of

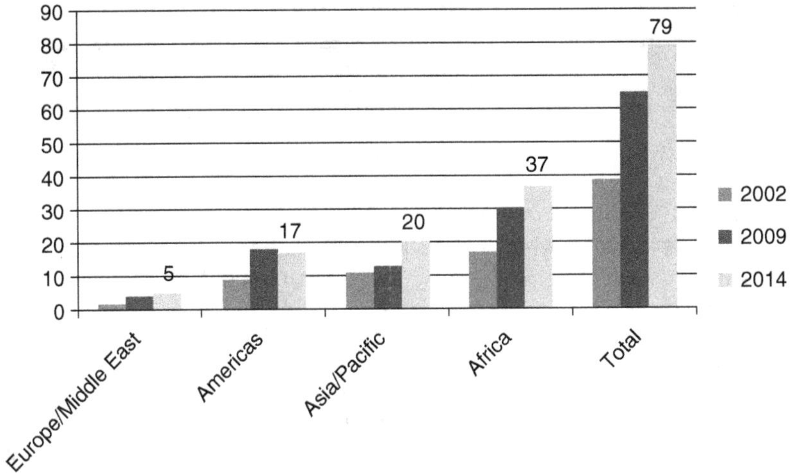

Figure 2.1 Geographical distribution of WTO Members applying export duties
Source: OECD, 'Analysis of Non-Tariff Barriers: The Case of Export Duties', p. 13; Kim, 'Recent Trends in Export Restrictions', pp. 15–6; and the review of 2010–2014 TPRs.

The use of export taxes predominates in developing countries (forty-seven WTO Members in 2014 compared with twenty-six in the period 1995–2002). Least developed countries (LDCs) have also increasingly resorted to export duties (twenty-seven WTO Members in 2014 compared with ten in 2002). Advanced economies, on the other hand, only marginally use these instruments.[45] Figure 2.1 illustrates that almost half of the countries applying export duties are located in the African continent, with eighteen countries imposing export taxes in 1995–2002 out of twenty-six countries reviewed, and thirty-seven out of forty-two countries in 2003–2014. The Asia/Pacific region has also intensified the use of export taxes, with the number of countries applying these instruments almost doubling between 2002 and 2014 and accounting for more than two-thirds of the twenty-nine countries reviewed. In the American continent, seventeen countries adopted export taxes between 2003 and 2014 compared with nine countries in the period 1995–2002.[46]

the Congo, Gabon, Guinea Bissau, Former Republic of Yugoslavia, Kuwait, Myanmar, Nepal, Papua New Guinea, Saudi Arabia, Tonga, Vietnam and Zimbabwe.

[45] OECD, 'Analysis of Non-Tariff Barriers: The Case of Export Duties'; Kim, 'Recent Trends in Export Restrictions', pp. 15–16; and review of 2010–2014 TPRs.

[46] Kim, 'Recent Trends in Export Restrictions', pp. 36–45. Among the WTO Members reviewed in the period 2010–2014, the following additional countries have introduced

Table 2.1 *Newly introduced export duties (2008–2014)*

	2008–2009	2009–2010	2010–2011	2011–2012	2012–2013	2013–2014
Export-restrictive measures	20	25	64	32	27	36
Export duties	8	10	13	8	4	12

Source: WTO Docs. WT/TPR/OV/13–17 (various years) and WT/TPR/OV/W/1–8 (various years).

The use of export taxes has also increased since the outbreak of the global financial and economic crisis. The WTO trade monitoring reports have registered the introduction of new export duties and/or the upward adjustment of existing export duties by WTO Members in each reporting period. As shown in Table 2.1, the number of reported export duty measures has varied throughout the years, as has the number of new trade-restrictive measures imposed on exportation. Yet, export duty measures have consistently figured among the most frequently resorted to measures restricting exports in each monitoring period, and in some cases the most recurring type of export restrictions.[47]

Moreover, the incidence of export taxes is particularly high in the case of industrial raw materials and, in particular, the minerals and metals sector. According to the OECD Inventory on export restrictions, 'export taxes [we] re by far the leading type of export restrictions' between 2009 and 2010,[48] and the second most recurring type of export restriction in the years 2009–2012.[49] In this period, twenty-three out of the seventy-two countries surveyed applied export duties on mineral raw materials, on fifty-five types or groups of minerals and metals.[50] In 2012, the OECD recorded 144 measures at the HS 6-digit level of product classification affecting minerals and metals, either in their primary or semi-processed forms, and another 141 affecting metal waste and scrap.[51] Significantly, the OECD noted that the majority of

export duties: Cambodia, Cote d'Ivoire, Democratic Republic of the Congo, Mauritius, Gabon, Guinea Bissau, Honduras, Mauritania, Myanmar, Kyrgyz Republic, Papua New Guinea, Saudi Arabia, Vietnam and Zimbabwe. Benin, Burundi, Paraguay and St Kitts and Nevis, on the other hand, have removed previously reported export duties.

47 See, e.g. WTO Doc. WTO/TPR/OV/W/3, 14 June 2012, p. 15.

48 Fliess *et al.*, 'Taking Stock of Measures', p. 13.

49 Fliess *et al.*, 'Recent Developments', pp. 11–13. 50 Ibid., p. 11. 51 Ibid., p. 9.

such measures were introduced after 2009 and that the range of mineral products affected by such restrictions has not significantly changed since 2009.[52] Among the countries resorting most heavily to export duties on minerals and metals are Argentina, India, Vietnam and the Russian Federation.[53]

The vast majority of export duties (86 per cent) are *ad valorem* and generally higher than corresponding import tariffs applied on such raw materials: the average export tax in 2012 was 10.9 per cent for primary minerals and metals, and approximately 11 per cent for waste and scrap products. For primary minerals and metals, the average tax rate is frequently higher on semi-processed forms than on raw minerals and metals, while the opposite is generally true for the maximum tax rates.[54] The OECD also noted that in the minerals and metals sector, the new reported export taxes (introduced or tightened) were rarely of a temporary nature, and that the number of removed or relaxed export taxes was also relatively limited, mainly resulting from the implementation of country-specific obligations assumed by selected newly acceded WTO Members under their post-1994 accession protocols.[55]

Finally, the reported export taxes are normally applied on a most-favoured nation (MFN) basis.[56] In a few cases, however, exemptions from export taxes for selected trading partners may be granted on the basis of regional trade agreements (e.g. the Customs Union of Belarus, Kazakhstan and the Russian Federation).[57]

2.4.3 *Export duties or taxes applied to critical mineral and metals*

The trend of proliferation of export taxes affecting raw materials and, in particular, the minerals and metals sector is confirmed when analysing the duties introduced and/or maintained on critical minerals and metals, especially since the outbreak of the global financial and economic crisis in

52 Ibid., p. 7.

53 See K. C. Fung and J. Korinek, 'Economics of Export Restrictions as Applied to Industrial Raw Materials', OECD Trade Policy Papers, No. 155 (OECD Publishing, 2014); OECD Inventory, 'Restrictions on Exports of Raw Materials', http://qdd.oecd.org/subject.aspx?subject=8F4CFFA0-3A25-43F2-A778-E8FEE81D89E2# (accessed December 2014); European Commission, 'Eleventh Report', pp. 13–14, 97–106.

54 Fliess *et al.*, 'Recent Developments', p. 29.

55 Ibid., p. 8. In the case of China, the removal of export taxes on various mineral raw materials followed the two WTO disputes *China–Raw Materials* and *China–Rare Earths*. For a full account see below, Chapter 4.

56 See below, Chapter 4, Section 4.3.4.

57 Fliess *et al.*, 'Recent Developments', p. 32.

2008. Appendix 1 reports the information on the export taxes in place up to 31 December 2014, collected from a survey of the OECD Inventory, the WTO TPRs, the WTO trade monitoring reports and the reports of the Directorate-General for Trade of the European Commission.

The Appendix shows a very high incidence of export taxes applied to critical minerals and metals. Out of the thirty identified materials, all have been subject to export duties in at least one form and in at least one country since 2008.[58] Overall, fourteen countries have imposed taxes on exports of critical minerals and metals. In most cases, more than one country applies export duties on various forms of each considered material. Exceptions are limited to a few cases: borates (including tellurium) and lithium are solely restricted by Argentina, whereas fluorspar, magnesite, phosphates, rare earth elements and silicon are taxed on exportation only by China.[59] In both cases, however, the restricting country is a top producer. China is by far the world's leading mine producer of all such materials, while Argentina is the third largest global producer of both borates and lithium.[60] In contrast, copper is the critical material which, in its various forms, is taxed on exportation by the highest number of countries (eight countries).

The bulk of the measures in place are applied by a relatively small number of countries. China, Vietnam, Argentina and the Russian Federation have taxed the highest number of critical minerals and metals on exportation (respectively, 25, 11, 10, and 5 types or groups of minerals and metals). China's regime of export taxes is particularly pervasive. The Chinese government yearly reviews and publishes in December a Tariff

[58] In some cases, the measures have been in place for many years and sometimes even decades. However, the majority were either introduced or reviewed on or after 2008. If reviewed after 2008, they were mostly left unchanged or, if adjusted, they were increased compared with previous levels. See Appendix 1.

[59] Following the *China–Raw Materials* dispute, China removed the export duties applied to fluorspar with effect from 1 January 2013. Fluorspar is thus the only critical material not reportedly subject to export duties at the time of writing. The WTO Appellate Body has also condemned the wide range of export duties imposed by China on various forms of molybdenum, rare earth elements and tungsten in a second WTO dispute, *China–Rare Earths*. The report was adopted on 29 August 2014 and China is expected to bring its export regime into compliance with the Appellate Body's findings by 2 May 2015. WTO Docs. WT/DS431/16, WT/DS432/14 and WT/DS433/14, 10 December 2014. As China is currently the only country imposing export duties on rare earth elements, similar to the case with fluorspar, these materials will not be subject to any taxes on exportation after China has implemented the Appellate Body decision.

[60] World Mining Congress, 'World Mining Data Report 2014', www.wmc.org.pl/sites/default/files/WMD2014.pdf (accessed December 2014).

Implementation Plan, which contains a list of commodities to be subject to a new and/or adjusted export tax rate, in accordance with the Regulations on Import and Export Tariffs.[61] Export taxes are thus systematically applied to a wide range of commodities. Since 2008, the number of HS 8-digit tariff lines subject to statutory export duties has been increasing, going from 95 HS 8-digit tariff lines in 2009 to 102 in 2013.[62] Moreover, China also applies *interim* export duties (i.e. duties applied for a limited period). In 2013, 320 HS 8-digit tariff lines were subject to *interim* export duties.[63] When imposed on items which are already subject to a statutory rate, *interim* export duties prevail and are normally lower.[64] In most cases, however, *interim* export taxes are applied on additional tariff lines items which are not subject to statutory export taxes.[65] Based on this system, China still cumulatively applied export duties on 346 HS 8-digit tariff lines in 2014 (including both statutory and *interim*).[66]

China has imposed export taxes on a wide range of critical minerals and metals in their primary state. In a number of cases, those are the materials for which it enjoys a dominant and/or a quasi-monopolistic position in world mine production (e.g. antimony, graphite, magnesite, molybdenum, rare earth elements, tungsten and vanadium); in others (e.g. copper, manganese, silver and titanium), China is still among the four top producers, accounting for no less than 10 per cent of global primary production.[67] Relevant exceptions are the export taxes applied on ores and concentrates of niobium, platinum-group metals (PGMs)

[61] At the time of writing, the latest Regulations on Import and Export Tariffs came into force on 1 January 2004. See WTO Doc. WT/TPR/S/300/Rev.1, 7 October 2014, p. 79.

[62] WTO Doc. WT/TPR/S/264/Rev.1, 20 July 2012, p. 58; WTO Doc. WT/TPR/S/300/Rev.1, p. 80.

[63] Ibid.

[64] There are a few exceptions to this trend, which, however, do not concern critical minerals and metals. Ibid.

[65] Overall, *interim* duty rates are lower than statutory rates, ranging from 0 to 35 per cent compared with 20–50 per cent of statutory rates. Ibid. Yet, the former may be flanked by 'special' duty rates which are mainly seasonal and substantially higher than *interim* duty rates. For instance, in 2013, thirteen HS 8-digit tariff lines were subject to a 75 per cent special duty rate. Interim duty rates are revised from time to time, but are normally in place for many years and sometimes even decades. WTO Doc. WT/TPR/S/199/Rev.1, 12 August 2008, p. 74.

[66] T. P. Stewart, E. J. Drake and P. Gong, 'WTO Rules against China's Export Restraints on Rare Earths: A Step Forward in Levelling the Playing Field', 26 March 2014, Stewart and Stewart, www.stewartlaw.com/Article/ViewArticle/989 (accessed December 2014).

[67] World Mining Congress, 'World Mining Data Report 2014'.

and tantalum. Along with ores and concentrates, China also taxes exports of oxides and semi-refined/processed minerals and metals. The export tax rates range between 5 per cent and 40 per cent, with the taxes on ores and concentrates generally higher than those on metal chemical compounds, alloys and metals. Importantly, however, China imposes higher rates on iron and ferro-alloy metals that are fundamental for steelmaking, and on 'technology metals', indispensable for high-tech applications (e.g. various forms of rare-earth-related chemical compounds and non-ferrous minor metals).[68]

The case of Argentina differs from the Chinese model in that the former has horizontally taxed all exports since 2002. The introduction of this instrument followed the devaluation of the peso at the beginning of the 2000s and was established by the former Ministry of the Economy and Infrastructure as a price policy tool aimed at contrasting exchange rate fluctuations on domestic prices and as a fiscal measure aimed at restoring public finances.[69] Currently, a general tax rate of 5 per cent applies to all commodities (with the exception of some dairy products), while higher export duties apply to specific products or categories of products.[70] Metal minerals and their concentrates are subject to a 10 per cent export duty rate.[71] The Argentinean system is based on the principle of differential taxation, as the export duty rate varies according to the degree of processing and is systematically different for primary and semi-processed forms of each exported material – higher for the former and lower for the latter.[72] Although originally introduced on a declaratorily temporary basis as per Resolution No. 11/2002, neither the resolution itself nor its complementary or amendatory regulations have so far contained any indication about the timetable for the elimination of such instruments.[73] Export duties have thus become a major source of tax revenue for the Argentinean government, accounting for more than 20 per cent of total returns.[74]

Vietnam is another example of a relatively well-endowed country resorting to progressive taxation on exports from the mining sector. Ores and concentrates are taxed at a rate that ranges from 22 to 30 per

[68] See Appendix 1.

[69] Resolution No. 11/2002, cited in WTO Doc. WT/TPR/S/176, 8 January 2007, p. 67.

[70] For instance, fuels are taxed at higher rates. WTO Doc. WT/TPR/S/277, 13 February 2013, p. 96. See also above, Introduction, Section I.2.2.

[71] WTO Doc. WT/TPR/S/176, p. 68.

[72] Fliess *et al.*, 'Recent Developments', p. 13.

[73] WTO Doc. WT/TPR/S/277, p. 97.

[74] Ibid.

cent, which has frequently been increased in the last few years (e.g. in the cases of PGMs, tantalum and titanium).[75] Semi-processed materials, on the other hand, are generally taxed at 5–15 per cent.[76] Among the restricted minerals and metals are titanium and tungsten, for which Vietnam is the fifth largest mine producer worldwide.[77]

The Russian Federation, by contrast, reportedly imposes export taxes on fewer mineral commodities in their primary and/or semi-processed forms, namely copper, molybdenum, nickel, PGMs and tungsten. In the cases of copper and nickel, it applies taxes on the exportation of a wide range of compounds, alloys and semi-processed metal forms but not on ores and concentrates. Yet, the Russian Federation is a top mine producer of all these materials. In fact, it is the world's second largest primary producer of PGMs and tungsten, the third largest for nickel, the sixth for copper and the ninth for molybdenum (accounting for around 27 per cent, 5 per cent, 14 per cent, 4 per cent and 2 per cent, respectively).[78] The export duty rates are however lower than those applied by other leading producers such as China, and never exceed 10 per cent. In some cases (e.g. certain forms of copper and nickel), the existing export duty rates were replaced after the creation of the Customs Union with Belarus and Kazakhstan and are thus common to the three countries.[79]

Other countries applying export restrictions on more than one type or group of critical minerals and metals include the Democratic Republic of the Congo (DRC), Malaysia, India and Indonesia. DRC has traditionally taxed exports from the mining sector at a rate of 10 per cent as per its Customs Code.[80] Export duties are therefore horizontally imposed on mineral ores and concentrates, including on those critical minerals and metals that are abundant in DRC, such as cobalt, copper and tantalum.[81]

[75] The most recent legislation on export duties consists of the Ministry of Finance Circular 157/2011/tt-BTB of 14 November 2011 and the Ministry of Finance Circular 193/2012/TT-BTC of 15 November 2012. See European Commission, 'Eleventh Report', p. 106.

[76] Fliess *et al.*, 'Recent Developments', p. 13.

[77] Vietnam is also among the top twenty mine producers of antimony, ranking fourteenth in 2012. See World Mining Congress, 'World Mining Data Report 2014'.

[78] Importantly, the cases of lowest share of global primary production coincide with situations of heavy Chinese concentration (e.g. in the cases of molybdenum and tungsten). See World Mining Congress, 'World Mining Data Report 2014'.

[79] WTO Doc. WT/TPR/OV/W/5/Rev.1, 7 September 2011, p. 22.

[80] WTO Doc. WT/TPR/S/240, 20 October 2010, p. 42.

[81] DRC is the leading mine producer of cobalt, accounting for approximately 67 per cent of total mine production, and it is among the top ten producers of copper and tantalum. World Mining Congress, 'World Mining Data Report 2014'.

India taxes exports of chromium and manganese in their ore and concentrate forms, and is also a top producer of such materials, ranking fourth (with around 11 per cent of world mine production) and seventh (with approximately 5 per cent) based on its mine output, respectively.[82] In contrast, Malaysia applies export duties on copper, nickel, PGMs and silver, and copper and manganese, mainly in semi-processed forms, whereas Indonesia has recently introduced export taxes on ores and concentrates of copper and manganese, among others.[83] Yet, neither Malaysia nor Indonesia enjoy a leading position in global mine production of the critical materials concerned.[84]

Finally, another group of countries imposes export duties on selected critical minerals and metals. For instance, Gabon taxes solely exports of manganese ores and concentrates, Zambia imposes an export duty on copper ores and concentrates and Zimbabwe on chromium ores and concentrates. In all cases, the countries are among the top ten leading mine producers.[85]

When considering waste and scrap forms of critical materials, the landscape of export duties appears somewhat different. Countries such as Argentina, Vietnam, the Russian Federation and China are still widely making use of such instruments (respectively, on 20, 18, 14 and 12 types or groups). In the case of the Russian Federation, the average export duty rate is considerably higher than the rates imposed on primary and semi-processed forms of critical minerals and metals. Yet, export duties are also extensively used by other WTO Members, such as Sri Lanka, Pakistan and Ukraine. These differences reflect, on the one hand, the different stages of the supply chain where countries are placed and, on the other hand, the relatively less intensive mining activities in these territories.[86] In such cases, waste and scrap metals are generally taxed at a flat rate, which tends to be higher than the average rate.[87] Some Candidate WTO Members such as Afghanistan and Kazakhstan have also applied taxes on exports of waste and scrap forms of specific critical metals (respectively, antimony, beryllium and chromium; and copper).

In light of the above, it is possible to qualify the proliferation of export taxes on critical minerals and metals. The first feature is that each type or group of minerals and metals has been subject to export taxes since 2008 by a total of fourteen countries. While some of these measures pre-dated

[82] Ibid. [83] European Commission, 'Eleventh Report', pp. 101–2.
[84] World Mining Congress, 'World Mining Data Report 2014'. [85] Ibid. [86] Ibid.
[87] For instance, Sri Lanka and Pakistan apply a flat rate of 25 per cent on waste and scrap metals. Ukraine, on the other hand, has a tax of 21 per cent. See Appendix 1.

2008, the bulk of them were introduced and/or extended on or after that date. In some cases, the introduction of the measures was intended to be temporary, but they were often maintained longer than anticipated (e.g. the export duty on copper waste and scrap applied by Pakistan) and sometimes for even more than a decade (e.g. in the case of Argentina). Moreover, when subject to a revision or renewal, whether temporary or not, the export duty rates have tended to be increased (e.g. in the cases of nickel and copper, as well as rare earth elements), always depending on whether they are applied on ores and concentrates, semi-processed forms or waste and scrap. Export duty rates, moreover, generally decrease at higher levels of processing (progressive taxation on exportation).

There are, however, some exceptions to these trends. Countries such as Ukraine and Vietnam have gradually decreased the export duty rates applied on various forms of non-ferrous scrap, while China has eliminated the taxes applied on exports of various forms of critical minerals and metals, namely coking coal, fluorspar, magnesium, manganese and silicon. Significantly, these developments were instigated as a result of the implementation of the individual obligations on the use of export taxes assumed by these countries under their respective WTO accession protocols, mostly consisting of the obligation to remove and/or progressively phase down the export duties applied to a certain (country-specific) number of products.[88] As shown in Appendix 1, however, the use of these instruments often exceeds, sometimes widely, the undertaken commitments, as countries extensively resort to export taxes on unbound items (e.g. Vietnam and Russian Federation) or to export duties which are in violation of their WTO commitments.[89] The most evident case is China's regime of export duties, which still encompasses a wide range of measures in breach of China's WTO obligations. Among them, the export duties applied to various forms of rare earth elements, molybdenum and tungsten were condemned by the Appellate Body in the *China–Rare Earths* Report adopted on 29 August 2014. China is expected to bring its export regime in compliance with the Appellate Body's findings by 2 May 2015.[90]

[88] For a full account see below, Chapter 4.

[89] The violation may arise out of two possible scenarios: a country imposes an export duty on a specific product notwithstanding the obligation to refrain from using such an instrument (e.g. China); or, a country imposes an export duty rate which is in excess of the transitional rate agreed upon in a specific phasing-down timetable (e.g. Ukraine). See below, Chapter 4.

[90] WTO Docs. WT/DS431/16, WT/DS432/14 and WT/DS433/14, 10 December 2014.

In conclusion, when restricting the analysis to critical minerals and metals, the phenomenon of 'proliferation' of export duties does not refer to the expansion of the number of countries resorting to such measures, as has occurred on a more general level, but rather to the multiplication of the number of tariff lines subject to export duties by a limited number of countries. These countries are often dominant world mine producers and large exporters.[91] In some important cases, including China, Vietnam, the Russian Federation and Ukraine, they also have undertaken individual obligations on the use of export duties. Relevant exceptions include Argentina, and a few relatively well-endowed countries such as the DRC, Malaysia and Indonesia, Gabon and Zambia, which either impose export taxes on specific critical minerals and metals or mainly apply them to waste and scrap forms.

2.4.4 *Other price or tax measures imposed on exportation*

In addition to export duties or taxes, some WTO Members make use of a number of other price or tax measures to restrict the exportation of raw materials – in particular, critical minerals and metals. One of the instruments used is the so-called fiscal tax on exports or export royalty, which is 'a tax not paid at the border, but that only applies to, or discriminates against, goods or commodities intended for export'.[92] According to the OECD, four countries have applied mining royalties on exports of various minerals and metals since 2008 (Afghanistan, Colombia, the Dominican Republic and Guinea). Afghanistan and the Dominican Republic have also used fiscal taxes on the exportation of, respectively, rare earth elements and chromium.[93] Exports of critical minerals and metals have also been subject to fiscal taxes. Colombia has collected fiscal taxes on ores and concentrates of exported copper, silver and PGMs since 1996. The Dominican Republic has maintained a fiscal tax of 5 per cent of the free on board (FOB) value of exports of antimony, copper, chromium, cobalt, manganese, molybdenum, nickel, tungsten, titanium, niobium, tantalum, vanadium, silver and PMG ore and concentrates since the 1970s. In September 2013, Rwanda introduced mining royalties of 4 per cent of the norm value on exports of niobium, tantalum and vanadium ores and concentrates.[94]

[91] For instance, China's exports of rare earth elements account for more than 60 per cent of world exports. WTO Doc. WT/TPR/S/300/Rev.1, p. 81.

[92] Fliess *et al.*, 'Recent Developments', p. 40. [93] Ibid., p. 13. [94] Ibid.

Another type of export restriction that intervenes at the level of prices is the 'export surtax', which is a 'tax collected on goods or commodities when they leave a customs territory, and which is applied in addition to the normal export tax rate'.[95] Based on OECD data, two countries have resorted to export surtaxes since 2008. The first is Bolivia, which has applied such instruments on a number of base and minor metals.[96] Specifically, among the critical minerals and metals under review, Bolivia has imposed an export surtax on various forms of antimony and tungsten, from ores and concentrates to waste and scrap forms. China has also imposed particularly high export surtaxes (75 per cent) on various forms of phosphates and potash. The measures on phosphates, however, were eliminated in July 2009.[97]

Other tax measures include, for instance, the 'congestion surcharge' of 20 per cent applied by India to railway traffic directed to Pakistan and Bangladesh.[98] Among the materials affected by such a measure are a number of critical minerals and metals, including various forms of antimony, beryllium, cobalt, coke, copper, chromium, gallium, germanium, graphite, indium, niobium, rhenium, manganese, magnesium, molybdenum, nickel, PGMs, silver, rare earth elements, tantalum, titanium and tungsten, including waste and scrap.[99]

Finally, another type of export restriction that is conventionally classified as a price measure is the value-added tax (VAT) rebate reduction or withdrawal. It technically consists of 'denying VAT reimbursement to exports in whole or part [so that] it is less advantageous to export a product than to sell it domestically'.[100] Although not an export tax *stricto sensu*, the reduction of VAT rebates essentially constitutes a levy on exports consisting of the difference between the VAT rate actually paid and the rebate rate. In other words, it raises the cost of exports analogous to an export tax, and thus may be adjusted by national governments as a means to 'control, restrict, or otherwise "manage" the export of certain products'.[101] China is currently the only country using this instrument both with respect to agricultural commodities and industrial raw materials.[102] According to the WTO Secretariat, more than 400 HS 8-digit tariff

[95] Ibid., p. 40. [96] Ibid., p. 13. [97] OECD Inventory.
[98] Fliess *et al.*, 'Recent Developments', p. 15. [99] OECD Inventory.
[100] Fliess *et al.*, 'Recent Developments', p. 40.
[101] WTO Doc. WT/TPR/S/264/Rev. 1, p. 59; WTO Doc. WT/TPR/OV/W/5/Rev.1, p. 17.
[102] According to the OECD, China has refused to reimburse VAT for exports of ninety-three agricultural items since 2007 (HS 6-digit). Fliess *et al.*, 'Recent Developments', p. 14.

lines have been affected by this type of measure since July 2010.[103] Among them, thirty HS 6-digit tariff lines in the minerals and metals sector have been subject to VAT rebates withdrawal.[104] In particular, China has refused VAT reimbursement to exporters of many critical minerals and metals. According to the OECD Inventory, China applies such instruments on various forms of antimony, beryllium, cobalt, copper, gallium, germanium, indium, manganese, magnesium, molybdenum, niobium, phosphates, rhenium, silicon, silver, tantalum, tungsten and vanadium.[105] These measures intervene exclusively at the semi-processed stage,[106] and have been criticized for unfairly assisting domestic downstream processing by 'encourag[ing] exports of products produced locally that use the input to produce downstream products'.[107]

2.5 Recent trends in the use of QRs on exports

In contrast to price measures, QRs are measures which operate by limiting the quantity of exports that may be exported. Based on the terminology used in Article XI:1 GATT, this category of export restraints includes export 'prohibitions' and 'restrictions', where the latter may be made effective through export quotas, licences and any other measures capable of having a limiting effect on export volumes. Such a broad definition allows the inclusion in this category of different types of 'classic' quantity measures, such as export bans and export quotas, administrative measures such as export licensing schemes, and other measures such as minimum export prices, qualified exporters lists, captive mining and domestic market obligations that may result in a limitation of exports.

2.5.1 Export prohibitions or bans

Export prohibitions or bans are 'absolute restriction[s] on exports'.[108] In certain cases, they may also be conditional, in the sense that an outright prohibition may apply only where certain conditions are not met.[109]

[103] See WTO Doc. WT/TPR/S/264/Rev.1, p. 59.
[104] Fliess *et al.*, 'Taking Stock of Measures', p. 12.
[105] Fliess *et al.*, 'Recent Developments', p. 40.
[106] Fliess *et al.*, 'Taking Stock of Measures', p. 12.
[107] See WTO Doc. WT/TPR/S/264/Rev.1, p. 59; European Commission, 'Annex V', p. 130.
[108] Fliess *et al.*, 'Recent Developments', p. 40.
[109] Van den Bossche and Zdouc, *The Law and Policy*, p. 481.

A plurality of WTO Members resort to export bans in order to comply with the obligations contained in a number of multilateral agreements.[110] In such cases, the prohibition of exports corresponds to non-economic public policy goals such as security, life, public health, safety and environmental protection.[111] Among the international instruments most frequently cited in justification of export prohibitions are multilateral environmental agreements (MEAs) such as the Basel Convention on the Control of Transboundary Movements of Hazardous Wastes and their Disposal, the Convention on International Trade in Endangered Species of Wild Fauna and Flora, and the Montreal Protocol on Substances that Deplete the Ozone Layer, as well as various conventions on the non-proliferation of weapons (e.g. the Convention on Chemical Weapons and the Treaty on Nuclear Non-Proliferation) and multilateral export control arrangements (e.g. the Australia Group, the Missile Technology Control Regime, the Nuclear Suppliers Group and the Wassenaar Arrangement).[112] The restrictions implemented in pursuance of such conventions are therefore usually excluded from discussions concerning the challenges posed by the proliferation of export restrictions.[113]

However, the use of export prohibitions has increased in the last decade. Export bans have extensively been used in the agricultural sector to prevent food shortages[114] and severe price volatility of essential foodstuffs.[115] Such instruments peaked during the food crises of 2007–2008 and 2010–2011, and have almost always been announced as temporary measures although often being extended longer than anticipated.[116] A few countries (namely, India, Indonesia, Nigeria and the United States) regularly prohibit exports of certain wood products.[117]

In the minerals and metals sector, export bans are rarely reported.[118] Only a few countries actually ban exports of critical minerals and

[110] Kim, 'Recent Trends in Export Restrictions', p. 16. [111] Ibid., p. 17.

[112] Ibid. See also OECD, 'Analysis of Non-Tariff Measures: The Case of Export Restrictions', p. 17.

[113] Fliess *et al.*, 'Recent Developments', p. 29.

[114] Recent examples include temporary export bans on wheat, rice, onions and sugar. WTO Docs. WT/TPR/OV/W/1-8 (various years).

[115] See above, Introduction, Section I.2.1.

[116] Liapis, 'How Export Restrictive Measures'. According to Piermartini, the temporariness of export bans undermines their effectiveness as a long-term policy. Piermartini, 'The Role of Export Taxes', p. 3.

[117] Fliess *et al.*, 'Recent Developments', p. 13. Indonesia is the only country that has reviewed, and tightened, its export policy on wood products in recent years. Ibid., p. 22.

[118] Fliess *et al.*, 'Taking Stock of Measures', p. 12; Fliess *et al.*, 'Recent Developments', p. 9.

metals. Zimbabwe, the tenth biggest world mine producer of chromium, has prohibited the exportation of ores and concentrates since 2007, with an interruption between 2010 and 2011 during which it substituted the measure with an export tax.[119] The Philippines, moreover, has banned exports of PGM ores and concentrates since the 1970s, while Venezuela has not allowed exports of ash and residues containing antimony, beryllium, chromium, copper and precious metals since 2005.[120] In 2012, the Indonesian Ministry of Energy and Mineral Resources adopted a regulation on Increasing Value-Added Minerals Through Processing and Refining, which introduces an export ban on unprocessed minerals, except coal, from 12 January 2014.[121]

Lastly, export prohibitions are increasingly being used in the waste and scrap metal sector. According to the OECD, 'export prohibitions account for a much higher share of total export restrictions in this industry (8% in 2009 and 23% in 2012) than in the primary minerals and metals industry (5% in 2009 and 14% in 2012)'.[122] While these data take into account both ferrous and non-ferrous scrap, it should be noted that the use of export bans is intensifying with respect to a number of critical base and minor scrap metals. One group of countries (Azerbaijan, Burundi, Guyana, Jamaica, Kenya, Rwanda, Tanzania, Uganda and Venezuela) has regularly prohibited the exportation of a wide range of materials, including antimony, beryllium, chromium, cobalt, copper, gallium, germanium, indium, niobium, rhenium, magnesium, manganese, molybdenum, nickel, tantalum, titanium, tungsten and vanadium. Precious metals such as PGMs and silver have also been subject to export bans by Azerbaijan and Venezuela in their waste and scrap forms.[123] In most cases, these measures are still in place, although a few countries applying them on a temporary basis, such as Guyana and Jamaica, have reportedly abandoned them completely or partly after some years.[124] Copper waste and scrap has also been subject to temporary export bans by countries such as Argentina and Uruguay in the last few years.[125]

[119] See OECD Inventory and Appendix 1. [120] See OECD Inventory.

[121] European Commission, 'Eleventh Report', p. 101.

[122] Fliess *et al.*, 'Recent Developments', p. 19.

[123] See OECD Inventory. Burundi, Kenya, Rwanda, Tanzania and Uganda introduced export prohibitions in 2010 within the framework of the East African Community Customs Union.

[124] Ibid.

[125] Ibid. Argentina maintained the export ban from 4 September 2012 to 30 August 2013; Uruguay from 24 November 2006 to 30 April 2009.

2.5.2 *Export quotas*

Export quotas are numerical ceilings imposed by an exporting country in order to cap the total volume of exports of a certain product.[126] Similar to export prohibitions, export quotas are also widely used by WTO Members to achieve various non-economic goals pursued in a number of MEAs and multilateral export control arrangements.[127] In these cases, the use of export quotas is not controversial and therefore will not be scrutinized in this study.

The incidence of export quotas on commodity trade is relatively limited compared with export taxes or other types of QRs such as export prohibitions. In the agricultural sector, export quotas were applied by only three countries, Argentina, China and Ukraine, on a total of seven tariff lines related to bulk commodities (HS 6-digit) between 2009 and 2012.[128] Likewise, two countries, Canada and the Russian Federation, imposed quotas on the exportation of various wood products (eight HS 6-digit tariff lines) in the same period.[129] In the minerals and metals sector, three countries have reportedly resorted to export quotas since 2008, namely China, Ukraine and Belarus.[130] The latter two countries, however, imposed export quotas on a narrow range of products. Since 2010, Belarus has imposed export quotas on various forms of ferrous waste and scrap used in steelmaking as well as on copper, nickel and aluminium waste and scrap.[131] Ukraine has used export quotas on copper ash and residues, and on gold and silver waste and scrap forms.[132]

The bulk of export quotas on mineral raw materials have been applied by China. According to the OECD Inventory, China maintained export quotas on forty-four types or groups of minerals in the period 2009–2012.[133] The WTO Secretariat reports that in 2013, these accounted for 193 tariff lines at HS 8-digit level.[134] Among the restricted materials are critical minerals and metals, including various forms of antimony, coke, fluorspar, magnesite, magnesium, molybdenum, niobium, phosphates, rare earth elements, rhenium, silicon, silver, tungsten and vanadium. Waste and scrap forms are also generally capped.[135] The OECD has calculated that

[126] Fliess *et al.*, 'Recent Developments', p. 40. [127] See above, Section 2.5.1.
[128] Fliess *et al.*, 'Recent Developments', pp. 9, 29. [129] Ibid., pp. 9, 14.
[130] See OECD Inventory. [131] Ibid. [132] Ibid.
[133] Fliess *et al.*, 'Recent Developments', p. 13.
[134] WTO Doc. WT/TPR/S/300/Rev.1, p. 81.
[135] A single shared quota applies to the group of minerals encompassing gallium, germanium, indium, niobium, rhenium and vanadium (together with hafnium). See OECD Inventory.

'the share of world exports subject to quantitative restrictions was highest for antimony (50% of world exports), tungsten (45%), magnesite (44%) … and rare earths (39%)'.[136]

China's regime of export quotas is based on a series of measures consisting of basic framework legislation (the Foreign Trade Law), a set of implementing regulations on administration of imports and exports (the Regulation on the Administration of the Import and Export of Goods), as well as specific annual measures indicating all products subject to export quotas for each year (the annual Export Licensing Catalogue) and the related annual quota amounts (the annual Export Quota Amounts).[137] Both the Catalogue and the annual quota amounts are formulated and announced on 31 October of each year for the following year, coming under the responsibility of the Chinese Ministry of Commerce (MOFCOM).[138] Exporters can apply for an export quota until 15 November of each year. Annual quotas are then generally allocated by 15 December and, in some cases, twice a year through a 'first batch' and a 'second batch' occurring in December and July, respectively. The allocation of quotas is done either directly or through a quota bidding system.[139] Among the critical minerals and metals subject to export quotas by China, MOFCOM directly allocates those for antimony and articles thereof, tungsten and articles thereof, silver, indium and articles thereof, molybdenum and phosphorus ore, whereas export quotas for magnesium ore are allocated through bidding.[140]

In the former case, MOFCOM determines the total ceiling for quotas based on a series of factors which include: '(i) the needs of guaranteeing the safety of the national economy; (ii) the needs of protecting the limited domestic resources; (iii) development planning, objectives and policies of the State on the relevant industries; and (iv) demands on the international and domestic markets, and the production and sales status'.[141] Upon receipt of the applications, MOFCOM allocates the quotas directly or through local administrative authorities guided by the following criteria: '(i) the export performance of the particular good; (ii) the utilization ratio of the export quota; (iii) the business management/operation capacity of the applicant; (iv) the production scale and resource status of the applicant

[136] Fliess *et al.*, 'Recent Developments', p. 14.

[137] For a thorough account of the functioning of China's export quota regime see *China–Raw Materials*, Panel Reports, paras. 7.172–201.

[138] WTO Doc. WT/TPR/S/300/Rev.1, p. 81. [139] Ibid. [140] Ibid.

[141] See China's Export Quota Administration Measures, Article 10, cited in *China–Raw Materials*, Panel Reports, para. 7.177.

enterprise or area during the previous three years'.[142] In the latter case, any interested enterprise approved to participate in export quota bidding submits a bidding price and a bidding quantity to China's Export Quota Bidding Committee, which is established within MOFCOM and governs each step of the bidding procedure.[143] China's Bidding Office determines the winning bidders by ranking all bids received from applicants in descending order (i.e. from the highest bid prices to the lowest) and matching them with the related bidding quantities until the total bidding quantity equals the annual quota amount available for each product.[144] An enterprise must thus be awarded a portion of the export quota as part of the bidding process in order to export.

In both cases, MOFCOM sets out general eligibility requirements for enterprises, which frequently vary depending on the nationality of the enterprises. Sector-specific eligibility requirements are also established and have often included prior export experience or export performance requirements.[145] Often, foreign-invested enterprises are required to satisfy additional requirements.[146]

MOFCOM yearly adjusts the annual quota amounts for each product subject to export quotas. According to the OECD Inventory, since 2008, China has tightened the ceiling imposed on a number of critical minerals and metals, not only in terms of absolute value but also as a percentage of total Chinese production.[147] A striking example in this

142 See China's Export Quota Administration Measures, Article 19, cited in *China–Raw Materials*, Panel Reports, para. 7.178. Moreover, MOFCOM can also adjust its decisions provided that major changes occur in the international market or domestic resources, or there are obvious imbalances in the quota-use pace between enterprises or areas. Ibid., para. 7.179, citing Article 20 of China's Export Quota Administration Measures.

143 See China's Export Quota Bidding Measures, Article 26, cited in *China–Raw Materials*, Panel Reports, para. 7.199.

144 See China's Export Quota Bidding Measures and Export Quota Bidding Implementation Rules, cited in *China–Raw Materials*, Panel Reports, paras. 7.187–200.

145 See, e.g. the requirements established for exporters of bauxite, coke, fluorspar, silicon carbide and zinc, detailed in *China–Raw Materials*, Panel Reports, paras. 7.177–86 and paras. 7.198–201. See also the requirements imposed on exporters of rare earth elements, molybdenum and tungsten described in *China–Rare Earths*, Panel Reports, paras. 7.204–35.

146 W. M. Morrison and R. Tang, 'China's Rare Earths Industry and Export Regime: Economic and Trade Implications for the United States', US Congressional Research Service (CRS) Report for Congress, 30 April 2012, http://china.usc.edu/App_Images//crs-2012-rare-earth-industry.pdf (accessed December 2014).

147 J. Korinek and J. Kim, 'Export Restrictions on Strategic Raw Materials and their Impact on Trade and Global Supply', in OECD, *The Economic Impact of Export Restrictions on Raw Materials* (OECD Publishing, 2010), pp. 123–30.

respect is the evolution of the export quotas imposed on rare earth elements. In 2008, the allocated quotas were set at 49,990 metric tons, significantly reduced from 59,643 metric tons in 2007. In 2010, the annual quota amount was further decreased to 30,259 metric tons.[148] For the next 4 years, the total allocated quality did not vary much, always remaining within the range of 30,000–31,000 metric tons until 2014.[149] However, from 2011, the list of products covered by the quota was updated to include various ferro-alloys that were not covered before.[150] Hence, although left almost unchanged at a formal level, it has been estimated that 'by including this new category of products in the quota, China in fact reduced the 2011 rare earth export quota by as much as 30%, while passing the message that the 2011 quota was not reduced when compared to that of 2010'.[151] Moreover, the allocation of the quotas was reduced more significantly for foreign-invested than Chinese-owned companies.[152]

There are however some exceptions to this general trend. China has eliminated the export quotas applied to various forms of fluorspar since 2012, as well as a number of other minerals and metals including coke.[153] Once again, the removal of these measures was instigated by the need to implement the Appellate Body's recommendations in the *China–Raw Materials* dispute.[154] A second Appellate Body ruling, *China–Rare Earths*, condemned China's export quotas on various forms of rare earth elements, tungsten and molybdenum. Accordingly, at the end of 2014 MOFCOM announced that China would eliminate the export quotas for such materials as of 2015.[155]

148 Morrison and Tang, 'China's Rare Earths Industry', p. 17.

149 G. Hatch, 'The Second Round of Chinese Rare-Earth Export-Quota Allocations for 2014', *Technology Metal Research*, 19 July 2014, www.techmetalsresearch.com/2014/07/the-second-round-of-chinese-rare-earth-export-quota-allocations-for-2014/ (accessed December 2014).

150 WTO Doc. WT/TPR/OV/16, 31 January 2014.

151 European Commission, 'Ninth Report', p. 119.

152 Ibid.; Morrison and Tang, 'China's Rare Earths Industry', p. 18.

153 WTO Doc. WT/TPR/S/300/Rev.1, p. 81.

154 For a detailed account see below, Chapter 6.

155 Sinocast Metals, Mining and Chemicals Beat, 'China Cancels 2015 Tungsten, Molybdenum Export Quotas', *Kitco News*, 30 December 2014, www.kitco.com/news/2014-12-30/China-Cancels-2015-Tungsten-Molybdenum-Export-Quotas.html (accessed December 2014); Backer and McKenzie, 'Changes to Export Requirements for Rare Earths from China', 9 January 2015.

2.5.3 *Export licensing schemes or requirements*

An export licensing requirement is '[t]he requirement to obtain prior approval, in the form of a license, to export a good or a commodity'.[156] In other words, enterprises interested in exporting must submit an application for a licence. The applications are then reviewed by the governments, which can thus monitor export flows. Export licensing schemes can be automatic or non-automatic. In the case of automatic licensing, 'approval is granted in all cases, usually immediately upon standardised application'.[157] Automatic schemes therefore do not impose any significant extra burden or transaction costs on exporters and are mainly implemented for statistical purposes.[158] Non-automatic licences, on the other hand, are released only upon submission of an application or other documentation to the relevant administrative bodies as a prior condition for exportation. In these cases, exporters must fulfil all the necessary requirements in order to be granted approval in the form of a licence.[159] Depending on how they operate, export licensing schemes may entail significant additional costs for exporting enterprises and thus have a negative impact on export volumes. For instance, excessively long processing times and opaque and/or discretionary rules for the review of applications could *de facto* negatively impact on export volumes.[160] Excessively stringent export licensing schemes may even result in an export ban.[161]

Non-automatic licensing schemes are frequently used to administer export quotas. Accordingly, a large number of WTO Members implement such schemes in combination with those export quotas imposed in pursuance of MEAs and other international agreements.[162] Countries applying export quotas on industrial raw materials, including critical minerals and metals, also utilize non-automatic licensing schemes in conjunction with such quotas.[163] In the case of China, for instance, enterprises that are allocated a quota are issued with a certificate of quota that entitles them to apply for an export licence. Different application procedures are in place and a different

156 Fliess *et al.*, 'Recent Developments', p. 40. 157 Ibid. 158 Ibid. 159 Ibid.
160 Korinek and Kim, 'Export Restrictions on Strategic Raw Materials', p. 109.
161 European Commission, 'Tenth Report on Potentially Trade-Restrictive Measures Identified in the Context of the Financial and Economic Crisis', 2013, http://trade.ec.europa.eu/doclib/docs/2013/september/tradoc_151703.pdf (accessed November 2014), pp. 131–2.
162 See OECD, 'Analysis of Non-Tariff Barriers: The Case of Export Restrictions', pp. 15–17.
163 See OECD Inventory.

export licence (either an 'export quota licence' or an 'export quota bidding licence') is released, depending on whether the quota is awarded directly by MOFCOM or through bidding.[164]

While raw materials capped on exportation are often subject to non-automatic export licensing schemes, countries far more extensively resort to such schemes as an instrument in and of itself. According to the OECD Inventory, export licensing requirements have been the most frequently used type of export restriction applied to industrial raw materials, by a total of twenty-nine countries since 2008.[165] In the minerals and metals sector, recourse to such instruments has been particularly extensive, with 240 HS6 tariff lines affected by non-automatic export licensing requirements in 2012.[166] Overall, China, the Dominican Republic, Malaysia, the Philippines and the Russian Federation reportedly apply non-automatic export licensing schemes on the highest number of products.[167]

Almost the entire range of critical minerals and metals has been subject to non-automatic export licensing requirements by at least one country and in at least one form. Appendix 2 details the countries using export licensing schemes for each material in unprocessed, semi-processed and waste and scrap forms based on OECD Inventory data, WTO trade monitoring reports and various TPRs. Those most frequently applying such instruments to ores and concentrates represent a relatively small group of WTO Members composed of China, Grenada, Malaysia, the Philippines, the Russian Federation, South Africa and, to a limited extent, Argentina and India. China and the Russian Federation use export licensing schemes for the highest number of critical materials in their semi-processed forms, although the Russian Federation reportedly removed these measures in 2012 for most of the products concerned, with the notable exceptions of some molybdenum and PGM products.[168] Other countries using export licensing schemes on semi-processed products include South Africa (chromium), Malaysia (magnesite), Brazil (magnesium), Indonesia (silver), and the Philippines (titanium). A few materials such as copper, manganese and PGMs have comparatively been more exposed to export licensing schemes in their semi-processed forms. In these cases, the measures are used by numerous WTO Members including, again, South Africa, Malaysia and

[164] WTO Doc. WT/TPR/S/300/Rev.1, pp. 81–2.

[165] Fliess *et al.*, 'Recent Developments', p. 9. Export licensing schemes were, in contrast, hardly used in the agricultural sector, with only Argentina and Indonesia reportedly using them in 2009. Ibid., p. 11.

[166] These include semi-processed and waste and scrap forms of minerals and metals. Ibid.

[167] Ibid. [168] See OECD Inventory.

Indonesia, and also Japan, Ukraine, Colombia, Paraguay and Zambia. Other countries include WTO Candidate Members such as Belarus. Thus, many of the world's leading producers are applying export licensing schemes on critical minerals and metals.

Scrap and waste metals are frequently subject to export licensing requirements. Most countries apply such measures simultaneously on many base and minor metals.[169] A number of developing countries as well as LDCs systematically subject a wide range of critical metals to export licensing schemes in their waste and scrap forms, including Afghanistan, Algeria, Guyana, Jamaica, Mauritius, Morocco, Sri Lanka, Tajikistan, Trinidad and Tobago, Turkmenistan and Zambia. Some of these countries appear to apply them because of difficulties in determining the origin and/or the purity of the materials,[170] or to prevent theft and smuggling induced by increased scrap prices.[171] In addition, non-automatic export licences are also required by China, South Africa and the Russian Federation to sell metal scrap abroad.[172]

Most countries have introduced and/or adjusted their export licensing schemes in the last few years, although in a number of cases (e.g. Malaysia, Japan, Brazil and the Russian Federation) the systems date back to the 1990s. Yet, difficulties in acquiring consistent detailed information on the functioning of these measures make it hard to review them in a systematic way.[173] The latest trade monitoring reports and TPRs show that a number of countries have introduced more stringent procedures and requirements for enterprises to obtain an export licence in the past few years.[174] Turkey, for instance, has recently tightened the export licensing scheme applicable to copper scrap. As of 21 May 2010, exporting firms need to fulfil three conditions: '(i) copper scrap which will be exported shall be pre-investigated on site by supervisors from the Standardisation Department of the Foreign Trade; (ii) submission of written confirmation received from at least three domestic producers showing that copper scrap would not be used for their production; (iii) submission of a contract that shows export

[169] Fliess *et al.*, 'Recent Developments', p. 11.

[170] Korinek and Kim, 'Export Restrictions on Strategic Raw Materials', p. 110.

[171] Fliess *et al.*, 'Recent Developments', p. 25.

[172] The Russian Federation reportedly eliminated these instruments for most forms of critical metal scrap since 28 January 2012. See OECD Inventory.

[173] According to the OECD, 'little is known about the consistency and even-handedness with which government agencies administer export licences'. Fliess *et al.*, 'Recent Developments', p. 33.

[174] European Commission, 'Tenth Report', pp. 125–33; European Commission, 'Eleventh Report', pp. 97–106.

connection'.[175] In South Africa, a 5-year policy directive dated 10 May 2013 prescribed that exports of non-ferrous waste and scrap metal must first be 'offered to domestic users for a period determined by the International Trade Administration Commission of South Africa (ITAC) and at a price discount or other formula determined by ITAC intended to facilitate local rather than export sale'.[176] On 9 May 2014, moreover, South Africa further amended the export licensing scheme in place for non-ferrous scrap metals, introducing the additional condition that scrap metals should have 'previously – and unsuccessfully – been offered to domestic consumers at a price 20% below international spot prices'.[177] Similarly, Indonesia recently introduced a government regulation concerning the implementation of, among others, mineral mining business activities, which indicates the minimum levels of domestic processing and refining for metallic and non-metallic minerals prior to obtaining permission to sell them abroad in specific amounts.[178] In Mauritius, only Mauritian-owned companies are allowed to apply for a licence if they wish to export scrap metals and, if they fulfil the eligibility requirements, they have to pay for the licence, which is only valid for six months.[179] Moreover, following elimination of the export quotas on rare earth elements, China announced stricter licensing rules for the exportation of such materials, including a further reduction in the number of licensed companies.[180] However, against the tide, Tajikistan eliminated all export licensing schemes in place in the country in 2012 in the context of its accession to the WTO.[181]

2.5.4 *Minimum export prices and other types of export restrictions*

A 'minimum export price' (MEP), also known as 'administered pricing', is the 'minimum permitted price for a good being exported'.[182] A MEP

[175] European Commission, 'Eleventh Report', p. 105. [176] Ibid., p. 104.

[177] Ibid., p. 105. The European Commission has noted that such a measure 'will affect the quantity of scrap metal available for export as well as the price at which scrap metal is sold to all market actors'.

[178] Ibid., pp. 101–2. [179] Fliess *et al.*, 'Recent Developments', p. 19.

[180] Sinocast Metals, Mining and Chemicals Beat, 'China Cancels'; Backer and McKenzie, 'Changes to Export Requirements'; T. Weslowsky, 'China Tightens Control on the World's Supply of Rare Earths in 2015', Interview with Jack Lifton (InvestorIntel), 10 January 2015. The number of licensed companies has been progressively reduced by China in the last few years. See Morrison and Tang, 'China's Rare Earths Industry', p. 18.

[181] WTO Docs. WT/ACC/TJK/3, WT/ACC/TJK/24, and WT/ACC/TJK/7/Rev.7, cited in the OECD Inventory.

[182] Fliess *et al.*, 'Recent Developments', p. 40.

may not necessarily be binding but can be used as a reference price.[183] Although MEPs are often associated with export taxes or used to calculate export taxes,[184] they are conventionally categorized as export restrictions due to their potentially negative impacts on export volumes.[185]

Administered pricing is not a frequently used type of export restriction. According to the OECD Inventory and the latest WTO trade monitoring reports, only seven countries have resorted to MEPs since 2008: Argentina, China, India, Indonesia, Pakistan, the Russian Federation and Vietnam. Pakistan and Vietnam used minimum export prices for rice during the period 2008–2010, while India reintroduced minimum export prices for agricultural commodities such as onions and potatoes in 2014.[186] Indonesia and the Russian Federation still use these measures for wood products and diamonds, respectively.[187] However, China dismantled its system of coordinated export prices – which it had imposed on a number of raw materials, including various critical minerals and metals – in July 2010.[188] Finally, Argentina has applied precautionary price references for a wide range of semi-refined/processed articles of copper exports since 2010, which it regularly reviews.[189]

Other types of export restriction include the so-called domestic market obligation, that is, 'the requirement for producers … to allocate a proportion of their annual production output to the domestic market'.[190] The Russian Federation, for instance, applies such instrument along with export taxes on various forms of precious metals, including PGMs.[191] Other countries restrict exports by requiring exporting firms to register with the government authorities. Indonesia has 'qualified exporters lists', whereby the government allocates the right to export precious metals to those companies that have successfully applied for registration.[192] Similarly, China's MOFCOM formulates and periodically adjusts the list of companies that are allowed to export specific

[183] Ibid. [184] Ibid.

[185] See, e.g. Van den Bossche and Zdouc, *The Law and Policy*, p. 444; WTO Doc. WT/TPR/OV/W/5/Rev.1, p. 17.

[186] WT/TPR/OV/17, p. 158. [187] Fliess *et al.*, 'Recent Developments', p. 15.

[188] See China's Chamber of Commerce of Metals, Minerals and Chemicals Importers and Exporters (CCCMC), Resolution on Abolition of Coordination and Administration of Export Commodities, cited in *China–Raw Materials*, Panel Reports, paras. 7.1012–26.

[189] See OECD Inventory.

[190] OECD Inventory and Fliess *et al.*, 'Recent Developments', p. 41.

[191] See OECD Inventory.

[192] Fliess *et al.*, 'Recent Developments', pp. 14, 40. Indonesia also applies this instrument on certain wood products.

commodities.[193] In some cases, exports of specific products are subject to state trading. State trading itself is considered an implicit form of export restriction inasmuch as state trading enterprises (STEs) are capable of controlling exports and thereby impacting domestic supply, demand and prices.[194] Among the critical raw materials, antimony ore and antimony products, silver, and tungsten ore and tungsten products are subject to export state trading and thus are not exportable by foreign-invested companies.[195] The Chinese authorities maintain that the allocation of export quantities among STEs is governed by market principles and that STEs determine the prices at which they export based on domestic prices plus transportation and storage costs, interest rates, inspection fees and international market prices.[196]

Lastly, another type of export restriction is known as 'captive mining'. The expression refers to the practice whereby 'a processing company is required to own the mine that produces its inputs, or has been awarded mining rights with the intention that the company will mine the commodity for use in its own domestic processes and not trade it'.[197] India uses the policy of captive mining concessions on a wide range of mineral products, including various forms of coke and manganese.[198] Indonesia has also recently implemented a comprehensive scheme of mining concessions aimed at ensuring that holders of mining rights process and refine products domestically instead of trading them in their concentrate forms.[199]

2.6 Conclusions

The recent proliferation of export restrictions has affected various categories of primary products, from food commodities to agricultural raw materials, from wood products to minerals and metals. Minerals and metals have, in particular, been severely hit by the new wave of export restrictions that started with the commodity boom of the 2000s. The

[193] Ibid., p. 14. [194] WTO Doc. WT/TPR/OV/W/5/Rev.1, p. 17.

[195] WTO Docs. WT/TPR/S/264/Rev. 1, p. 207; WT/TPR/S/300/Rev.1, pp. 103–4. Other products subject to export state trading in China are rice, maize, cotton, coal, crude oil and processed oil, as well as tobacco. For agricultural commodities, the Chinese authorities maintain that state trading responds to the need to ensure food security, whereas in the case of industrial raw materials the main goals are resource conservation and environmental protection. WTO Doc. G/STR/N/9/CHN/Add.1, 14 July 2003.

[196] WTO Docs. WT/TPR/S/264/Rev. 1, p. 61; WT/TPR/S/300/Rev.1, p. 103.

[197] Fliess *et al.*, 'Recent Developments', p. 41. [198] See OECD Inventory.

[199] European Commission 'Eleventh Report', pp. 101–2.

outbreak of the global financial and economic crisis in 2008–2009 marked the introduction and/or the adjustment of most of these measures, with a peak between 2010 and 2011 and, unlike other sectors, a new phase of intensification in 2013–2014.

All of the critical minerals and metals under scrutiny in this study have been affected by export restrictions. The most frequently used types of export restrictions are export licensing schemes, export taxes, export prohibitions and export quotas. Commonly, each type or group of minerals is subject to various categories of export restrictions. Often, the considered materials are taxed in more than one form (unprocessed, semi-processed and/or waste and scrap). The highest number of export restrictions has been applied to products used in steel-making and technology metals.

Within this context, the bulk of export restrictions are applied by a relatively limited number of countries, including several leading producers. In many cases, these countries individually command a substantial share of world exports. In others, export restrictions are cumulatively applied by those countries which jointly control a large share of world exports.[200] This is not only a reflection of the high geographical concentration of critical minerals and metals but also means that the proportion of total exports subject to export restrictions is particularly high.[201] China, in particular, maintains the most comprehensive export regime on minerals and metals, affecting almost the entire range of critical minerals and metals, and in any case all critical materials for which it enjoys a dominant position in the world markets. Other leading producers widely using export restrictions on several critical minerals and metals are the Russian Federation, Argentina, Indonesia, India and Ukraine. The majority of these countries often use more than one type of export restriction. A number of other countries, however, apply a single type of export restriction to various critical minerals and metals, or various types of export restriction to just a few of such materials (e.g. Vietnam, South Africa, Democratic Republic of the Congo, Ukraine and Venezuela).

These countries have rarely removed or relaxed existing export restrictions, mostly tending to maintain them and often even tightening them (e.g. by increasing export duty rates or reducing the annual quota amounts). Relevant exceptions to this trend are the measures that China implemented following the WTO Appellate Body's recommendations in the *China–Raw Materials* and *China–Rare Earths*

[200] Fliess *et al.*, 'Recent Developments', p. 27. [201] Ibid., pp. 13, 20.

disputes. Yet, in some cases China has compensated for the removal of such measures with the reinforcement of other types of export restrictions (e.g. it cancelled the export quotas on rare earth elements but concurrently announced tighter export licensing rules). Other countries such as the Russian Federation, Vietnam and Ukraine have also relaxed or eliminated a number of export restrictions, mostly export duties, in accordance with the specific obligations assumed under their post-1994 WTO accession protocols. In contrast, a number of original WTO Members have introduced new or stricter measures limiting exports of critical minerals and metals in the last few years. For example, South Africa and Indonesia have explicitly used export restrictions to support a more comprehensive policy of downstream processing promotion.

In light of all the foregoing, the proliferation of export restrictions imposed on critical minerals and metals may be seen as a 'vertical' phenomenon, inasmuch as it results from the multiplication, and increased restrictiveness, of the number of measures introduced or maintained by a nucleus of leading producers rather than from the multiplication of the number of countries resorting to such measures at a 'horizontal' level. Insofar as countries implementing export restrictions control large shares of world exports, the vertical proliferation of export restrictions on critical minerals and metals may influence world prices of and negatively affect access to such resources. Depending on the use that imposing countries make domestically of the restricted materials, these effects may trigger or accelerate major transformations in the composition of international trade patterns. It is therefore particularly important that the analysis of the policy goals attached to export restrictions in the sector complement the study of the economic impacts of such measures on world supply and demand patterns as well as on international prices.

3

Economic effects and policy goals of mineral export restraints

3.1 Introduction

The recrudescence of export restrictions on raw materials has generated growing concern among World Trade Organization (WTO) Members, as import-dependent countries have started to fear the effects of supply distortions in the international markets. In the minerals and metals sector, in particular, the proliferation of export restrictions has caused great alarm due to the indispensability of critical minerals and metals, the heavy geographical concentration of mine production of critical materials and the sector-specific dynamics of the mining industry, typically exposing import-dependent countries to higher supply risks in the event of trade disruptions (e.g. lagged response of supply, low demand elasticity). The invocation of non-economic concerns on the part of restricting leading producers has, moreover, further fuelled the debate; the revival of industrial policies in most of the countries concerned has in particular been scrutinized as rather being instrumental to the use of export restrictions as economic development tools, in apparent contradiction with their declared goals. Export restrictions have thus become an easy instrument to blame for the major transformations occurring in international trade patterns.

This chapter completes the empirical analysis of the phenomenon of export restrictions on critical minerals and metals by giving an overview of the economic implications and rationales attached to export restrictions in the minerals and metals sector. Section 3.2 explains the effects of export restrictions predicted by standard economic theory, followed by an account of the sector-specific factors that, in the case of minerals and metals (and particularly in the case of critical minerals and metals), may amplify the restrictiveness of export restrictions. Section 3.3 gives an account of the policy goals most frequently invoked by countries restricting exports of minerals and metals and discusses the core development question at the heart of the debate over the policy space to be left to WTO

Members to implement export restrictions in the sector. Section 3.4 illustrates some examples of the practical effects of export restrictions on critical minerals and metals drawn from empirical evidence. Finally, Section 3.5 provides some general conclusions on the effects of the vertical proliferation of export restrictions on critical minerals and metals as emerged in the first part of this study.

3.2 The economic effects of export restrictions

An increasing body of economic literature has investigated the economic effects of export restrictions in the last few years. The bulk of recent economic analyses have focused on the impact of export taxes, as the most transparent and directly measurable type of export restriction.[1] Some studies have also attempted to assess the impacts of quantitative export restrictions, such as export quotas and bans (modelled as a quota set to zero).[2] Yet, in these cases, the theoretical models are generally based on the assumption that the effects of an export quota are similar to the effects of a corresponding export tax. In principle, an export quota may in fact be converted into its equivalent export tax, that is, there exists a level of export tax that would produce the same impacts.[3] These types of models, however, fail to take into account factors such as the level of stringency and/or discretion in the administration of the quota.[4] The economic effects do in fact vary, the level of restrictiveness being equal[5], depending on how strict the export licensing requirements imposed in connection with the quotas may be.

[1] J. Bonarriva, M. Koscielski and E. Wilson 'Export Controls: An Overview of their Use, Economic Effects, and Treatment in the Global Trading System', US International Trade Commission, Office of Industries Working Papers, August 2009, p. 16.

[2] K.C. Fung and J. Korinek, 'Economics of Export Restrictions as Applied to Industrial Raw Materials', OECD Trade Policy Papers, No. 155 (OECD Publishing, 2014), pp. 18–20.

[3] S. Mitra and T. Josling, 'Agricultural Export Restrictions: Welfare Implications and Trade Disciplines', International Food and Agricultural Trade Policy Council (IPC) Position Paper, Agricultural and Rural Development Policy Series, 2009, p. 24. This is generally true in a model of perfect competition. Bhagwati, however, shows that tariffs may not have their quota-equivalent in situations of imperfect competition. See J. Bhagwati, 'On the Equivalence of Tariffs and Quotas', in R. Caves, H. G. Johnson and P. B. Kenen (eds.), *Trade, Growth, and the Balance of Payments – Essays in Honor of Gottfried Haberler* (Rand McNally, 1965), pp. 53–67.

[4] Fung and Korinek, 'Economics of Export Restrictions', p. 18.

[5] Intuitively, the tighter the level of the ceiling (and, thus, the higher the level of the corresponding export duty rate), the more the distortive effects of the measure increase. Mitra and Josling, 'Agricultural Export Restrictions', p. 24.

In extreme cases, licensing requirements may be so stringent as to amount to a *de facto* export ban.[6]

Bearing these factors in mind, the economic literature has devoted considerable attention to the analysis of the impact of export taxes and attributed similar effects to other types of export restrictions in general.[7] Special attention has been given to the economics of agricultural export restrictions and, in particular, food commodities in the context of the food crises.[8] A relatively smaller, but expanding, body of literature has started to investigate the economic implications of export restrictions applied specifically to industrial raw materials such as minerals and metals.[9]

3.2.1 *Effects predicted by theoretical economic models*

The literature on the economics of export restrictions applied to mineral resources has mainly used static models of perfect competition derived from standard economic theory.[10] In these studies, the effects of an export tax are normally assessed against two different scenarios: in the first one, the country imposing the restriction is 'small', in the sense that its share of exports is not large enough to influence world prices; in the second one, the restricting country is 'large', that is, with a share of

[6] European Commission, 'Eleventh Report on Potentially Trade-Restrictive Measures Identified in the Context of the Financial and Economic Crisis', 2014, http://horizon.hozint.com/2014/11/dg-trade-11th-report-on-potentially-trade-restrictive-measures/ (accessed December 2014), p. 105.

[7] Fung and Korinek, 'Economics of Export Restrictions', p. 18.

[8] See, e.g. A. Anderson and N. Nelgen, 'Trade Barrier Volatility and Agricultural Price Stabilization', *World Development* 40 (2011), 36; A. Bouet and D. Laborde Debucquet, 'The Economics of Export Taxes in the Context of Food Security', in OECD, *The Economic Impact of Export Restrictions on Raw Materials* (OECD Publishing, 2010); P. Giordani, N. Rocha and M. Ruta, 'Food Prices and the Multiplier Effect of Export Policy', WTO Economic Research and Statistics Division, Staff Working Paper ERSD-2012-08, https://www.wto.org/english/res_e/reser_e/ersd201208_e.pdf (accessed December 2014); R. Piermartini, 'The Role of Export Taxes in the Field of Primary Commodities', WTO Discussion Paper, Economic Research and Statistics Division, 2004, www.wto.org/english/res_e/booksp_e/discussion_papers4_e.pdf (accessed May 2014). For a detailed account see above, Introduction, Section I.2.1.

[9] See, in particular, J. Korinek and J. Kim, 'Export Restriction on Strategic Raw Materials and their Impact on Trade and Global Supply', in OECD, *The Economic Impact of Export Restrictions on Raw Materials* (OECD Publishing, 2010); J. Latina, R. Piermartini and M. Ruta, 'Natural Resources and Non-Cooperative Trade Policy', *International Economics and Economic Policy* 8 (2012), 177; Fung and Korinek, 'Economics of Export Restrictions'.

[10] A static analysis considers only 'changes to production, consumption and prices that result from the export tax alone'. Fung and Korinek, 'Economics of Export Restrictions', p. 18.

exports capable of affecting world market prices.[11] In both cases, export taxes induce a contraction of exports in the imposing country due to the fact that domestic producers will prefer to sell the taxed products at home rather than abroad to avoid bearing the cost of the tax.[12] The domestic supply of the taxed product will thus increase, thereby generating a reduction in the domestic price and, in turn, a decline in domestic production.[13]

When a country is large, however, the effects of an export tax also impact the world market. This is because the diversion of reduced exports onto the domestic market due to the price effect of export taxes (i.e. the fact that exporting the taxed product is more expensive) induces a contraction of world supply and a consequent increase in world prices.[14] In the case of a large country, therefore, export taxes will create a wedge between the domestic price of the taxed good and its international price up to the point where the differential between the latter and the former is equal to the tax.[15] Given the geographical concentration of mineral resources, the model that best describes the effects of export taxes on minerals and metals is that of the large country. The same holds true in the case of critical minerals and metals, whose geographical distribution and mine production are particularly uneven.

In a model of perfect competition, moreover, export quotas are conventionally considered as equivalent to export taxes.[16] The major difference between the two is that an export tax generates government revenues, whereas the recipient of the quota rent varies depending on how the quota is administered. In the case of a quota bidding system, for instance, the quota rent is still taken by the government, as if it were an export tax of equivalent level.[17] In the case where the quota is attributed by national authorities, depending on the allocation criteria adopted, the quota rent may be distributed to the exporting firms.[18]

Based on the peculiarities of the mining sector, the Organisation for Economic Co-operation and Development (OECD) recently developed a theoretical model of international oligopoly based on the Cournot-Nash

[11] See Piermartini, 'The Role of Export Taxes', pp. 3–4.
[12] Fung and Korinek, 'Economics of Export Restrictions', p. 15.
[13] Piermartini, 'The Role of Export Taxes', p. 3. [14] Ibid.
[15] Fung and Korinek, 'Economics of Export Restrictions', p. 17. [16] Ibid., p. 18.
[17] Ibid.
[18] The exporting firms may decide to act as 'single sellers' in order to raise the price of the restricted good in the market of the importing countries. D. Appleyard, A. Field and S. Cobb, *International Economics*, 7th edn (Boston, MA: McGraw Hill, 2010), cited in Fung and Korinek, 'Economics of Export Restrictions', p. 18.

model of oligopolistic (quantity-setting) firms.[19] Importantly, the effects on domestic and foreign markets predicted under this static model following the introduction and/or increase of an export tax are similar to those described in the models of perfect competition. In the domestic market (i.e. the market of the country imposing a restriction), the prices of the taxed good will decline, inducing the domestic firm to decrease production and accordingly reduce its exports of the product. Correspondingly, the foreign firm will be pushed to produce greater quantities of the taxed good due to its increased price in the world market. Demand for the taxed product will thus fall in the world market and increase domestically.[20] The Cournot-Nash duopoly model also predicts similar impacts on producers and consumers in the domestic and foreign markets when used to assess the effects of export quotas.[21] However, in a situation of imperfect competition, an export quota is considered more distortionary than an export tax yielding the same level of exports. In particular, due to the distribution of the quota rent, an export quota encourages collusive behaviours on the part of oligopolistic firms so that export quotas generally lead to higher prices in the world markets and a higher downward shift in the demand of the importing countries.[22] Moreover, export quotas do not allow for an increase in exports in response to an unexpected surge in demand for the restricted product in the importing countries.[23] These properties make export quotas (and, thus, export bans) welfare inferior to export quotas.[24]

3.2.2 *Changes in welfare and redistributional effects linked to export restrictions*

The distortionary effects of export restrictions translate into changes in welfare and related redistributional effects. When a large country imposes an export tax in a perfectly competitive market, the decline in domestic prices induces producers to produce and sell less, while domestic consumers

[19] According to the OECD, an international oligopoly would better reflect the reality of the mining sector, characterized by a relatively small number of mine producing and exporting countries, as well as a limited number of companies enjoying substantial market power. For simplicity, the OECD considered a duopoly, with two international firms – one domestic and one foreign – producing identical products. Ibid., p. 20.

[20] Ibid., p. 21. [21] Ibid., pp. 24–5. [22] Ibid., pp. 25–6. [23] Ibid., p. 25.

[24] Ibid., p. 26. It should be noted, however, that excessively high levels of export taxes have the same effect as an export ban. Piermartini, 'The Role of Export Taxes', p. 8. The rate at which export taxes act as export bans depends on the elasticity of demand for a specific raw material. Fung and Korinek, 'Economics of Export Restrictions', p. 26.

benefiting from lower prices are pushed to consume too much. These production and consumption distortions are known as the efficiency losses or social costs of export restrictions.[25] At the same time, the exporting country benefits from the terms of trade gain induced by the increase in the world price of the exported (taxed) product.[26] Depending on whether this gain is greater than the social costs associated with the export tax, the net welfare effect in the exporting country may be positive or negative.[27] On a global level, however, the terms of trade gain experienced in a large exporting country is counterbalanced by the deterioration of the terms of trade in the importing country, while the latter also bears the social costs of the export tax.[28] Thus, although capable of ameliorating the welfare of a large exporting country, an export tax always unambiguously reduces global welfare.[29] In other words, an export tax is a 'beggar-thy-neighbour' policy, in the sense that it benefits the imposing country by making other countries worse off.[30]

An export tax also has redistributional effects in both the exporting and the importing countries. In the exporting country, producers will suffer from the decline of domestic prices, whereas consumers will benefit from cheaper inputs as if they were indirectly subsidized.[31] In other words, as noted by Fung and Korinek, an export tax 'actually transfers welfare from the sector producing the raw commodity to the processing industry that uses it'.[32] The opposite is true for the importing country.[33] The public sector will also benefit from the revenues generated by the introduction or increase of an export tax.[34]

25 Piermartini, 'The Role of Export Taxes', p. 4; Fung and Korinek, 'Economics of Export Restrictions', p. 17.

26 The expression 'terms of trade' refers to the ratio between the price of a country's exports and the price of its imports. A terms of trade gain indicates an increase in a country's welfare, since it implies an improvement in its real income. P. Krugman and M. Obstfeld, *International Economics: Theory and Policy*, 6th edn (Pearson Addison-Wesley, 2003), p. 101.

27 In the case of a small country, there is no terms of trade gain because there is no variation in the world price of the taxed good. Hence, the welfare implications of an export tax are unequivocally negative. Piermartini, 'The Role of Export Taxes', p. 4.

28 In the importing country, producers produce too much as a result of higher world prices, whereas consumers reduce consumption due to a loss in income. Ibid.

29 Fung and Korinek, 'Economics of Export Restrictions', p. 17.

30 Krugman and Obstfeld, *International Economics*, p. 544.

31 Piermartini, 'The Role of Export Taxes', pp. 4–5.

32 Fung and Korinek, 'Economics of Export Restrictions', p. 17.

33 Piermartini, 'The Role of Export Taxes', p. 5.

34 This is indeed one main difference between an export tax and an export quota. Fung and Korinek, 'Economics of Export Restrictions', p. 25.

Similarly, the net welfare effect of an export tax in a situation of international oligopoly may differ for the exporting country depending on whether there are producers and consumers in both the exporting and importing countries or, instead, production and consumption only take place, respectively, in the exporting country and in the importing country. In the first scenario, the export tax may be beneficial for the country that imposes the export tax inasmuch as its rate is set at an optimal level. In the second scenario, the introduction or increase of an export tax always improves the domestic welfare of the exporting (producing) country.[35] In both cases, however, the benefits generated domestically by the export tax will be more than offset by the welfare losses experienced in the importing country, so global welfare will deteriorate.[36]

The Cournot-Nash oligopoly model developed by the OECD also helps to elaborate on the redistributional effects of export taxes when applied in the minerals and metals sector. Fung and Korinek adapted this model to the distinctive features of the international mineral industry. According to their analysis, assuming that a raw material firm and a downstream firm exist in both the exporting and importing countries, an export tax will create two distinct effects: a profit-shifting or rent-shifting effect and a job-shifting or employment-shifting effect.[37] The profit-shifting effect means that, by imposing an export tax, the government 'shifts rent away from its own [upstream] firm to the foreign [raw materials] firm'.[38] This is because the domestic firm reduces its output due to the decline in the domestic price of the taxed good, while the foreign firm increases its production as prices go up.[39] The job-shifting effect refers to the redistributional effects generated by an export tax at the level of downstream firms. In the domestic country, the downstream sector consuming the product that is subject to an export tax will benefit from cheaper inputs and accordingly expand its output, employment and profits, thanks to a lower marginal cost of production. In the importing country, in contrast, the downstream firm will reduce its output, employment and profits due to the higher marginal cost induced by the higher world price of the taxed material.[40] Overall, an export tax will thus benefit foreign firms operating in the upstream mineral sector of the importing country and domestic downstream firms in the exporting country.

Lastly, winners and losers in both the exporting and the importing countries are not any different in the case of export quotas. However, for

[35] Ibid., p. 23. [36] Ibid. [37] Ibid., pp. 22–4. [38] Ibid., p. 22.

[39] The model assumes that 'profits are positively related to outputs and market shares'. Ibid.

[40] Ibid., pp. 23–4.

the reasons mentioned above, export quotas lead to greater welfare losses than the corresponding level of export taxes.[41]

3.2.3 Sector-specific factors influencing the effects of export restrictions on minerals and metals

The minerals and metals sector presents distinct characteristics and market conditions that may influence the level of restrictiveness of export restrictions and the magnitude of their impact on specific economic actors in the exporting and importing countries. In the case of critical minerals and metals, these characteristics are, importantly, accentuated.

Generally, the restrictiveness of export restrictions is higher the more concentrated the production and exports are. As minerals and metals are often produced and exported by a limited number of large countries, export restrictions are likely to impact significantly on world prices and thus have greater welfare implications for the global economy. High geographical concentration also influences the elasticity of supply in the mineral sector which, in turn, is an important factor in determining the magnitude of the effects attributable to export restrictions.[42] In the short term, in particular, supply elasticity tends to be very low if mine production is heavily concentrated in the country imposing the export restriction. Higher world prices would in principle push mining companies to develop additional production capacity in other countries that have reserves but, due to decision and implementation lags, increased levels of output may require some years to materialize.[43] At the same time, the uncertainty of the policy is likely to affect the medium- to long-term investment response due to the high capital costs and long-term payback perspectives characterizing the mineral sector. In other words, mining companies will tend to 'engage in less investment than they normally would if they were responding to changing market conditions than a policy that may be altered'.[44]

[41] Ibid., pp. 24–6.

[42] Supply elasticity is defined as 'the percentage increase in the quantity of a good produced following a one per cent increase in its price'. Similarly, demand elasticity indicates 'the percentage fall in the demanded quantity of a good as its price increases by one per cent'. The higher the elasticity the higher the responsiveness of the demand and supply to price variations. Piermartini, 'The Role of Export Taxes', p. 6.

[43] See above, Chapter 1, Section 1.3.1.

[44] Fung and Korinek, 'Economics of Export Restrictions', p. 27. In this respect, export restrictions may be used to rationalize entrepreneurial decisions and discourage the overexpansion of exploration and extraction activities in a period of sharp increases of

The impact of export restrictions will also depend on the elasticity of demand. The impact of an export quota is bigger in the case of inelastic demand, whereas an export tax will produce greater welfare losses when demand is responsive to price variations.[45] In the mineral industry, elasticity tends to be low in the short-run when minerals and metals are difficult to substitute, especially if they are used in trace amounts or in the manufacture of strategic products.[46] The use of export restrictions will, however, foster technological innovation so that the substitutability rate will increase and/or smaller quantities of the restricted minerals will be needed in the medium to long term.[47] Yet, in the short to medium term, the structural characteristics of the mineral industry may likely contribute to amplify the welfare losses induced by export restrictions.[48]

Other sector-specific factors may influence the precise impact on and reaction of upstream and downstream producers affected by export restrictions, as well as the choice of the type of export restriction on the part of governments. Fung and Korinek have examined these factors in detail.[49] A main factor is the ownership of the raw materials and downstream firms. In the mineral sector, it is frequently the case that the upstream firm is foreign owned, and often a multinational affiliate.[50] In these circumstances, as Fung and Korinek note, 'the multinational raw materials producers may react by shifting investment from the country imposing the restriction to its mining operations elsewhere in order to benefit from higher world prices and avoid the restrictive measures'.[51] Under this scenario, the government would prefer export taxes over quantitative restrictions in order to collect revenues from foreign-owned companies who would have employed their profits abroad.[52] Yet, the welfare gain produced by government revenues will decline over time, as companies decrease investment in mining and exploration activities in the exporting country.[53] In contrast, when mineral raw materials producers are state-owned enterprises, the government may gain from imposing an export quota, while the level of production may

world prices deemed to be temporary in the international markets. J.C. Gomes-Sabaini, 'Role of Export Taxes', in V. Tanzi (ed.), *Fiscal Policy in Open Developing Economies* (International Monetary Fund, 1998), p. 46.

45 Fung and Korinek, 'Economics of Export Restrictions', p. 19.

46 Korinek and Kim, 'Export Restrictions on Strategic Raw Materials', p. 104.

47 Fung and Korinek, 'Economics of Export Restrictions', p. 27.

48 Gomes-Sabaini, 'Role of Export Taxes', p. 47.

49 Fung and Korinek, 'Economics of Export Restrictions', pp. 26–32. 50 Ibid., p. 23.

51 Ibid., p. 29. 52 Ibid., p. 23. 53 Ibid., p. 33.

not shrink in order to let domestic downstream firms benefit from access to cheaper inputs.[54]

The incentive for reallocating profits from the upstream mineral sector to the downstream sector by means of an export restriction may be particularly strong in those developing countries whose mineral exports make up a large proportion of gross domestic product (GDP). This is because of the job-shifting effect induced by these measures, [55] which leads to lower levels of employment in the upstream mineral sector but generates jobs in more labour-intensive downstream sectors.[56] Yet, in certain low-income countries, the downstream sector may not be developed enough or have the necessary technology to initiate or expand production and generate the higher profits associated with higher value-added mining production. In these cases, governments imposing export restrictions may match these instruments with policies aimed at creating a favourable climate for investment in their downstream industry. Downstream foreign firms may thus be encouraged to either sell their proprietary processes and/or know-how or directly transfer some of their production to the country imposing the export restrictions.[57] The potential benefits linked to the introduction of an export restriction for the domestic downstream industry will, however, be diminished if other countries, especially other large producers, react by introducing similar measures as a response. In this scenario, world prices will go up further, potentially encouraging escalation in the use of export restrictive measures and in any case creating greater uncertainty for markets and trading partners.[58]

Finally, in the minerals and metals sector, the imposition of an export restriction entails different effects for complementary and substitute products, and may produce complex dynamics when a mineral product can be extracted as a carrier metal in primary mines or as a by-product (such as in the case of minor metals) depending on whether the export restriction is imposed on the former or on the latter.[59]

[54] Ibid., p. 29. [55] Ibid., p. 24.

[56] However, Fung and Korinek warn that the drop in employment in the extraction of raw materials 'may bring particular challenges since many mines are located in geographically remote areas with few alternatives for employment'. Ibid., p. 27.

[57] Ibid., p. 27.

[58] B. Fliess, C. Arriola and P. Liapis, 'Recent Developments in the Use of Export Restrictions in Raw Materials Trade', in OECD, *Export Restrictions in Raw Materials Trade: Facts, Fallacies and Better Practices* (OECD Publishing, 2014), p. 31.

[59] Ibid., p. 29.

3.3 Main policy goals of export restrictions in the minerals and metals sector

Countries applying export restrictions declare a wide variety of public policy goals. Some of them are specific to a certain type of restrictive measure, such as the generation of government revenues for export taxes. Others are more typically cited to justify the export restrictions used in specific sectors. For instance, the control of food price inflation and food security are among the most commonly declared rationales for applying export restrictions on agricultural commodities.[60] A number of economic and non-economic goals also recur in the minerals and metals sector. Among the former, the most frequently reported are, in addition to the generation of government revenues, the safeguarding of domestic supply and the promotion of downstream processing.[61] Developing countries, in particular, restrict exports of mineral products in an attempt to move up the value chain in order to achieve economic diversification.[62] In these cases, export barriers are defended as economic development tools.[63]

The non-economic public policy goals most commonly cited to justify mineral export restraints are environment-related goals such as mineral resources conservation and public health and/or environmental protection.[64] The invocation of such goals draws on the externalities linked to mining exploration and exploitation. Yet, the effectiveness of export restrictions as environmental tools has increasingly been subject to debate, especially when such instruments are applied to semi-refined/processed mineral products.[65]

3.3.1 *Export restrictions as developmental tools*

Export restrictions in the minerals and metals sector are most commonly used to promote and/or protect local downstream processing industries.

[60] See above, Introduction, Section 3.2.1.

[61] Fliess *et al.*, 'Recent Developments', pp. 25–6.

[62] M. Radetzki, *A Handbook of Primary Commodities in the Global Economy* (Cambridge University Press, 2010), pp. 188–99.

[63] For a full account see below, Section 3.3.1.

[64] B. Fliess, H. Gou and T. Mard, 'Taking Stock of Measures Restricting the Export of Raw Materials: Analysis of OECD Inventory Data', OECD Trade Policy Working Papers, No. 140 (OECD Publishing, 2012), pp. 15–6; Fliess *et al.*, 'Recent Developments', p. 26.

[65] Ibid., p. 25. For a detailed account see below, Section 3.3.3.

The general goal is to encourage economic diversification through the exportation of greater volumes of higher value-added products.[66] Among the countries turning to export restrictions in an attempt to subsidize the local downstream sector are a wide range of emerging economies, least developed countries and other low-income mineral-based economies. The *trait d'union* among them is the intention to move away from commodity dependence and a model of resource-led development.[67] The correlation between higher value-added exports and economic growth is in fact well established in a substantial body of economic literature.

On the one hand, as noted by Fliess *et al.*, 'prices of processed products usually are higher and do not fluctuate as much as primary commodities, processed products can be used locally as well as exported, the local workforce acquires new skills and more links are created with other sectors of the economy'.[68] On the other hand, based on the 'resource curse' theory elaborated by Sachs and Warner, mineral-dependent countries may experience slower economic performances due to the inverse association between natural resource abundance and economic growth.[69] Moreover, countries that excessively rely on mineral export earnings risk being affected by the so-called Dutch disease in periods of high

66 Piermartini, 'The Role of Export Taxes', p. 9; Fliess *et al.*, 'Taking Stock of Measures', pp. 15–16; Fliess *et al.*, 'Recent Developments', p. 26.

67 As Edward Leamer notes, international trade patterns have throughout most of the twentieth century followed a two-tier track: on the one hand, developed countries have traditionally imported raw materials and exported sophisticated manufactured goods; on the other hand, developing countries have exported raw materials and imported higher value-added products. E. E. Leamer, *Sources of International Comparative Advantage: Theory and Evidence* (MIT Press, 1984), Tables 4.4 and 4.5.

68 Fliess *et al.*, 'Recent Developments', p. 25. For an account of the importance of diversifying the export base see R. Hausmann and D. Rodrik, 'What You Export Matters', *Journal of Economic Growth* 12 (2007), 1.

69 See J. Sachs and A. Warner, 'Natural Resources and Economic Development: The Curse of Natural Resources', *European Economic Review* 45 (2001), 827. It should be noted, however, that the resource curse theory is not undisputed. A whole body of economic literature has provided numerous examples of successful stories of resource-led development. A. Maddison, 'Explaining the Economic Performance of Nations', in W. Baumol, R. Nelson and E. N. Wolff (eds.), *Convergence of Productivity: Cross-National Studies and Historical Evidence* (Oxford University Press, 1994); P. Maxwell, 'Chile's Recent Copper Driven Prosperity', *Minerals and Energy* 19 (2004), 16; G. Wright and J. Czelusta, 'The Myth of the Resource Curse', *Challenge* 47 (2004), 6; R. Findlay and M. Lundhal, 'Resource-Led Growth – A Long-term Perspective: The Relevance of the 1870–1914 Experience for Today's Developing Economies', World Institute for Development Economics Research Working Paper No. 162/1999.

commodity prices.[70] The expression refers to the process by which mineral abundance may lock countries into a pattern of mineral-dependence. The process develops as follows: during times of booming mineral prices, the large external surpluses generated from mineral export earnings induce the appreciation of a country's currency; as a result of the appreciation, other exports become more expensive and, therefore, the manufacturing sector of the mineral-abundant country loses competitiveness; investment and output will thus decline, ultimately leading to the shrinkage of the manufacturing sector up to the point where the country becomes a mono-economy.[71]

Drawing on this paradigm, developing countries have increasingly purported to use export restrictions to foster the expansion of their industrial structure and thus reap the benefits of mineral abundance exploitation domestically. In the case of low-income mineral-based countries, dependent on the exportation of a few mineral commodities for a large share of their GDP and with a weak industrial base, export restrictions are mainly applied on unprocessed mineral products in order to promote the development of infant downstream firms.[72] The upgrade of the trade and export structure is considered as a means of reducing vulnerability to the commodity price cycle.[73] In the case of resource-abundant emerging economies with a more diversified export base, export restrictions have been regarded as instrumental in consolidating

[70] The expression was introduced in the late 1970s to describe the shrinkage of the manufacturing sector experienced by the Netherlands throughout the decade as a consequence of the excessive exploitation of natural gas earnings. See G. Davies, 'Learning to Love the Dutch Disease: Evidence from the Mineral Economies', *World Development* 23 (1995), 1765. However, other notable examples occurred more than a century ago, led by the boom of gold, guano and sugar exports in Australia, Chile and Peru, and Cuba, respectively. Radetzki, *A Handbook of Primary Commodities*, p. 207.

[71] Ibid., pp. 207–10. According to Radetzki, the exploitation of a mineral boom would not be harmful *per se* if it was not for the fact that 'the bonanza often ends with a bang'. Similarly, Davies posits that 'the Dutch disease truly becomes a problem if there exists some market failure inhibiting an appropriate structural adjustment or if there is some existing distortion in the economy which is intensified by the mineral export boom'. See G. Davies, 'Trade in Mineral Resources', WTO Economic Research and Statistics Division, Staff Working Paper ERSD-2010-01, www.wto.org/english/res_e/reser_e/ersd201001_e.pdf (accessed December 2014), p. 16.

[72] According to the infant industry argument, newly established industries will benefit from temporary protection until they can compete with foreign industries and potentially develop a comparative advantage. Piermartini, 'The Role of Export Taxes', p. 10.

[73] H. Bargawi, E. Bova, B. Ferrarini and S. Newman, 'Low-income Countries and Commodity Price Volatility', in T. Cottier and P. Delimatsis (eds.), *The Prospects of International Trade Regulation* (Cambridge University Press, 2010), pp. 464–71.

(and at times accelerating) an ongoing process of economic transition and thus have also more consistently been applied to semi-refined/processed mineral products.[74]

Not surprisingly, the use of mineral export restraints on the part of countries such as China, the Russian Federation, India, Argentina and, most recently, South Africa and Indonesia has caused the greatest alarm among net importing industrialized countries. Advanced economies have not only objected to the proliferating use of export restrictions in these countries, but have noted with concern that such measures were adopted within the context of comprehensive industrialization plans aimed at strengthening key manufacturing sectors. According to the European Commission, recent years have shown 'a consolidation of trade measures as part of industrial policy',[75] while 'the conditions of access to natural resources can and are deteriorating quickly, through easily imposable border measures'.[76] The Commission has further posited that the unprecedented increase in the use of export restrictions 'represents a great danger to the effective functioning of value chains, and can eventually harm its own perpetrators, as no country is self-sufficient in terms of resources needs'.[77]

Yet, emerging economies have tightened and/or expanded the reach of their industrial strategies in the last few years. Paradigmatic of this trend is, once again, the case of China. In 2011, China launched its Twelfth National Five-Year Plan (2011–2015). The plan aimed at strengthening various traditional industries, including iron/steel and non-ferrous metals, construction, equipment manufacturing, shipbuilding and automotive, as well as fostering the expansion of seven leading-edge manufacturing sectors (energy conservation and environmental protection, new generation information technology, biotechnology, high-end manufacturing equipment, new energy and materials and new-energy vehicles).[78] As a follow-up to the National Plan, China has also elaborated a number of sector-specific plans envisaging targeted objectives for the reinforcement of 'national champions' in different industries, including the raw materials industry. Such plans encompass the identification of strategic industrial development policies, and generally match these

[74] Fliess *et al.*, 'Recent Developments', p. 25.

[75] European Commission, 'Ninth Report on Potentially Trade-Restrictive Measures Identified in the Context of the Financial and Economic Crisis', 2012, http://trade.ec.europa.eu/doclib/docs/2012/june/tradoc_149526.pdf (accessed December 2014), p. 9.

[76] European Commission, 'Eleventh Report', p. 13. [77] Ibid., p. 38.

[78] European Commission, 'Ninth Report', pp. 9–10.

policies with trade-restrictive measures such as export restrictions. For instance, the export restrictions on rare earth elements were tightened as part of the latest Five-year Rare Earths Industry Development Plan drafted by the country's Ministry of Industry and Information Technology.[79]

Other countries using mineral export restraints along with comprehensive industrial programmes include India, South Africa and Indonesia. India, for instance, approved the National Manufacturing Policy 2011 on 25 November 2011. The policy was the first of its kind in fostering high economic growth rates in the country throughout the 2010s.[80] South Africa is currently implementing the Second Industrial Policy Action Plan (2013–2016) with the goal of supporting a wide range of industrial sectors, including metals, transport equipment, green energy, ships and automotive, aerospace, defence and electrotechnical.[81] Lastly, Indonesia has recently launched the regulation on Increasing Value-Added Minerals Through Processing and Refining, which greatly relies on the use of export restrictions.[82]

Significantly, the recent revival of industrial policy in emerging economies has focused on supporting the upgrade of strategic manufacturing sectors that make intensive use of critical minerals and metals. Depending on the stage of development of each economy, the focus has either been on downstream sectors that directly process mineral raw materials (e.g. the metals industry) or on manufacturing sectors further down the value chain, which use critical minerals and metals in their semi-refined/processed forms in final products.

In as much as the use of export restrictions has proved instrumental in ensuring the success of these comprehensive industrialization plans, export restrictions have become a major point of contention between the Western world and the most advanced developing countries. The emerging economies have repeatedly defended these measures as legitimate developmental tools,[83] while net-importing industrialized countries have firmly contested mineral export restraints, both politically and legally

[79] European Commission, 'Fifth Report on Potentially Trade-Restrictive Measures Identified in the Context of the Financial and Economic Crisis', 2009, http://trade.ec.europa.eu/doclib/docs/2009/november/tradoc_145270.pdf (accessed November 2014), p. 52.

[80] European Commission, 'Ninth Report', p. 10. [81] Ibid.

[82] European Commission, 'Eleventh Report', pp. 101–2.

[83] See, e.g. WTO Doc. WT/MIN(11)/ST/19, 16 December 2011, cited in WTO Doc. WT/TPR/S/277, 13 February 2013, p. 96; WTO Doc. WT/TPR/M/249/Add.2, 28 October 2011, p. 45.

before the WTO dispute settlement bodies, as beggar-thy-neighbour instruments.[84] The question of the policy space left to WTO Members to impose export restrictions for development purposes has thus gained prominence in the international trade debate while at the same time actualizing the long-standing contrast between import-dependent industrialized economies and resource-abundant developing countries in the multilateral trading system.[85]

3.3.2 *Export restrictions as environmental tools*

As mentioned, among the non-economic public policy goals cited in justification of export restrictions applied to minerals and metals, the most frequently invoked rationales are environment-related. Countries often declare that export restrictions are meant to either conserve exhaustible mineral resources, or protect public health or the environment.[86] In other words, export restrictions would serve to correct the negative environmental externalities associated with mining activities.[87] Indeed, because mineral resources are non-renewable, quantitative export restrictions may be used in certain cases to 'achieve optimal rates of extraction in the case that private sector producers would have an incentive to extract minerals faster than it is optimal'.[88] The same holds true for export taxes of equivalent level inasmuch as they are expected to lead to the same reduction in domestic production.[89] A slower pace of extraction would, in turn, mitigate the environmentally harmful effects linked to the exploitation of mining products, including air, water and land pollution, energy and water consumption, land alteration, deforestation and consequent damage to public health and ecosystems.[90] Ultimately, the use of export restrictions as environmental tools purportedly serves to ensure the achievement of sustainable development.[91]

[84] For a full account see below, Chapter 6. [85] See below, Chapter 7.

[86] Korinek and Kim, 'Export Restrictions on Strategic Raw Materials', p. 110; Fliess *et al.*, 'Recent Developments', pp. 25–6.

[87] An externality occurs when an economic activity impacts on a party that is not directly involved in such activity. It follows that the price does not reflect the full costs or benefits in the production or consumption of a certain good or service. WTO, 'World Trade Report 2010: Trade in Natural Resources', www.wto.org/english/res_e/booksp_e/anrep_e/world_trade_report10_e.pdf (accessed December 2014), p. 88.

[88] Fung and Korinek, 'Economics of Export Restrictions', pp. 19–20.

[89] See above, Section 3.2.1.

[90] OECD, 'Minerals and Pro-Poor Growth: The Economics and Politics', DAC Guidelines and Reference Series (OECD Publishing, 2008), pp. 138–9.

[91] For a full account see below, Chapter 6.

Among the countries applying mineral export restraints, those restricting exports on environmental grounds are China, India, Indonesia, South Africa and Malaysia.[92] In some cases, such instruments have been used in conjunction with domestic measures aimed at tightening control over exploration and exploitation of mineral resources.[93] For instance, China has lately adopted a series of measures with the stated goal of rationally and sustainably managing the supply and use of its mineral resources.[94] Accordingly, it has consistently reiterated the centrality of environmental concerns in its export regime on minerals and metals, and accordingly put environment-related goals such as resource conservation and public health protection at the heart of its defending strategy before the WTO dispute settlement bodies in two disputes, *China–Raw Materials* and *China–Rare Earths.*[95]

All this notwithstanding, the use of export restrictions as genuine environmental tools has been subject to increasing controversy. Objectors have in particular argued that the supposed contraction in domestic production predicted by theoretical economic models may in practice not materialize if the restricted mining products are sold to expanding downstream industries in the domestic market.[96] This scenario is likely to occur in situations where the imposing country has a fast-developing manufacturing sector with large raw material-using downstream firms and, more so, in those countries that implement comprehensive industrialization programmes targeting the expansion of industries that avail themselves of restricted minerals and metals.

In light of this, evaluating the capacity of export restrictions to authentically respond to environmental concerns requires a case-by-case assessment. Yet, a whole body of literature has suggested that export restrictions may not be the most appropriate tools to achieve environment-related goals

[92] Fliess *et al.*, 'Recent Developments', p. 26.

[93] For an overview, see the latest TPRs of countries such as China, India, South Africa and the Democratic Republic of the Congo, particularly Section 3.4 on the topic of 'Mining Sectors'.

[94] W. M. Morrison and R. Tang, 'China's Rare Earths Industry and Export Regime: Economic and Trade Implications for the United States', US Congressional Research Service (CRS) Report for Congress, 30 April 2012, p. 16; B. Gu, 'Mineral Export Restraints and Sustainable Development – Are Rare Earths Testing the WTO's Loopholes?', *Journal of International Economic Law* 14 (2011), 765, p. 774.

[95] See Declaration by the Spokesman of China's Ministry of Commerce, 'China's Rare Earths Policy Justified', 16 March 2012, http://english.mofcom.gov.cn/aarticle/counselorsreport/europereport/201203/20120308027910.html (accessed January 2015). For a more detailed account see below, Chapter 6.

[96] Fung and Korinek, 'Economics of Export Restrictions', p. 34.

when compared with alternative options directly discouraging total production (or pollution) levels domestically (e.g. mineral royalties, resources taxes, volume restrictions on mining and production and pollution taxes).[97]

3.3.3 *Other policy objectives associated with export restrictions*

In addition to the above-mentioned goals, export restrictions are also used to achieve a wide range of economic and non-economic policy objectives. One of the most frequently cited motives for governments to resort to export taxes is the generation of government revenues.[98] Generally, countries using export taxes as a source of public finances are developing countries with relatively underdeveloped tax administration systems.[99] In a few other cases, including Argentina, export taxes are horizontally applied to all exports, thereby making up a large proportion of total tax revenues and total export value.[100] In these cases, however, export taxes may produce large inefficiencies, inasmuch as governments tend to overestimate expected revenues by considering constant levels of production.[101] In the case of Argentina, Noguès estimated that the removal of export taxes would increase its GDP by 2–4 per cent.[102] The maintenance of such instruments in these types of scenarios may respond to complementary policy objectives. For instance, the OECD has noted that '[a]lthough not stated as the rationale for the tax structure used, Argentina's export taxes favour exports of processed products over primary materials, and hence are supportive of local processing activities'.[103] Other economic goals frequently cited by low-income mineral-based economies include the control of exchange flows and inflation.[104]

[97] See, e.g. Korinek and Kim, 'Export Restrictions on Strategic Raw Materials', p. 119; M. Ruta and A. J. Venables, 'International Trade in Natural Resources: Practice and Policy', *Annual Review of Resource Economics* 4 (2012), 331, p. 346; Fung and Korinek, 'Economics of Export Restrictions', p. 34.

[98] Fliess *et al.*, 'Recent Developments', p. 26.

[99] Fung and Korinek, 'Economics of Export Restrictions', p. 34. In these cases, export taxes are second-best alternatives to more efficient systems of taxation based on the collection of income taxes or land taxes. Piermartini, 'The Role of Export Taxes', p. 14.

[100] In Argentina, export duties revenues amount to 20.5 per cent of total tax revenue and 15.6 per cent of total export value. See WT/TPR/S/277, p. 97.

[101] Fung and Korinek, 'Economics of Export Restrictions', p. 33.

[102] J. Noguès, 'The Domestic Impact of Export Restrictions: The Case of Argentina', IPC Position Paper, Agricultural and Rural Development Policy Series, 2008, cited in Fung and Korinek, 'Economics of Export Restrictions', p. 16.

[103] Fliess *et al.*, 'Recent Developments', p. 13.

[104] Bargawi *et al.*, 'Low-income Countries', pp. 468–71.

Exchange rate management may however run counter to the fulfilment of other development objectives.[105]

Finally, among the non-economic policy goals most commonly reported by countries using export restrictions is the need to control illegal export activity.[106] Many developing countries applying export licensing schemes and/or export bans have cited the need to monitor exports of metals, particularly waste and scrap, in order to prevent theft and smuggling induced by high scrap prices[107] or to check shipments and verify the origin or the purity of the materials.[108]

3.4 Empirical evidence on the effects of export restrictions on critical minerals and metals

Theoretical economic models describing the effects of export restrictions and declared policy objectives pursued by governments applying these instruments are important guiding elements that inform the empirical analysis of the impact of specific export restrictive measures. Yet, the precise effects of export restrictions may not always adhere to those predicted by static economic models, and case-by-case assessments may also contradict stated rationales. Furthermore, the analysis of the empirical impacts of export restrictions may be a complex exercise owing to several factors: (i) effects on world prices linked to export restrictions may be difficult to disentangle from more general price increases driven by growing demand from emerging economies and/or for new applications;[109] (ii) more than one type of export restriction is often applied simultaneously;[110] (iii) depending on the specific type of export restriction in use, several factors directly influencing its level of restrictiveness may be difficult to ascertain (e.g. the way an export quota is administered);[111] (iv) the reaction of other countries may directly jeopardize predictions and/or the achievement of pursued policy objectives.[112]

105 Fung and Korinek, 'Economics of Export Restrictions', p. 35. 106 Ibid., pp. 34–5.
107 Fliess *et al.*, 'Recent Developments', p. 25.
108 Korinek and Kim, 'Export Restrictions on Strategic Raw Materials', p. 110.
109 Radetzki, *A Handbook of Primary Commodities*, p. 66; Ruta and Venables, 'International Trade in Natural Resources', p. 338; G. Peeling, P. Stothart and B. Toms, 'Increasing Demand for and Restricted Supply of Raw Materials', in OECD, *The Economic Impact of Export Restrictions on Raw Materials* (OECD Publishing, 2010), p. 156.
110 Gomes-Sabaini, 'Role of Export Taxes', p. 46.
111 Fung and Korinek, 'Economics of Export Restrictions', pp. 9, 18.
112 Ibid., pp. 35–6.

Against this backdrop, an increasing body of economic literature has attempted to assess the precise impacts of specific export restrictions, either in multi-country and multi-product settings or focusing on single case studies.[113] Among the latter, efforts have mainly been directed to assessing the effects of export restrictions used by large countries and, in particular, those emerging economies experiencing high growth rates and adopting aggressive industrial policy strategies. A number of studies have also investigated the impact of export restrictions applied to critical minerals and metals.

The OECD has specifically assessed the impact of various types of export restrictions applied by China on critical materials, including coke, molybdenum and rare earth elements. In the case of coke, China's export restrictions created a substantial contraction of Chinese exports and a corresponding price differential of 241 USD per metric ton between world prices and Chinese domestic prices.[114] By 2008, Chinese steel producers had a cost advantage of almost 145 USD per metric ton over foreign companies, equal to more than 20 per cent of the international market price for carbon steel.[115]

China's export regime on certain unprocessed and semi-processed molybdenum products also induced a significant decline in Chinese exports. However, the drop in exports of the restricted materials was accompanied by a sharp increase (by 120 per cent) of exports of molybdenum articles further down the value chain.[116] Importantly, the rise in Chinese exports of intermediate molybdenum products was possible through the use of domestic below-the-world-price inputs on the part of Chinese downstream firms – an effect in apparent contradiction to the goals of environmental protection and resource conservation reiterated by the Chinese government, in pursuance of which domestic production of molybdenum would have had to decline.[117]

[113] See, e.g. Piermartini, 'The Role of Export Taxes'; Ruta and Venables, 'International Trade in Natural Resources'; O. Solleder, 'Trade Effects of Export Taxes', Graduate Institute Working Paper No. 8/2013, http://graduateinstitute.ch/webdav/site/international_economics/shared/international_economics/publications/working%20papers/2013/HEIDWP08-2013.pdf (accessed January 2015).

[114] A. H. Price and D. Scott Nance, 'Export Barriers and the Steel Industry', in OECD, *The Economic Impact of Export Restrictions on Raw Materials* (OECD Publishing, 2010), p. 91.

[115] Ibid. [116] Korinek and Kim, 'Export Restrictions on Strategic Raw Materials', p. 112.

[117] Fung and Korinek, 'Economics of Export Restrictions', p. 19.

China's export restrictions on rare earth elements have also created great distortions in the world market.[118] The use of export taxes and other tax measures such as the withdrawal of value-added tax (VAT) rebates determined a situation whereby Chinese companies were paying 31 per cent less for rare earth raw materials than their foreign competitors in 2008.[119] At the same time, access to supply was compromised for foreign consumers after China tightened annual quota amounts below global demand levels in 2009. The Chinese government justified the measures by stating that rare earth elements would be exhausted within twenty to thirty years with an unchanged rate of extraction.[120] Yet, China's export regime on rare earth elements *de facto* created a dual pricing regime, leaving Chinese rare earth processors with a competitive advantage.[121] The same was true for Chinese downstream producers of tungsten articles.[122]

Korinek and Kim have also studied the impact of India's export tax on chromium ore between 2007 and 2008. Total chromite production in India did not significantly decrease but was diverted to the domestic market, thereby helping Indian ferro-chrome producers who had been penalized until then by high chromite prices. Indian chromite exports decreased from 1,432,730 tons in 2006 to 550,532 tons in 2008.[123] China – the biggest importer of chromite – reacted by changing its source of imports from India to South Africa. As a result, South African exports of chromite increased by 200 per cent between 2006 and 2008, almost causing the South African government to introduce an export tax of its own in order to stop the Chinese surge of imports and keep exporting higher value-added chromium articles.[124] According to the OECD, 'had South Africa applied an export tax, it would have offset the impact of the Indian measure by reducing the price gap in the chromite exports of India and South Africa. Furthermore, such measures, by further reducing international supply, would have led to an even higher international price of chromite'.[125] The reaction of other countries and, in particular, other

[118] Korinek and Kim, 'Export Restrictions on Strategic Raw Materials', pp. 13, 19–22.

[119] Fung and Korinek, 'Economics of Export Restrictions', p. 31.

[120] MOFCOM Declaration, 'Ten Rare Metals Are Put Into Consideration for Strategic Stockpiles', cited in Gu, 'Mineral Export Restraints', p. 768.

[121] European Commission, 'Annex V', p. 209. [122] Ibid.

[123] Korinek and Kim, 'Export Restrictions on Strategic Raw Materials', pp. 114–15.

[124] Ibid., pp. 115–16. China's share of world chromite production is negligible (1 per cent in 2008). However, China accounts for around 20 per cent of world's ferro-chrome production.

[125] Ibid., p. 116.

large producers to the introduction of an export restriction may thus greatly influence the effectiveness of such instruments in ensuring the fulfilment of intended policy objectives.

3.5 Conclusions

The analysis of the economic effects of mineral export restraints and the rationales most commonly cited in justification of these instruments sheds light on the reasons why the proliferation of export restrictions in the minerals and metals sector has become a major point of contention between net importing Western economies and resource-abundant developing countries. The vertical proliferation of export restrictions applied on critical minerals and metals, in particular, presents at its extreme all the controversial elements raised by the use of export barriers. First, the relatively limited number of large countries mainly responsible for such proliferation control significant (if not dominant) export market shares and are thus capable of influencing world prices and availability of world supply. Second, these countries are in most cases emerging economies experiencing high levels of growth and rapid industrialization. Some of them have actively encouraged the development and expansion of key manufacturing sectors traditionally dominated by Western countries through the launch of comprehensive industrial policy programmes. Export restrictions are usually part of such programmes, although many countries have frequently invoked non-economic goals in justification of these measures, namely resource conservation and public health or environmental protection. Empirical evidence shows, however, that the precise impact of export restrictions often contradicts or impairs the declared environmental objectives.

Thus, the ambivalence of export restrictions – on the one hand, industrial policy instruments and, on the other hand, sustainable development tools – partially explains why the issue of export restrictions has exacerbated the classic divide between Western economies and resource-rich developing countries. The latter defend the right to use export restrictions as developmental tools in order to accelerate or consolidate a path of sustainable industrialization that would definitely 'unlock' them from a pattern of commodity dependence and resource-led development. Net importing industrialized countries endeavour to keep their dominant positions in the international markets in the midst of a global economic crisis and the emergence of new global players progressively challenging their manufacturing sectors. Inevitably, these dynamics have generated a certain degree of rhetoric in the debate over mineral export restraints.

As the first part of this study shows, however, the reality is much more complex and nuanced. The ongoing transformations occurring in the international minerals and metals industry and, more generally, in world trade patterns are the result of epochal geopolitical and geoeconomic changes that could not have been instigated solely by the use of export restrictions. Yet, the vertical proliferation of export restrictions has certainly been instrumental in the achievement of such transformations.

PART II

Existing WTO disciplines on export restrictions

Part I has explored the magnitude of the proliferation of export restrictions applied to minerals and metals and, in particular, critical minerals and metals. The objective of Part II is to identify World Trade Organization (WTO) rules that are relevant and applicable to all types of export restrictions in place in the minerals and metals sector, especially those imposed on critical minerals and metals. These include WTO disciplines on export duties (Chapter 4), WTO rules on export prohibitions and other quantitative restrictions on exports (Chapter 5), and the available exceptions for WTO-inconsistent export restrictive measures (Chapter 6). In Part I, cyclical and structural driving forces behind the proliferation of export restrictions on minerals and metals were identified. These revealed the existence of a certain degree of rhetoric over the use of mineral export restraints, which could arguably have a bearing on the emergence of a conventional wisdom on the presumed deficiency of WTO disciplines on the export side and the escalation of tensions concerning the employment of export restrictions on critical minerals and metals.

The analysis of existing WTO rules on export restrictions is the second step of the two-tiered analysis aimed at investigating whether and to what extent the proliferation of export restrictions on minerals and metals, as qualified in the first part of this work, could be deemed to result from a regulatory deficiency in the WTO legal framework. In particular, Part II aims to test the prevailing conventional wisdom for the identified narrow segment of critical minerals and metals, based on the peculiar features of the proliferation of export restrictions currently affecting such materials.

4

WTO disciplines on export duties

4.1 Introduction

The empirical analysis of the proliferation of export restrictions on raw materials revealed that export taxes or duties are among the most common types of export restriction. In particular, export duties are the second most frequently reported export restrictive measures in the minerals and metals sector. They are also used extensively for critical minerals and metals. As captured by the idea of vertical proliferation, the bulk of export taxes in place in this segment have been introduced and/or maintained by a relatively small number of large exporters. Significantly, among them are WTO Members that acceded to the Organization after its entry into force and assumed specific obligations on the use of export duties in their accession protocols in addition to those arising out of the General Agreement on Tariffs and Trade (GATT). While the negotiation of so-called WTO-plus obligations has become a common practice in the context of WTO accession, the export duty commitments undertaken by newly acceded WTO Members greatly vary in scope and coverage.

This chapter gives an account of the array of WTO obligations binding upon WTO Members on the use of export duties. Section 4.2 explains the historical reasons behind the 'asymmetry' in the treatment of import and export tariffs in the GATT. Section 4.3 illustrates the general GATT rules relevant and applicable to export duties. Section 4.4 gives an overview of the existing export duty commitments assumed in post-1994 accession protocols. Section 4.5 discusses transparency issues in the use of export duties. This is followed by some conclusions on the reach of existing WTO disciplines on export duties when seen through the lens of the vertical proliferation of export taxes currently applied to critical minerals and metals.

4.2 Asymmetry in the WTO treatment of import and export tariffs: origins and rationale

Since its early decades, the multilateral trading system has strongly focused on how to improve market access for foreign exporters.[1] The focus on import tariff reduction naturally developed as a result of the peculiar historical circumstances under which the GATT was designed. The GATT drafters were animated by the belief that the escalation of import tariffs triggered by the financial crisis of the 1930s had played a major role in pushing countries into the disaster of the Second World War.[2] Export taxes had been sparingly used until that point.[3] Moreover, the severe balance of payments deficits created by the conflict seemed to reasonably rule out countries having any interest in intensifying the use of export tariffs as a protectionist tool.[4]

Against this backdrop, the GATT resulted from an import tariff reduction negotiation concluded among twenty-three countries. Originally, it was conceived as part of a much more ambitious project aimed at creating the International Trade Organization (ITO).[5] The incorporation of many of the commercial policy provisions contained in the draft Havana Charter was intended to 'ensure that the agreed tariff reductions were not diluted or undercut by other trade measures'.[6] It was only after the definitive failure of the ITO project that the commercial policy role originally intended for the newly established organization was taken on by the GATT.[7] Yet, in its

[1] For a detailed account of the history of the GATT see J. H. Jackson, *World Trade and the Law of GATT* (Bobbs-Merrill Co., Inc., 1969).

[2] L. Ehring and G. F. Chianale, 'Export Restrictions in the Field of Energy', in Y. Selivanova (ed.), *Regulation of Energy in International Trade Law: WTO, NAFTA and Energy Charter* (Kluwer Law International, 2012), p. 112.

[3] P. C. Mavroidis, *Trade in Goods: The GATT and the Other Agreements Regulating Trade in Goods*, 2nd edn (Cambridge University Press, 2012), pp. 59–60, citing the Report of the US Committee on Relaxation of Trade Barriers of 8 December 1943, International Trade Files, Lot File 57D-284. Export duties were, however, used much more intensively during wartime. Ibid. See also N. J. Schrijiver, *Sovereignty Over Natural Resources: Balancing Rights and Duties*, 2nd edn (Cambridge University Press, 2008), pp. 4-5.

[4] See R. Gardner, *Sterling-Dollar Diplomacy in Current Perspective* (Columbia University Press, 1980).

[5] For an account of the ill-fated Havana Charter establishing the ITO see G. Sacerdoti, 'Havana Charter (1948)', in R. Wolfrum (ed.), *Max Planck Encyclopedia of Public International Law* (Oxford University Press, 2013), Vol. IV, pp. 718–725.

[6] WTO, 'World Trade Report 2012: Trade and Public Policies: A Closer Look at Non-Tariff Measures in the 21st Century', www.wto.org/english/res_e/booksp_e/anrep_e/world_trade_report12_e.pdf (accessed December 2014), p. 39.

[7] Ibid.

early stages, the GATT mainly continued to serve 'as a negotiating forum in which reciprocal, voluntary and non-discriminatory (MFN) tariff bargaining among member governments would lead to tariff "binding"'.[8]

Originally neglected due to the inherent dynamics leading to the GATT creation,[9] the issue of export tariffs was doomed to remain marginal to the agenda of the GATT contracting parties, the symmetry between import and export tariffs notwithstanding.[10] The negotiation of export duty concessions was considered for a long time 'a less pressing issue for the world trading system' compared with import tariff reduction.[11] The limited prevalence of export duties in the early decades of the GATT provides a partial explanation.[12] Political economic interests have also been instrumental in perpetuating the marginality of export taxes to the agenda of the GATT contracting parties.[13] As a result, as noted by Ehring and Chianale, 'scheduled import duty commitments nowadays cover practically all products in the case of most WTO Members, whereas this is by far not the case for export duties'.[14]

The central attention on import tariff reduction in the GATT, moreover, has induced a polarization in the way trade measures have been

[8] R. W. Staiger, 'Non-Tariff Measures and the WTO', WTO Economic Research and Statistics Division, Staff Working Paper ERSD-2012-01, www.wto.org/english/res_e/reser_e/ersd201201_e.pdf (accessed December 2014), p. 8.

[9] During the GATT negotiations, the United States had recommended the abolishment of export taxes, most probably as a way of preventing other countries from using an instrument that was already prohibited in the country as per Article I, §9, cl.5 of the Constitution (see E. M. Jensen, 'The Export Clause', *Florida Tax Review* 6 (2003), 6). See Report of the US Committee on Relaxation of Trade Barriers, pp. 35 ff. cited in Mavroidis, *Trade in Goods*, p. 59; D. A. Irwin, P. C. Mavroidis and A. O. Sykes, *The Genesis of the GATT* (Cambridge University Press, 2008), pp. 69–70. Commenting on the lack of success of the US negotiating position, Mavroidis notes that 'significant lack of practice in the field of export taxes [. . .] is probably the single most important reason explaining why the founding fathers did not spend time and effort designing a mechanism for negotiation of export tariffs à la Article II of the GATT'. Mavroidis, *Trade in Goods*, p. 59.

[10] The symmetry in the trade effects of import and export tariffs was first articulated by Lerner. According to the so-called Lerner symmetry, the social costs of an export tax are symmetrical to the social costs of an import duty. See A. P. Lerner, 'The Symmetry between Import and Export Taxes', *Economica* 3 (1936), 306.

[11] Staiger, 'Non-Tariff Measures', p. 8.

[12] Ibid.; Ehring and Chianale, 'Export Restrictions', p. 112.

[13] Mavroidis explains: 'exporters would not be necessarily interested in seeing their revenues divided between them and their government (that would pocket the export tax). Assuming that their lobby power matters to the government, one would expect few export taxes. The producers' interests are of course the exact opposite when it comes to import tariffs and this probably explains proliferation of import and almost total absence of export duties'. Mavroidis, *Trade in Goods*, p. 60.

[14] Ehring and Chianale, 'Export Restrictions', p. 112.

conceptualized in the multilateral trading system, so that import tariffs have long been considered separately from all other 'non-tariff barriers' (NTBs) affecting trade in goods.[15] The rationale behind this dualism is effectively explained by Staiger: 'it is import tariffs alone that are the policy measure with which negotiated market access commitments are made through negotiated tariff bindings and in this way, tariffs have a special place relative to all non-tariff measures in the GATT/WTO'.[16] The asymmetrical attention paid to export duties has thus contributed to dissociate such measures from their analogous ones on the import side, as if export duty concessions could not be negotiated following the same procedures governing the reduction of import tariffs.[17] Export duties or taxes have indeed consistently been categorized as non-tariff measures since the GATT Secretariat compiled the first Non-Tariff Measure Inventory in 1968.[18] This classification is however in apparent contradiction to the existence of a few scheduled export duty concessions negotiated in the GATT history and, more recently, in the Uruguay Round.[19] Moreover, it has determined a situation whereby export taxes have not, until very recently, been strongly perceived as a category in its own right, with peculiar trade effects that make it much more similar to import tariffs than to other non-tariff barriers. Rather, they have generally been considered as another type of export restriction.[20]

15 WTO, 'World Trade Report 2012', p. 38. 16 Staiger, 'Non-Tariff Measures', p. 6.

17 Export taxes have been considered non-tariff barriers since the time of the negotiation of the GATT. See E/PC/T/C.6/9, 23 January 1947/24 January 1947, and E/PC/T/C.6/W.4, 3 February 1947, cited in Irwin *et al.*, *The Genesis of the GATT*, p. 136. For a discussion on the possibility of scheduling export duty commitments and the enforceability of such commitments see below, Section 4.3.1.

18 See, for the GATT era, the Inventory on Non-Tariff Measures of the Committee on Industrial Products (GATT Doc. COM.IND/6, 11 December 1968 and various Addenda); the Inventory of Non-Tariff Measures compiled by the Secretariat (GATT Doc. COM. IND/W/102, 11 April 1973); and the Analysis of the Group's [Technical Group on Quantitative Restrictions and other Non-Tariff Measures] Documentation (GATT Doc. NTM(TG)/W/5, 28 February 1989, Annex 12). Moreover, the Indicative List of Notifiable Measures annexed to the 1993 Decision on Notification Procedures adopted at the Uruguay Round includes 'export taxes' in the category of non-tariff measures. See below, Section 4.5.

19 In 1947, the United Kingdom made a concession to the Malayan Union on export duties on tin ore and concentrates. In the Uruguay Round, Australia committed to refrain from imposing export duties on various raw materials (iron ore, titanium ore, zirconium ore, coal, peat, coke, refined copper, unwrought nickel, nickel oxide, and lead waste and scrap) in exchange for certain concessions made by the European Communities. The concessions were scheduled in both cases. See J. Y. Qin, 'Reforming WTO Discipline on Export Duties: Sovereignty over Natural Resources, Economic Development and Environmental Protection', *Journal of World Trade* 46 (2012), 1147, p. 1152.

20 In this sense, see also WTO Doc. WT/TPR/OV/W/5/Rev.1, 7 September 2011, pp. 17–8.

The concurrence of these factors has not only had an impact on the drafting of the GATT disciplines on export duties, but has also affected the interpretation and the implementation of these disciplines right up to the present WTO era.

4.3 Export duties or taxes in the GATT

4.3.1 Enforceability of export duty concessions included in GATT schedules

Tariff reduction is the cornerstone of the multilateral trading system. Customs duties, on either importation or exportation, are not prohibited *per se* in WTO law, but WTO Members are encouraged to negotiate their progressive reduction.[21] Article XXVIII:1(*bis*) GATT, labelled 'Tariff Negotiations', posits:

> The contracting parties recognize that custom duties often constitute serious obstacles to trade; thus, negotiations on a reciprocal and mutually advantageous basis, directed to the *substantial reduction of the general level of tariffs* and other charges *on* imports and *exports* ... are of great importance to the expansion of international trade. The contracting parties may therefore sponsor negotiations from time to time (emphasis added).

According to Article XXVIII:2(a)(*bis*), furthermore, '[s]uch negotiations may be directed towards the reduction of duties, the binding of duties at then existing levels or undertakings that individual duties or the average duties on specified categories of products shall not exceed specified levels'.

Article II GATT, labelled 'Schedules of Concessions', gives legal effect to the tariff concessions negotiated by WTO Members. Tariff binding commitments are set forth in so-called schedules of concessions, which are annexed to the GATT and incorporated into the GATT legal framework through Article II:7 GATT. As per Article II:1(b), in particular, WTO Members shall refrain from imposing ordinary custom duties in excess of scheduled tariff bindings. Yet, this Article expressly refers to duties imposed solely on importation.[22] No other GATT provision

[21] For a detailed account of the legal framework for the negotiation of tariff concessions in the GATT see Jackson, *World Trade and the Law of GATT*, pp. 201–47.

[22] More precisely, Article II:1(b) GATT gives legal effect to import duty commitments included in Part I of the WTO Members' GATT schedules of concessions ('MFN Concessions'). Moreover, Article II:1(c) GATT concerns import duty commitments inserted in Part II of GATT schedules of concessions ('Preferential Concessions'). The other paragraphs of Article II also only apply to importation. See Ehring and Chianale, 'Export Restrictions', p. 112.

explicitly envisages a parallel mechanism for scheduling and binding export duties.

Despite the formal lack of a legal framework, nothing in principle precludes that WTO Members negotiate export duty commitments in accordance with Article XXVIII(*bis*) GATT using a scheduling and binding procedure *à la* Article II:1(b) GATT.[23] Whether these commitments are enforceable or not in a manner similar to import duty bindings is, however, still subject to debate. Jackson argues that export duty commitments listed in GATT schedules cannot be considered tariff concessions within the meaning of Article II:1(b) GATT as this Article explicitly mentions import tariffs only. He concludes that export duty concessions cannot bind WTO Members as per Article II:1(b) GATT.[24] While Jackson's view has not remained isolated,[25] other authors have argued that export duty commitments are enforceable inasmuch as WTO Members list such commitments in their GATT schedules of concessions.[26] Interestingly, however, different views have been expressed as to the legal basis for enforcing export duty concessions.

Matsushita argues that export duty concessions are enforceable through Article II:1(b) GATT itself, the absence therein of an express reference to export notwithstanding. According to the author, the language of Article II:1(b) 'is merely indicative of the fact that the negotiators of the GATT 1947 were preoccupied with the elimination and reduction of import

[23] See also, among others, Irwin *et al.*, *The Genesis of the GATT*, p. 180; Mavroidis, *Trade in Goods*, p. 61; M. Matsushita, 'Export Control of Natural Resources: WTO Panel Ruling on the Chinese Export Restrictions of Natural Resources', *Trade Law and Development* 3 (2011), 267, pp. 273–4; Ehring and Chianale, 'Export Restrictions', pp. 113–14; D. Crosby, 'WTO Legal Status and Evolving Practice of Export Taxes', *Bridges* 12, No. 5 (October–November 2008), 3.

[24] Yet, the conclusion of John Jackson was based on the assumption that no WTO Member had ever listed export duty commitments in its GATT schedule. Jackson, *World Trade and the Law of GATT*, p. 499. See also Matsushita, 'Export Control', pp. 273–4; and Ehring and Chianale, 'Export Restrictions', p. 116. As mentioned above, however, export duty concessions have been granted at least twice (by the United Kingdom in 1947 and by Australia in 1994) since the creation of the GATT. Qin, 'Reforming WTO Discipline', p. 1152.

[25] See, e.g. M. G. Desta, 'OPEC, the WTO, Regionalism and Unilateralism', *Journal of World Trade* 37 (2003), 523, p. 533; R. Bhala, *Modern GATT Law: A Treatise on the General Agreements on Tariffs and Trade* (Sweet & Maxwell, 2005), p. 287.

[26] See, among others, F. Roessler, 'GATT and Access to Supplies', *Journal of World Trade* 9 (1975), 25; United Nations Conference on Trade and Development (UNCTAD), 'Trade Agreements, Petroleum and Supply Policies' (Geneva: United Nations, 2002), p. 2; Matsushita, 'Export Control', p. 274; Ehring and Chianale, 'Export Restrictions', pp. 112–17.

tariffs, and ... cannot be interpreted to necessarily imply that their intention was to exclude export duties altogether from the scope of Article II:1 (b)'.[27] On the other hand, Ehring and Chianale point to the language of Article II:1(a) GATT to support the enforceability of export duty commitments inserted into GATT schedules. Article II:1(a) GATT posits that 'each contracting party shall accord to the *commerce* of the other contracting parties treatment no less favourable than that provided for *in the appropriate Part of the appropriate Schedule* annexed to this Agreement' (emphasis added). The use of the word 'commerce' has been interpreted to leave Members free to negotiate commitments to be included in their GATT schedules both on the import and on the export side. Accordingly, the authors argue that Article II:1(a) GATT, as a self-standing legal obligation,[28] gives legal effect to such export duty concessions, while maintaining that 'it is beyond question that the wording of Article II of the GATT 1994 is precisely not symmetric, in that the long provisions in Article II:1(b) and II:1(c) speak about imports only and not exports'.[29]

According to both views, scheduled export duty concessions are incorporated into the GATT Agreement as per Article II:7 GATT. Yet, the enforceability through Article II:1(b) would automatically open the door to the applicability of the GATT-specific adjustment procedures contemplated for import duty concessions falling within the remit of Article II:2(b) GATT (e.g. the renegotiation procedure under Article XXVIII GATT or the requirements for the revocation of concessions under Article XVIII GATT) to scheduled and bound export duties.[30] In the latter view, however, these provisions 'either do not exist for export duties, or it is difficult to operationalize them in that context'.[31]

The question of whether export duty concessions are legally enforceable under the GATT, and under what basis, has not yet been brought before the WTO disputes settlement bodies. So far, only Australia has negotiated export duty concessions, and included them in Section 2 of Part I of its Schedule on 'MFN import tariff commitments on non-agricultural products' by means of a note which states: '[t]here shall be no export duty on

[27] Matsushita, 'Export Control', pp. 273–4. In this sense, Matsushita espoused the arguments of Roessler. See Roessler, 'GATT and Access to Supplies', pp. 33–5.

[28] According to the authors, the independent normative nature of Article II:1(a) GATT was confirmed by the Panel in *Korea–Various Measures on Beef*, paras. 774–9.

[29] Ehring and Chianale, 'Export Restrictions', pp. 115–17.

[30] See Matsushita, 'Export Control', p. 273. For a detailed account of the GATT provisions on the adjustment of import duty concessions see below, Section 4.3.2.

[31] Ehring and Chianale, 'Export Restrictions', p. 116.

this product'.[32] As export duties proliferate, however, the issue is likely to gain prominence on the international trade agenda, and countries may want to seek clarity before deciding to schedule and bind export duties. In this respect, the author shares the view that export duty concessions are enforceable through Article II:1(a) GATT. The capability of Article II:1(a) to provide a legal basis for giving legal effect to scheduled and bound export duties seems not only confirmed by the use of the word 'commerce', but also by its independent normative character[33] and the consequent broad scope which has already been attributed to it.[34] In particular, it should be noted that WTO Members can negotiate concessions relating to non-tariff barriers in Part III of their GATT schedules of concessions.[35] The possibility left to Members to schedule NTB concessions in a Part not yet expressly envisaged in the GATT 1947, that is, an 'appropriate Part of the appropriate Schedule', is clearly a product of the implementation of Article II:1(a) GATT. The non-enforceability of export duty concessions would thus be in apparent contradiction to the ample room of manoeuvre left to WTO Members to negotiate measures as diverse as the macro-category of non-tariff barriers may encompass under Article II:2(a) GATT, not to mention the fact that export duties have also – albeit wrongly in our opinion – been quite consistently classified as non-tariff barriers in the WTO.[36] Finally, the

[32] Australia committed to eliminate export duties on eleven HS 8-digit tariff lines (iron ore, titanium ore, zirconium ore, coal, peat, coke, refined copper, unwrought nickel, nickel oxide and lead waste and scrap). Initially given in exchange for certain import commitments undertaken by the European Communities, the concessions were then extended to all WTO Members under the most favoured nation clause. Australia's Uruguay Schedule, AUS1-201 to AUS1-204, cited in Qin, 'Reforming WTO Discipline', p. 1152. See also Ehring and Chianale, 'Export Restrictions', p. 114.

[33] The negotiating history of Article II also confirms the independent nature of Article II:1(a) GATT. Roessler, 'GATT and Access to Supplies', pp. 28–9, cited in Ehring and Chianale, 'Export Restrictions', p. 116.

[34] See GATT Panel Report, *United States–Restrictions on Imports of Sugar*, paras. 5.2 and 5.3, confirming the ample margin of flexibility left to WTO Members to decide what commitments to include in their GATT schedules (cited in Ehring and Chianale, 'Export Restrictions', p. 113).

[35] See WTO 'Goods Schedule: Members Commitments', www.wto.org/english/tratop_e/schedules_e/goods_schedules_e.htm (accessed January 2015). Lacking a clear definition of the non-tariff barriers schedulable under Part III, however, this mechanism has rarely been used except for some commitments concerning import licensing. Organization for Economic Co-operation and Development (OECD), 'Analysis of Non-Tariff Barriers: The Case of Export Duties', 31 January 2003, http://search.oecd.org/officialdocuments/displaydocumentpdf/?doclanguage=en&cote=td/tc/wp(2002)54/final (accessed December 2014), p. 7.

[36] See above, Section 4.2. Interestingly, Australia's Consolidated Loose-leaf Schedule includes the export duty commitments negotiated by the country in the Uruguay Round in Part III on 'Non-Tariff Barriers'. See Australian Government, Department of Foreign Affairs and

language of Article XVIII:1(*bis*) as relevant context for Article II GATT seems also to support this interpretative outcome.

In light of all the foregoing, WTO Members exclusively bound by GATT obligations as to their use of export duties remain free to voluntarily negotiate export duty commitments and list them in their GATT schedules of concessions. Once inscribed in GATT schedules, such concessions are incorporated into the GATT framework as per Article II:7 GATT and are therefore binding and legally enforceable by virtue of Article II:1(a) GATT. Among the WTO Members solely bound by standard GATT rules on export duties, Australia is to date the only one having assumed export duty commitments in the form of concessions included in its GATT Schedule.

4.3.2 *Differences in the treatment of import and export duty concessions in the GATT*

The GATT contains a number of provisions on the adjustment of scheduled concessions. These provisions, however, are mostly formulated in a way so as to specifically target import duty concessions. Their applicability to export duty commitments, albeit included in GATT schedules of concessions in the same way as import duty concessions, is therefore uncertain.

Article XXVIII GATT, labelled 'Modification of Schedules', sets forth a specific procedure for renegotiating scheduled tariff bindings, the so-called deconsolidation procedure. According to Article XXVIII:1 GATT, WTO Members may 'modify or withdraw a *concession* included *in the appropriate schedule annexed to this Agreement*' (emphasis added) on the first day of each three-year ('open season') renegotiation timeframe.[37] The modification or withdrawal should be negotiated and agreed:

Trade, 'Australia's WTO Tariffs Schedule: Consolidated Loose-leaf Schedule', www.dfat.gov.au/trade/negotiations/schedule/schedule.pdf (accessed January 2015). Yet, Australia's Consolidated Loose-leaf Schedule has not been certified at the request of Australia. See WTO Doc. G/MA/TAR/RS/85, 5 July 2002, cited in WTO, 'Current Situation of Schedules of WTO Members', www.wto.org/english/tratop_e/schedules_e/goods_schedules_table_e.htm (accessed January 2015).

37 The first open season period began on 1 January 1958 as per Article XXVIII:1 GATT. The latest one started on 1 January 2015. WTO Members are required to notify the intention to modify or withdraw no later than three months but not earlier than six months before the first day of the 3-year period. The agreement shall be reached before that date so that the modification or withdrawal can take effect on the first day of the open season timeframe. A. Hoda, *Tariff Negotiations and Renegotiations under the GATT and the WTO: Procedures and Practice* (Cambridge University Press, 2001), p. 15.

> with any contracting party with which such concession was initially negotiated and with any other contracting party determined by the Contracting Parties to have a principal supplying interest (which two preceding categories of contracting parties, together with the applicant contracting party, are in this Article hereinafter referred to as the 'contracting parties primarily concerned'), and subject to consultation with any other contracting party determined by the Contracting Parties to have a substantial interest in such concession.

As per Article XXVIII:2, WTO Members 'shall endeavour to maintain a general level of reciprocal and mutually advantageous concessions not less favourable to trade than that provided for in this Agreement prior to such negotiations', namely by conceding compensatory adjustment with respect to other products. Yet, Article XXVIII:3 clarifies that the failure to reach an agreement with the primarily concerned parties does not prevent a WTO Member from proceeding with the proposed modification or withdrawal, in which case the primarily concerned parties will also be free to withdraw substantially equivalent concessions. In this sense, Hoda suggests that 'the right to modify or withdraw a concession is absolute'.[38] Under the same substantive requirements, WTO Members can also enter into negotiations for the modification or withdrawal of concessions at any time in special circumstances authorized by the Ministerial Conference as per Article XXVIII:4 or before the end of the three-year period if any WTO Member elects to reserve the right to do so before the beginning of the period under Article XXVIII:5.[39]

The fact that Article XXVIII expressly refers to Members with 'a principal supplying interest' in a concession suggests at first glance an inherent 'bias' towards import duty concessions. Yet, many authors have advocated the applicability of Article XXVIII GATT to export duty concessions. As seen above, Matsushita assumes that any GATT adjustment procedures set forth for import duty concessions would be applicable to export duty commitments listed in GATT schedules as he considers the latter to fall within the scope of Article II:1(b) GATT.[40] Without addressing the question of whether export duty concessions are enforceable through Article II:1(a) or Article II:1(b) GATT, Qin infers from the fact that Article

[38] Ibid., p. 16.

[39] Approval of negotiation requests outside of the open season timeframe has become routine. Hoda, *Tariff Negotiations and Renegotiations*, p. 88. For an overview of the requests for modification or withdrawal of concessions in the WTO era see WTO, 'Current Situations of Schedules of WTO Members'. In the GATT era, 270 requests were filed under Article XXVIII. See Qin, 'Reforming WTO Discipline', p. 1159.

[40] Matsushita, 'Export Control', p. 273.

XXVIII is labelled 'Modification of Schedules' and refers to concessions 'included in the appropriate Schedule annexed to this Agreement' that Article XXVIII is 'equally valid and applicable to export concessions'.[41] This interpretation is not only supported by the wording of Article XXVIII (*bis*), which provides for relevant context, but is consistent with the interpretation of Article II:1(a) GATT given above. Moreover, it seems to be confirmed by the fact that the applicability of Article XXVIII was already explicitly envisaged for concessions relating to non-tariff measures as contained in Part III of GATT schedules, which are negotiated under Article II:1(a) GATT, as per Paragraph 6 of the Marrakesh Protocol.[42] As concessions contained in Part III of GATT Schedules may in principle incorporate export duty concessions, it seems reasonable to conclude that Article XXVIII GATT applies to export duty concessions, all the more considering that the unavailability of Article XXVIII GATT would leave WTO Members free to renegotiate measures as diverse as non-tariff barriers but not tariff-like measures such as export duties.

Article XXVIII is not the only GATT provision setting out an adjustment procedure for scheduled concessions. Other provisions, however, appear more specifically conceived for modifying or revoking import duty concessions under specific circumstances, which, *de facto*, makes them difficult to be operationalized on the export side. This is the case, for instance, with Article XVIII on 'Governmental Assistance to Economic Development'. The Article is designed to provide additional facilities for those countries 'the economies of which can only support low standards of living and are in the early stages of development'.[43] Article XVIII:2 states, in the relevant part:

> The contracting parties recognize further that it may be necessary for those contracting parties, in order to implement programmes and policies of economic development designed to raise the general standard of living of their people, to take *protective or other measures affecting imports*, and that such measures are justified in so far as they facilitate the attainment of the objectives of this Agreement. They agree, therefore, that those contracting parties should enjoy additional facilities to enable them (*a*) to maintain sufficient flexibility in their tariff structure to be able to grant the tariff protection required for the establishment of a particular industry . . . (emphasis added).

41 Qin, 'Reforming WTO Discipline', pp. 1160–1.

42 According to Paragraph 6 of the Protocol: '[i]n cases of modification or withdrawal of concessions relating to non-tariff measures as contained in Part III of the schedules, the provisions of Article XXVIII of GATT 1994 and the "Procedures for Negotiations under Article XXVIII" adopted on 10 November 1980 (BISD 27S/26–28) shall apply'.

43 See Article XVIII:1 and Article XVIII:4(a) GATT.

Under Article XVIII:7, one of these additional facilities envisaged to enable the countries referred to here to promote the establishment of a particular industry is the modification or the withdrawal of a concession. Article XVIII:7(a) reads:

> If a contracting party coming within the scope of paragraph 4 (*a*) of this Article considers it desirable, in order to promote the establishment of a particular industry with a view to raising the general standard of living of its people, to modify or withdraw a concession included in the appropriate Schedule annexed to this Agreement, it shall notify the Contracting Parties to this effect and enter into negotiations with any contracting party with which such concession was initially negotiated, and with any other contracting party determined by the Contracting Parties to have a substantial interest therein.

Article XVIII:7(a) is not import specific *per se*: similar to Article XXVIII, it refers to 'a concession included in the appropriate Schedule annexed to this Agreement' and does not even include the requirement to negotiate with Members 'having a principal supplying interest' as it is the case with Article XXVIII GATT. Yet, in contrast to Article XXVIII, whose relevant context suggests in favour of its applicability to export duty concessions, the relevant context for Article XVIII:7 seems to suggest otherwise for several reasons. First, Article XVIII:7 describes one of the additional facilities referred to under Article XVIII:2, which expressly limits its scope to measures affecting imports. Second, it is one of the two additional facilities identified under Article XVIII:3 and Article XVIII:4(a) as enabling a temporary deviation from other GATT provisions for countries with a low standard of living or at an early stage of development, the other one being the restriction of 'the quantity or value of merchandise permitted to be *imported*' to cope with balance of payment difficulties.[44] Moreover, Article XVIII also describes two 'special procedures' available to grant the governmental assistance required to promote the establishment of particular industries in those cases where no measure consistent with GATT provisions is practicable for countries with a low standard of living or at an early stage of development, or countries in the process of economic development. In both cases, those countries are required to 'indicate the specific measure *affecting imports* which [they] propos[e]'.[45] The non-applicability of Article XVIII to export duty concessions also

[44] See Article XVIII:9 GATT (emphasis added). See also Article XVIII:8–12. For a full account see below, Chapter 5, Section 5.3.2.

[45] See Article XVIII:13 (emphasis added). See also Article XVIII:22.

seems to be suggested by the language of Article XVIII:5. While this provision does specifically address the export side, it states:

> The contracting parties recognize that the export earnings of contracting parties, the economies of which are of the type described in paragraph 4 (*a*) and (*b*) above and which depend on exports of a small number of primary commodities, may be seriously reduced by a decline in the sale of such commodities. Accordingly, when the exports of primary commodities by such a contracting party are seriously affected by measures taken by another contracting party, it may have resort to the consultation provisions of Article XXII of this Agreement.

In other words, Article XVIII moves from the premise that countries at earlier stages of development may only lose from restricting primary exports and thus it contemplates enhanced tools to *encourage* export increases. From this perspective, it would seem logical to interpret the additional facilities meant to foster economic development under Article XVIII as targeting import commitments only.[46]

Finally, the GATT contains a so-called economic emergency exception allowing Members to withdraw or modify 'tariff concessions' as a way of implementing a safeguard measure under Article XIX. Yet, under Article XIX GATT and the related WTO Agreement on Safeguards[47], safeguard measures can be used as a safety valve in those cases where a surge of *imports* causes, or threatens to cause, serious injury to a domestic industry. Article XIX:1(a) GATT reads, in the relevant part:

> If, as a result of unforeseen developments and of the effect of the obligations incurred by a contracting party under this Agreement, including tariff concessions, any product is being *imported* into the territory of that contracting party in such increased quantities and under such conditions as to cause or threaten serious injury to domestic producers in that territory of like or directly competitive products, the contracting party shall be free, in respect of such product, and to the extent and for such time as may be necessary to prevent or remedy such injury, to ... withdraw or modify the concession (emphasis added).

[46] However, many countries resorting to export duties are least developed countries or low-income commodity-dependent countries aiming at fostering economic diversification and economic development. See above, Chapter 3.

[47] According to the existing WTO case law, safeguard measures need to comply with Article XIX GATT and the provisions of the WTO Agreement on Safeguards cumulatively. See, e.g., *Argentina–Footwear (EC)*, Appellate Body Report, para. 81.

Safeguard measures may thus only take the form of import tariff increases in derogation of Article II:1(b) GATT as per Article XIX GATT. In other words, assuming that a WTO Member scheduled and bound export duties for specific products as admitted by Article II:1(a) GATT, it would still be prevented from imposing on such products export duties in excess of those set forth in its GATT schedule even in the event that a sudden, sharp and substantial surge of exports of such products significantly impaired the position of a domestic industry.

In conclusion, inasmuch as export duty commitments listed in GATT schedules are enforceable through Article II:1(a) GATT, GATT adjustment procedures and exceptions relevant to import duty concessions may not always apply. While export duty concessions seem likely to fall within the scope of the renegotiation procedure described under Article XXVIIII GATT, the import-specific language of other GATT provisions such as Article XVIII and Article XIX renders their applicability to export duty concessions more difficult. This circumstance may well be the result of the fact that, at the time the GATT was drafted, import duty concessions were the only measures being negotiated. Yet, it introduces an element of asymmetry in the legal treatment of import and export duties under the GATT.

4.3.3 Admissibility of export duties under Article XI:1 GATT

Inasmuch as WTO Members have not negotiated export duty concessions in their GATT schedules, they remain free to introduce and/or maintain export duties as per Article XI:1 GATT. Article XI:1, labelled 'General Elimination of Quantitative Restrictions' reads:

> No prohibitions or restrictions *other than duties, taxes or other charges*, whether made effective through quotas, import or export licenses or other measures, shall be instituted or maintained by any contracting party on the importation of any product of the territory of any other contracting party or on the exportation or sale for export of any product destined for the territory of any other contracting party (emphasis added).

The exclusion of 'duties, taxes or other charges' on exportation (and importation) from the general elimination obligation imposed on the use of quantitative restrictions[48] reflects the preference for tariffs over

[48] For an account of the export quantitative restrictions covered, and thus prohibited, by Article XI:1 GATT see below, Chapter 5.

quantitative restrictions that is at the heart of the GATT architecture. Its broad formulation ('duties, taxes and *other charges*'), moreover, seems capable of encompassing other types of price measures applied on exports (e.g. export surtaxes).[49] Yet, Article XI:1 GATT does not admit export duties *ab absoluto.* As per a combined reading with Article II GATT, as interpreted above, Article XI:1 GATT can only allow WTO Members to use export duties on unbound tariff lines. In other words, a WTO Member that has scheduled and bound export duties for a certain number of tariff lines cannot apply to such products export duties that are in excess of the levels indicated in its GATT schedule. It does, however, remain free to introduce and/or maintain export taxes on any other product.

The margin of manoeuvre left to WTO Members to resort to export duties as per Article XI:1 GATT has not yet been tested in WTO jurisprudence. Many authors, including Matsushita[50] and Howse and Josling,[51] have suggested that prohibitive export duties (i.e. export duties *de facto* amounting to export bans) would run counter to Article XI:1 GATT. While predominant, this view is rejected outright by more recent literature. Ehring and Chianale, in particular, argue that 'the distinction between prohibitive and non-prohibitive restrictions is alien to Article XI:1 of the GATT 1994. The theory on prohibitive customs duties under Article XI:1 therefore departs from the basic design of that Article'.[52] Yet, it seems reasonable to assume that ban-equivalent export duties could not be used to circumvent the obligation to eliminate export prohibitions under Article XI:1 GATT.

Howse and Josling also suggest that the purpose and effect of an export tax should play a role in the evaluation of whether an export tax is admissible under Article XI:1 GATT. In their view, a combined reading of Article XI:1 and Article VIII 'clarifies that the disciplines in Article VIIII on fees and charges do not prevent the use of export taxes as permitted by Article XI:1, but it also suggests that what was understood by an export tax was a measure imposed for fiscal (e.g. revenue-raising) purposes'.[53] Moreover, they posit that such a purposive interpretation would be consistent with the general WTO case

[49] See above, Chapter 2, Section 2.4.4. [50] Matsushita, 'Export Control' p. 273.

[51] R. Howse and T. Josling, 'Agricultural Export Restrictions and International Trade Law: A Way Forward', International Food and Agricultural Trade Policy Council (IPC) Position Paper, 2012, http://www.agritrade.org/Publications/documents/Howse_Josling_Export_Restriction_final.pdf (accessed November 2014), p. 17.

[52] Ehring and Chianale, 'Export Restrictions', p. 120.

[53] Howse and Josling, 'Agricultural Export Restrictions', p. 17.

law on Article XI GATT.[54] While not addressing the question explicitly, Ehring and Chianale seem not to share the premise of such an interpretation when they argue that '[n]othing in the negotiating history or practice of the GATT suggests that the words "duties, taxes or other charges" in Article XI:1 of the GATT 1994 should be read narrowly' and that 'the text of Article XI:1 of the GATT 1994 seem[s] sufficiently clear and focused on form rather than effects, in that export duties are not covered by its disciplines'.[55] It should also be added that a purposive differentiation between permitted and non-permitted custom duties does arguably not exist for import tariffs, and Article XI:1 is structured entirely symmetrically on the import and export sides.[56] Moreover, distinguishing between an export tax imposed for fiscal purposes and an export tax applied for 'trade-restricting purposes' as the authors suggest may be practically very difficult.[57]

In conclusion, Article XI:1 GATT allows WTO Members to introduce and/or maintain export duties on any product, provided that for such product they have either not negotiated any export duty concessions in their GATT schedules or, if they have, the applied export duty rate does not exceed the level indicated in their GATT schedules. Excessively high export duties effectively amounting to an export ban are likely to fall foul of Article XI:1 GATT.

[54] According to the authors, '[t]he broad scope of Article XI as identified in the jurisprudence (*India–Autos*, report of the panel) and the application of purposive interpretation to Article XI (also endorsed in *India–Autos*) provides a basis for interpreting the meaning of export taxes narrowly, excluding those with predominantly trade-restricting as opposed to revenue-raising purposes'. Ibid. For a detailed account of WTO jurisprudence on Article XI see below, Chapter 5.

[55] Ehring and Chianale, 'Export Restrictions', p. 120.

[56] As per Article II:1(b), WTO Members are free to impose import duties, irrespective of their rationale, insofar as they are not in excess of levels set forth in their GATT Schedules. See above, Section 4.3.1. Moreover, on the import side, developing countries enjoy greater flexibility in the use of import duties in that they are not required to make concessions inconsistent with their 'development, financial and trade needs'. GATT Doc. L/4903, BISD 26S/203, Differential and More Favourable Treatment, Reciprocity and Fuller Participation of Developing Countries (The Enabling Clause), 28 November 1979, para. 5.

[57] See, among others, the case of Argentina described above in Chapter 3, Section 3.3.3. The view that export duties should be allowed only for fiscal purposes under Article XI:1 GATT would arguably raise strenuous objections on the part of those WTO Members that defend export duties as economic development tools.

4.3.4 *Other GATT provisions applicable to export duties*

A number of GATT provisions have a symmetrical structure in the sense that they apply to both imported and exported products equally. Some of these provisions are also applicable and relevant to export duties. According to the most favoured nation clause (MFN) embodied in Article I:1 GATT, WTO Members shall apply export duties on a non-discriminatory basis. Article I:1 GATT reads, in the relevant part:

> With respect to customs duties and charges of any kind imposed on or in connection with ... exportation, ... and with respect to the method of levying such duties and charges, and with respect to all rules and formalities in connection with ... exportation, any advantage, ... granted by any contracting party to any product ... destined for any other country shall be accorded immediately and unconditionally to the like product ... destined for the territories of all other contracting parties.

The MFN obligation covers any export duty applied by WTO Members, whether scheduled and bound or admissible within the scope of Article XI:1 GATT. Moreover, its broad formulation suggests that other types of tax measures imposed on exports, such as export surtaxes ('charges of any kind imposed on ... exportation'), shall also be applied on an MFN basis.[58] Importantly, the MFN requirement applies to 'like' products. Based on a consistent WTO jurisprudence, the like product analysis involves the assessment of four general criteria: physical characteristics, end uses, consumer tastes and habits and tariff classification of the products at issue.[59]

Article VIII GATT, labelled 'Fees and Formalities Connected with Importation and Exportation', is also symmetrically structured and, to a certain extent, relevant to export duties. According to Article VIII:1(a):

> All fees and charges of whatever character (other than import and export duties and other than taxes within the purview of Article III) imposed by contracting parties on or in connection with importation or exportation shall be limited in amount to the approximate cost of services rendered and shall not represent an indirect protection to domestic products or a taxation of imports or exports for fiscal purposes.

Article VIII distinguishes between export duties, on the one hand, and fees and charges levied at the border in connection with exportation, on

[58] See Ehring and Chianale, 'Export Restrictions', p. 110.

[59] *Japan–Alcoholic Beverages II*, AB Report, p. 114; *EC–Asbestos*, AB Report, para. 101. See also C. Won-Mog, *"Like products" in International Trade Law: Towards a Consistent GATT/WTO Jurisprudence* (Cambridge University Press, 2003).

the other hand. While the former are excluded from its scope of application (and therefore remain permitted under the conditions set forth in Article II and Article XI GATT), the latter shall not exceed the 'approximate cost' of the 'service' rendered to the exporter. According to WTO jurisprudence, a 'service' within the meaning of Article VIII refers to any 'government activities closely enough connected to the processes of customs entry that they might, with no more than the customary artistic licence accorded to taxing authorities, be called a "service" to the importer in question'.[60] In other words, Article VIII covers those fees and charges applied in connection with customs-related government activities such as 'consular transactions', 'statistical services', and 'analysis and inspection' indicated in the illustrative list contained in Paragraph 4 of Article VIII.[61] Importantly, Article VIII applies independently of whether the product in question is subject to tariff binding.[62] Moreover, the use of fees and charges in connection with exportation is permitted under Article VIII GATT to the extent that it does not 'represent ... a taxation of imports or exports for fiscal purposes'. As noted above, some have inferred from this requirement that export duties within the meaning of Article XI:1 would only be permitted when used for revenue-raising purposes.[63] Leaving the debate on the relationship between Article VIII and Article XI:1 GATT aside, this condition contributes to shedding light on the notion of fees and charges imposed in connection with exportation (and importation) admitted for the purposes of Article VIII. In this regard, the Article makes clear that only fees and charges corresponding to customs-related services fall within its scope of application. Such fees are to be considered unlawful if disproportionate to the cost of the services rendered, that is, they cannot be used for purposes other than 'repaying' the costs of such services (e.g. fiscal purposes). Accordingly, the other paragraphs of Article VIII GATT go in the direction of simplification of export fees and charges, recognizing 'the need for reducing the number and diversity of fees and charges' (Article VIII:1(b)), as well as 'minimizing the incidence and complexity of ... export formalities and ... decreasing and simplifying ... export documentation requirements' (Article VIII(c)). WTO Members are required 'to review the operation of [their] laws and regulations' upon request (Article VIII:2) and to abstain from imposing 'substantial

[60] Panel Report, *US–Customs Users Fee*, BISD 35S/245, para. 77. [61] Ibid., para. 76.
[62] See Ehring and Chianale, 'Export Restrictions', p. 118.
[63] Mitra and Josling, 'Agricultural Export Restrictions', p. 17. See above, Section 4.3.3.

penalties for minor breaches of customs regulations or procedural requirements' (Article VIII:3).[64]

Finally, Article X GATT covers the 'Publication and Administration of Trade Regulations' and contains a number of provisions relevant to export duties. Article X:1 states, in the relevant part:

> Laws, regulations, judicial decisions and administrative rulings of general application, made effective by any contracting party, pertaining to ... rates of duty, taxes or other charges ... shall be published promptly in such a manner as to enable governments and traders to become acquainted with them.

The requirement to publish all trade-related regulations in a prompt and accessible manner is complemented with the requirement to refrain from enforcing measures of general application prior to their publication in Article X:2 GATT. Yet, this provision is not structured symmetrically and refers only to 'measures ... effecting an advance in a rate of duty or other charge on imports'. However, Article X:3(a) GATT is also applicable to duties applied on exports, inasmuch as it requires WTO Members to 'administer in a uniform, impartial and reasonable manner' measures of 'the kind described in paragraph 1 of this Article'. No obligation of notification, however, exists under this Article.

4.4 Export duty commitments in post-1994 accession protocols: the emergence of WTO-plus obligations

In addition to standard GATT disciplines on export duties, a certain number of WTO Members that acceded to the WTO after its entry into force have assumed country-specific obligations limiting their use of export duties in their accession protocols and/or the related Working Party Reports. The negotiation of 'additional' obligations in the context

64 The provisions of Article VIII GATT are updated and enhanced in Article 6 of the WTO Agreement on Trade Facilitation approved at the 2013 WTO Ministerial Conference that took place at Bali from 3 to 6 December 2013 (see WTO Doc. WT/MIN(13)/36, WT/L/911, 11 December 2013). The Agreement has been annexed to the Protocol Amending the Marrakesh Agreement Establishing the World Trade Organization approved and opened for acceptance through the General Council Decision of 27 November 2014 (WTO Doc. WT/L/940, 28 November 2014). Upon the entry into force of the Protocol by acceptance of two-thirds of WTO Members in accordance with Article X:3 of the Marrakesh Agreement, the WTO Agreement on Trade Facilitation will be inserted into Annex 1A to the WTO Agreement, after the Agreement on Safeguards. See Paragraph 4 of the Protocol Amending the Marrakesh Agreement Establishing the World Trade annexed to the General Council Decision.

of the WTO accession process is not a new practice. It is based on Article XII:1 of the Marrakesh Agreement, according to which any country may accede to the WTO Agreement 'on *terms* to be agreed between it and the WTO' (emphasis added). Failing a precise delimitation of the 'terms' subject to negotiation, incumbent WTO Members have systematically requested candidate WTO Members to undertake commitments which either exceed the obligations arising out of Multilateral Trade Agreements (so-called WTO-plus obligations) or do not otherwise exist for WTO Members as they concern issues lying outside the WTO mandate (so-called WTO-extra obligations),[65] as a means to 'counterbalance' the benefits accruing from accessing already-liberalized markets of other WTO Members.[66] Despite the overall paucity of commitments in the area of export duties on the part of WTO Members exclusively bound by the GATT,[67] the export duty commitments undertaken by newly acceded WTO Members on the use of such instruments have been conventionally categorized as WTO-plus obligations due to the possibility for WTO Members to assume obligations in the area in the form of export duty concessions to be included in their GATT schedules.[68]

Not all the newly acceded WTO Members have undertaken specific export duty commitments in their accession protocols. In almost all cases, however, the Working Party Reports include a section on 'Export Regulation' where the export regime of the country in question is reviewed having regard to, to the extent relevant to the case, export quantitative restrictions (including prohibitions), export duties and export licensing. This section is often concluded with a standard formula according to which the country declares that, from the date of accession, it will apply its laws and regulations governing export measures

[65] The distinction between WTO-plus and WTO-extra obligations is borrowed from H. Horn, P. C. Mavroidis and A. Sapir, 'Beyond the WTO? An Anatomy of EU and US Preferential Trade Agreements', *The World Economy* 33 (2010), 1565, p. 1567.

[66] See, e.g. Qin, 'Reforming WTO Discipline', pp. 1153–4. For a comprehensive account of such obligations in the WTO legal framework see S. Charnovitz, 'Mapping the Law of the WTO Accession', in M. Janow *et al.* (eds.), *The WTO: Governance, Dispute Settlement and Developing Countries* (Juris Publishing, 2008).

[67] See above, Section 4.3.1.

[68] Horn *et al.*, 'Beyond the WTO?', p. 1571; J.Y. Qin, '"WTO-plus" Obligations and Their Implications for the WTO Legal System: An Appraisal of the China Accession Protocol', *Journal of World Trade* 37 (2003), 483. Many authors are critical of the process of WTO accession and argue that it often raises problems of 'political imbalance'. See, e.g. Qin, 'Reforming WTO Discipline', p. 1157; K. Jones, 'The Political Economy of WTO Accession: the Unfinished Business of Universal Membership', *World Trade Review* 8 (2009), 279.

(including export duties) in conformity with the relevant WTO provisions. The statement also often includes express reference to specific GATT provisions or to the GATT Agreement or to the WTO Agreement in general, thus leaving the acceding Members with a margin of discretion comparable to that of the original WTO Members bound exclusively by the GATT as to their use of export duties.[69]

Following the greater prevalence of export duties in the last decade, the number of post-1994 accession protocols specifically addressing the issue has also increased. Among the thirty-two countries that have acceded to the WTO after 1994 (as at 31 December 2014), eleven have agreed upon specific export duty commitments under their accession protocols: Mongolia, Latvia, Croatia, China, Saudi Arabia, Vietnam, Ukraine, Montenegro, the Russian Federation, Lao People's Democratic Republic, and Tajikistan (in order of accession). In all cases but the Russian Federation, such commitments are WTO-plus obligations assumed under individual accession protocol provisions. The Russian Federation, however, negotiated its commitments in the form of export duty concessions inscribed in its GATT Schedule. As shown below, this choice differentiates Russian obligations from the WTO-plus obligations assumed by other newly acceded WTO Members.

4.4.1 An overview of export duty commitments assumed by newly acceded WTO Members

The range of export duty commitments undertaken by newly acceded WTO Members to date varies significantly in terms of scope and coverage. Some countries have agreed to eliminate export duties, either integrally or on specific products. Others have agreed to phase down and bind the export duty rates to be applied on a more or less comprehensive list of products. In a number of cases, countries have abided by obligations that involve a combination of the two.

The legal techniques adopted to negotiate export duty commitments vary as well. WTO-plus obligations have to date generally been set out in individual accession protocol provisions or in specific provisions of the Working Party reports that are incorporated by reference to the related accession protocols. In either case, they are legally binding. Accession

[69] For a detailed account see B. Karapinar, 'Defining the Legal Boundaries of Export Restrictions: A Case Law Analysis', *Journal of International Economic Law* 15 (2012), 443, pp. 458–9.

protocols are considered 'an integral part of the WTO Agreement' and are thus enforceable under WTO law.[70] However, the Russian Federation has recently incorporated its export duty commitments in a specific new Part (Part V) of its GATT Schedule of Concessions.

4.4.1.1 Mongolia

Mongolia is among those newly acceded countries that have undertaken WTO-plus commitments on the use of export duties only on specific products. Paragraph 24 of its Working Party Report reads, in the relevant part:

> The representative of Mongolia also stated that his government would maintain the prohibition on the export of raw cashmere only until 1 October 1996, when an export duty at the rate of not more than 30 per cent ad valorem would be introduced. That export duty would be phased out and eliminated within 10 years of the date of Mongolia's accession to the WTO.[71]

Mongolia acceded to the WTO on 27 June 1996 and accordingly eliminated its export duties on raw cashmere in 2006. The formulation of its commitment, however, leaves it, in principle, free to impose export taxes on any other products, as it is allowed under the terms of Article XI:1 GATT. Currently, however, Mongolia does not impose export duties on any raw materials.

4.4.1.2 Latvia

Latvia committed to abolish all the export duties that it maintained at the time of accession (over fifty products), with the sole exception of antiques. According to Paragraph 69 of Latvia's Working Party Report:

> The representative of Latvia confirmed that present export tariff rates related only to the goods listed in Annex 3 Export Duty Tariffs ... Latvia would abolish all export duties listed in Annex 3 by 1 January 2000 with the exception of the duty on antiques. The timetable for elimination of

[70] The WTO Agreement is a covered agreement and thus any violation of its provisions can be challenged before the WTO dispute settlement bodies. Accordingly, Members can initiate WTO dispute settlement proceedings on the basis of a claim of violation of accession protocols' provisions. See Qin, 'Reforming WTO Discipline', p. 1154.

[71] Report of the Working Party on the Accession of Mongolia to the World Trade Organization, WT/ACC/MNG/9, 27 June 1996. Paragraph 24 was incorporated into Mongolia's Accession Protocol as per Paragraph 61 of Mongolia's Working Party Report.

> export duties would be similar for regional trade agreement partners and partners to which MFN treatment was applied as indicated in Annex 3.[72]

Annex 3 lists various categories of products, namely wood products, various forms of ferrous waste and scrap, and antiques (printed books and works of art). With the exception of the latter, Latvia is thus prevented from reintroducing any export tax on the listed products. Nothing prevents it, however, from introducing taxes on exports of any non-listed products, including minerals and metals. Latvia is not anyway a significant mine producer of any critical mineral or group of minerals and does not resort to any export duties.[73]

4.4.1.3 Croatia

Croatia's export duty commitments do not tangibly limit its margin of discretion to resort to export duties any differently from original WTO Members. According to Paragraph 101 of its Working Party Report, Croatia simply agreed to apply export duties, after the accession to the WTO, 'only in accordance with the provisions of the WTO Agreement' and to publish any newly introduced export duty measures or changes in the application, level or scope of such measures in the Official Gazette Narodne Novine.[74]

Under Paragraph 100 of its Working Party Report, moreover, Croatia confirmed that it was not making use of export duties at the time of accession, but also clarified that:

> Croatia had not eliminated the authority to apply export tariffs as stipulated in Article 34(2) of the Customs Law, enabling his Government to impose export tariffs in exceptional cases for the protection of exhaustible natural resources, or to ensure essential materials to the domestic industry and to prevent shortages in domestic supply.

Paragraph 100 was not incorporated into the Accession Protocol and therefore is not legally binding. Yet, the enforceability of Paragraph 101 would neither have added nor diminished the rights of Croatia to use of export duties as arising out of Paragraph 100, inasmuch as the 'exceptional cases' evoked under Paragraph 101 are also recognized under

[72] Report of the Working Party on the Accession of Latvia to the World Trade Organization, WT/ACC/LVA/32, 19 February 1999. Paragraph 69 was incorporated into Latvia's Accession Protocol by virtue of Paragraph 131 of Latvia's Working Party Report.

[73] See above, Part I.

[74] Report of the Working Party on the Accession of Croatia to the World Trade Organization, WT/ACC/HRV/59, 29 June 2000. Paragraph 101 was incorporated into Croatia's Accession Protocol through Paragraph 225 of Croatia's Working Party Report.

GATT general exceptions.[75] Croatia does not in any case maintain export duties on any raw materials.

4.4.1.4 China

A particularly stringent regime on export duties has been undertaken by China by virtue of Paragraph 11.3 of its Accession Protocol[76] and the related provisions of the Working Party Report. According to Paragraph 11.3:

> China shall eliminate all taxes and charges applied to exports unless specifically provided for in Annex 6 of this Protocol or applied in conformity with the provisions of Article VIII of the GATT 1994.

Annex 6 to China's Accession Protocol, labelled 'Products Subject to Export Duty', lists eighty-four HS 8-digit products for which maximum levels of export duty are provided. With respect to these commitments, a Note to Annex 6 clarifies:

> China confirmed that the tariff levels included in this Annex are maximum levels which will not be exceeded. China confirmed furthermore that it would not increase the presently applied rates, except under exceptional circumstances. If such circumstances occurred, China would consult with affected Members prior to increasing applied tariffs with a view to finding a mutually acceptable solution.

Among the eighty-four products listed in Annex 6 are mostly mineral raw materials, from iron and ferro-alloy metals to non-ferrous metals and industrial minerals.[77] In particular, China has included commitments on numerous critical minerals and metals in various forms (ores and concentrates, semi-refined/processed products and/or waste and scrap): antimony, chromium, copper, manganese, nickel, niobium, phosphates, silicon, tantalum, tungsten and vanadium. The bound export duty rate ranges from 20 per cent to 40 per cent.[78] Appendix 1 shows that China not only maintains export duties on all types of listed critical products, but applies the maximum bound rate indicated as per Annex 6. Moreover, China has systematically applied export duties on a wide range of products which are not listed, thereby violating the general elimination obligation

[75] See below, Chapter 6.

[76] Protocol on the Accession of the People's Republic of China, WT/L/432, 23 November 2001.

[77] Out of eighty-four listed products, only four are fishery products or products of animal origin. See Annex 6 to China's Accession Protocol.

[78] See ibid.

assumed under Paragraph 11.3 of its Accession Protocol. Among them are several critical minerals and metals in various forms: beryllium, cobalt, coke, copper, fluorspar, gallium, germanium, graphite, indium, magnesite, magnesium, manganese, molybdenum, nickel, niobium, phosphates, platinum-group metals (PGMs), rare earth elements, silicon, silver, rhenium, tantalum, titanium, tungsten and vanadium.[79]

China's regime of export duties was accordingly challenged before the WTO dispute settlement bodies in two different disputes: in *China–Raw Materials*, the United States, the European Union and Mexico challenged the export duties that China maintained on various forms of bauxite, coke, fluorspar, magnesium, manganese, silicon metal, yellow phosphorous and zinc; in *China–Rare Earths*, the United States, the European Union and Japan brought complaints about Chinese export duties on different forms of rare earth elements, tungsten and molybdenum. In both cases, China's export taxes were found to be in breach of Paragraph 11.3 of China's Accession Protocol because the products in question were not listed in Annex 6 to China's Accession Protocol.[80]

In *China–Raw Materials*, the Appellate Body interpreted for the first time the scope of Paragraph 11.3 of China's Accession Protocol and the related provisions of China's Working Party Report.[81] In its view, a combined reading of Paragraph 11.3, Annex 6 and the Note to Annex 6

[79] For a detailed account see below, Appendix 1.

[80] *China–Raw Materials*, Panel Reports, paras. 7.75, 7.80, 7.84 and *China–Rare Earths*, Panel Reports, para. 7.48. In the case of yellow phosphorus, China had applied a special export duty of 50 per cent in addition to the statutory rate of 20 per cent provided for in Annex 6 until 1 July 2009 (i.e. before the establishment of the Panel). Thus, the Panel declared the statutory rate of 20 per cent in accordance with China's WTO obligations, but did not make any findings on the special rate in excess of the level indicated under Annex 6 because it considered that the measure fell outside of its terms of reference. *China–Raw Materials*, Panel Reports, paras. 7.69–71. Consequently, the question of whether the scope of the 'exceptional circumstances' requirement for Annex 6-listed products could be considered to overlap with Article XX GATT was not addressed. The Appellate Body merely noted that the reference to 'exceptional circumstances' could not be used to justify the use of export duties on non-listed products under Article XX GATT. *China–Raw Materials*, AB Reports, para. 282–4.

[81] In *China–Raw Materials*, the Panel and the Appellate Body identified Paragraphs 155 and 156 of the Working Party Report as the relevant context for Paragraph 11.3 of China's Accession Protocol. These paragraphs fall under Section C on 'Export Regulations' and deal solely with the commitment undertaken by China with respect to the elimination of export duties that does not otherwise exist in the GATT. Paragraph 155 reads: 'taxes and charges should be eliminated unless applied in conformity with GATT Article VIII or listed in Annex 6 to the Draft Protocol'. Paragraph 156 confirms: 'China noted that the majority of products were free of export duty, although 84 items, including tungsten ore, ferrosilicon and some aluminium products, were subject to export duties'. Panel Reports,

of China's Accession Protocol indicates that China must not impose export duties other than those falling within the scope of a specific set of exceptions: those covered by Article VIII GATT[82] and those applied in conformity with Annex 6. This implies that (i) China cannot apply export duties on products not listed in Annex 6;[83] (ii) the 'exceptional circumstances' provided for in the Note to Annex 6 cannot be invoked to impose export duties on non-listed products;[84] (iii) in the case of the eighty-four listed products, China could increase the applied export duties up to the maximum rate set out in Annex 6[85] by invoking the 'exceptional circumstances' exception provided for in the Note to Annex 6, but only insofar as it first fulfils the prior consultation requirement provided therein.[86]

4.4.1.5 Saudi Arabia

Saudi Arabia is among those newly acceded WTO Members that agreed not to impose export duties on a specific set of products. According to Paragraph 184 of its Working Party Report, Saudi Arabia refrained from using export duties on iron and steel scrap, but affirmed its right to impose such measures on any other products:

> Some Members of the Working Party requested information on export duties applied by Saudi Arabia. Those Members noted that the export duties applied by Saudi Arabia appeared to be imposed only for revenue purposes and would have trade-distorting effects. In response, the representative of Saudi Arabia stated that Article XI of the GATT 1994 expressly permitted the imposition of export duties, and did not restrict the right to impose such duties. Export duties applied only to un-tanned hides and skins, falling under HS Nos. 4101, 4102 and 4103. The rate of export duty was SAR 2000/ton (roughly 20 per cent). The representative of Saudi Arabia

China–Raw Materials, paras. 7.130–48. See also AB Reports, *China–Raw Materials*, paras. 294–9.

82 The Appellate Body clarified that the export duties regulated under Paragraph 11.3 of China's Accession Protocol do not fall within the scope of Article VIII GATT. AB Reports, *China–Raw Materials*, para. 290. For an account of the relationship between the export 'fees and charges' under Article VIII GATT and the export 'duties' as per Article XI GATT and Paragraph 11.3 of China's Accession Protocol see above, Section 4.3.4.

83 AB Reports, *China–Raw Materials*, para. 284.

84 Ibid.

85 Ibid., para. 285. This is because the word 'furthermore' in the second sentence of the Note to Annex 6 was interpreted by the Appellate Body to mean that the obligation contained in the second and third sentences of the Note (i.e. the exceptional circumstances requirement and the consultation requirement) are 'in addition to China's obligation under the first sentence not to exceed the maximum tariff levels provided for in Annex 6'. Ibid., para. 287.

86 The prior consultation requirement is not applicable to non-listed products. Ibid.

> confirmed that Saudi Arabia would not impose export duties on iron and steel scrap. The Working Party took note of this commitment.[87]

According to this commitment, Saudi Arabia remains free to introduce and/or maintain export duties on any product other than iron and steel scrap in accordance with Article XI:1 GATT. In this respect, should Saudi Arabia in the future negotiate export duty concessions on any product in its GATT Schedule, it would be bound by these commitments under Article II:1(a) GATT and could thus not exceed the maximum scheduled rates by invoking Paragraph 184 of its Working Party Report.

As noted in Chapter 2, Saudi Arabia does not impose export duties on minerals and metals. Since 10 September 2008, it progressively phased down the export duty rates on untanned hides and skins notified under its Working Party Report until complete elimination on 10 September 2013.[88]

4.4.1.6 Vietnam

Vietnam's WTO-plus obligations also concern a limited group of products, namely ferrous and non-ferrous scrap metals. Similar to Saudi Arabia, Vietnam reserved the right to impose export duties on any other products in accordance with relevant GATT rules. Paragraph 260 of Vietnam's Working Party Report reads:

> The representative of Viet Nam confirmed that Viet Nam would apply export duties, export fees and charges, as well as internal regulations and taxes applied on or in connection with exportation in conformity with the GATT 1994. With regard to export duties on ferrous and non-ferrous scrap metals, he confirmed that Viet Nam would reduce export duties in accordance with Table 17, and that Table 17 included all export duties that Viet Nam applied to ferrous and non-ferrous scrap metal. The Working Party took note of these commitments.[89]

In Table 17 of Vietnam's Working Party Report, Vietnam agreed to gradually reduce the export duty rates applied on ferrous and non-ferrous

[87] Report of the Working Party on the Accession of Saudi Arabia to the World Trade Organization, WT/ACC/SAU/61, 11 December 2005. Paragraph 184 was incorporated into Saudi Arabia's Accession Protocol by virtue of Paragraph 315 of Saudi Arabia's Working Party Report.

[88] WTO Doc. WT/TPR/S/256, 21 December 2011, p. 38.

[89] Report of the Working Party on the Accession of Viet Nam to the World Trade Organization, WT/ACC/VNM/48, 11 January 2007. Paragraph 260 was incorporated into Vietnam's Accession Protocol by virtue of Paragraph 527 of Vietnam's Working Party Report.

scrap metals at the time of the accession (which ranged from 35 to 45 per cent) each year, according to a specific timetable. Based on these commitments, Vietnam promised to limit to 17 per cent the export duty rate applied to iron and steel scrap, and to 22 per cent the export duty rate applied to a wide range of non-ferrous waste and scrap as from 2012 (i.e. after 4 years from the date of accession). Among the latter are many critical metal scrap forms, including antimony, beryllium, chromium, cobalt, copper, gallium, germanium, indium, magnesium, manganese, molybdenum, nickel, niobium, rhenium, tantalum, titanium, tungsten and vanadium. As shown in Appendix 1, Vietnam has followed the timetable set out in Table 17 of its Working Party Report and has thus abided by its WTO-plus commitments under Paragraph 260 of such Report.

Yet, Vietnam has also applied a wide range of export duties on products not listed in Table 17, including many unprocessed and semi-refined forms of critical minerals and metals.[90] These export duties are admitted by terms of Vietnam's WTO-plus obligations. Table 16 of Vietnam's Working Party Report, labelled 'Viet Nam's Export Duty Schedule', lists in fact forty-three products to which Vietnam applied export duties at the time of accession along with the corresponding duty rate. Among them are several mineral products, including various semi-processed and scrap forms of critical materials such as copper, nickel and other base metals. The indicated export duty rates, however, are not meant to be legally enforceable. This conclusion can be discerned by reading Paragraph 256 of Vietnam's Working Party Report, which states, in the relevant part:

> The representative of Viet Nam said that Viet Nam levied export duties on some minerals and natural resources exported in raw form ... Export duties were imposed in accordance with the Law on Import, Export Duties, and Decision No. 45/2002/QD-BTC of 10 April 2002. Goods subject to export duties are listed in Table 16.

Table 16 reproduces the list of export duties attached to Decision No. 45/2002/QD/BTC of 10 April 2002. In this respect, Vietnam seems to have reserved the right to increase the export duty rates applied to the products listed in Table 16 as well as to introduce export duties on products not listed therein. The legitimacy of Vietnam's unbound export duties is confirmed by the wording of Paragraph 257 of its Working Party Report, according to which the representative of Vietnam 'did not

[90] See Appendix 1.

consider the imposition of export duties inconsistent with WTO rules'.[91] Vietnam's proliferating use of export duties on products not listed under Table 17 is therefore not in breach of its WTO-plus obligations inasmuch as these measures comply with relevant GATT obligations.

4.4.1.7 Ukraine

Similar to Vietnam, Ukraine promised to gradually reduce the export duty rates applied on specific products listed in Table 20(b) of its Working Party Report according to an agreed timetable, but reserved the right to impose export duties on non-listed products. Paragraph 240 of Ukraine's Working Party Report reads, in the relevant part:

> The representative of Ukraine confirmed that at present export duties were applied only to the goods listed in Table 20(a). He further confirmed that Ukraine would reduce export duties in accordance with the binding schedule contained in Table 20(b). He also confirmed that *as regards these products*, Ukraine would not increase export duties, nor apply other measures having an equivalent effect, unless justified under the exceptions of the GATT 1994. He further confirmed that the current duties and any changes in their application would be published in the Official Gazette ... The Working Party agreed that these commitments do not constitute a reinterpretation of the GATT 1994, nor affect the rights and obligations of other members in respect of provisions on the application of export duties, that are measures in accordance with the GATT 1994 (emphasis added).[92]

Among more than seventy types of products bound under Table 20(b) are various categories of oilseeds, live cattle and hides, and ferrous and non-ferrous scrap. Under the heading 'Non-ferrous scrap', Ukraine committed to bind various critical mineral raw materials such as cobalt (various forms), copper (various forms), nickel (waste and scrap), titanium (waste and scrap) and tungsten (waste and scrap).[93] Accordingly, Ukraine has started to reduce the export duty rates applied on such products as from the date of accession. Based on the Organization for

[91] Vietnam's Working Party Report, para. 257. This interpretation holds true when considering that, in Table 16, are products such as copper waste and scrap that were also included in Table 17. Were the export duty rates indicated in Table 16 meant to be legally binding, there would have been no need to schedule maximum bound rates for the same materials in accordance with Table 17.

[92] Report of the Working Party on the Accession of Ukraine to the World Trade Organization, WT/ACC/UKR/152, 16 May 2008. Paragraph 240 was incorporated into Ukraine's Accession Protocol through Paragraph 512 of Ukraine's Working Party Report.

[93] Ibid., Table 20(b).

Economic Co-operation and Development (OECD) Inventory, however, it appears that the planned reductions have been delayed in relation to the timetable set out under Table 20(b). Moreover, Ukraine does not seem to apply export duties on unbound minerals and metals, including critical materials.[94]

Finally, Ukraine listed in Table 20(a) the export duties applied at the time of accession. It did not formally agree to bind the export duty rates applied to such products at a specific level in accordance with a binding schedule of the kind set out under Table 20(b). However, for each specific listed product, the column 'Date of elimination' reported the following statement: 'Ukraine will conduct a policy of gradually reducing the level of export duties. Some duties may be eliminated. Ukraine is not planning to bind export duties at "Zero"'.[95] Based on this statement, the export duty rates indicated in Table 20(a) do not seem legally enforceable. Accordingly, it would be difficult to argue that any of the duties maintained on the listed products run afoul of Ukraine's WTO obligations. Furthermore, Ukraine does not seem to be prevented from imposing export duties on products other those indicated under Table 20(a) and Table 20(b).[96]

4.4.1.8 The Russian Federation

The Russian Federation agreed to schedule and bind the export duty rates applied to over 700 tariff lines. It did so by explicitly creating a new Part in its GATT Schedule of Concessions – Part V – specifically devoted to 'Export Duties'. According to Paragraph 638 of the Russian Federation's Working Party Report:

> ... from the date of accession, ... products described in Part V of [the Schedule of Concessions and Commitments on Goods of the Russian Federation] would, subject to the terms, conditions or qualifications set-forth in that Part of the Schedule, be exempt from export duties in excess of those set-forth and provided therein. The representative of the Russian Federation further confirmed that the Russian Federation would not apply other measures having an equivalent effect to export duties on those products. He confirmed that, from the date of accession, the

[94] See OECD Inventory and Appendix 1.

[95] No critical minerals and metals are included in Table 20(a).

[96] According to Paragraph 230 of Ukraine's Working Party Report, 'the representative of Ukraine stated that export duties *per se* were consistent with the WTO Agreements, but he acknowledged that high export duties could act as trade barriers and hence needed reduction'.

> Russian Federation would apply export duties in conformity with the WTO Agreement, in particular with Article I of the GATT 1994.[97]

The 'terms, conditions, or qualifications' referred to under Paragraph 638 are set forth in the introductory statement to Part V of the Russian Federation's Schedule:

> [t]he Russian Federation undertakes not to increase export duties, or to reduce or to eliminate them, in accordance with the following schedule, except in accordance with the provisions of the GATT 1994.[98]

Part V of the Russian Federation's GATT Schedule contains export tariff bindings for a wide range of products, from fisheries to oilseeds, from mineral fuels to raw hides and skins, from wood products to non-energy minerals and metals. The Schedule indicates the rate applied by the government at the time of the accession and the rate which the Russian Federation committed to apply after an implementation period that varies from 3 to 5 years depending on the type of product. With the exception of certain energy commodities and certain items falling under the category of raw hides and skins, the export duty rates negotiated by the Russian Federation range from 0 to 20 per cent.[99]

Among the scheduled mineral products are also many critical minerals and metals in one or more forms (unprocessed, semi-refined/processed, and/or waste and scrap): coke, copper, germanium, magnesium, nickel, PGMs, silver, tantalum, titanium, tungsten and vanadium. The Russian Federation promised to eliminate most of them within 4 years of the date of accession. In other words, it committed to a 0 per cent export duty rate to be implemented by November 2015. At the time of accession, and until December 2014, the Russian Federation maintained export duty rates on such products generally varying between 5 and 10 per cent,[100] with the

[97] Report of the Working Party on the Accession of the Russian Federation to the World Trade Organization, WT/ACC/RUS/70, 17 November 2011. Paragraph 638 was incorporated into Russia's Accession Protocol by means of Paragraph 1450 of the Russian Federation's Working Party Report.

[98] See WTO Doc. WT/MIN(11)/2/Add.1, 17 November 2011.

[99] For crude oil, natural gas and various related products, a specific formula determines the maximum applicable export duty rate. It takes into account the world price of crude oil and the average LPG price at DAF Brest. See ibid., Part V, footnote # and footnote ##. For certain types of raw hides and skins, the Russian Federation committed to reduce its export duty from 500 Euro/1,000 kg to 200 Euro/1,000 kg within 5 years. In other cases, the Russian Federation negotiated compound export duties (e.g. certain oilseeds, certain raw hides and skins, certain wood products, and various forms of ferrous and non-ferrous waste and scrap).

[100] See ibid., Part V and Appendix 1.

exception of certain waste and scrap forms of cobalt, copper, nickel and titanium for which the initial level of export duty was considerably higher (up to 50 per cent for copper waste and scrap). In these cases, the Russian Federation committed to a compound export duty rate with an *ad valorem* component of 10 per cent and a specific component that varies depending on the type of product, and an implementation period ranging from 3 to 5 years.

As shown in Appendix 1, the Russian Federation does not maintain export taxes on all of the critical minerals and metals included in Part V of its GATT Schedule. Those that are maintained are not in breach of the Russian Federation's WTO obligations since the implementation period has not yet expired. The Russian Federation will, however, need to phase them out by November 2015. In a few cases (e.g. for certain forms of copper), however, the Russian Federation did at some point impose export duties that were not consistent with the commitments listed in its Schedule, albeit for a limited time period. At the same time, the Russian Federation has also applied export duties on non-scheduled critical minerals and metals.[101] These export duties, however, do not violate Russia's WTO obligations, as they are applied to products for which no commitment was undertaken.

In light of the foregoing, the obligations assumed by the Russian Federation on the use of export duties are export duty *concessions* within the meaning indicated above. In other words, the legal technique used by the Russian Federation corroborates the interpretation of Article II:1(a) GATT endorsed above. Based on such an interpretation, the Russian Federation negotiated export duty concessions and included them in 'an appropriate part' of its GATT Schedule as explicitly admitted in Article II:1(a) GATT and Article XXVIII(*bis*). Part V of the Russian Schedule of Concessions is thus incorporated into the GATT legal framework and enforceable through Article II:1(a) and Article II:7 GATT. In this respect, the export duty commitments negotiated by the Russian Federation are technically not WTO-plus obligations in the sense of the other country-specific obligations considered above. The Russian Federation has in fact used the same technique adopted by original WTO Members in the framework of the GATT (e.g. Australia) when assuming export duty commitments in the context of its accession. The chosen legal technique thus equalizes the legal treatment of Russia's specific obligations to the legal treatment prescribed for scheduled and bound export duty rates

[101] See Appendix 1.

negotiated in accordance with existing GATT provisions. As shown below, this circumstance allows the Russian Federation greater flexibility than any other newly acceded WTO Member undertaking WTO-plus obligations on export duties by virtue of individual accession protocol provisions.

4.4.1.9 Montenegro

Very stringent WTO-plus obligations on export duties were undertaken by Montenegro. According to Paragraph 132 of its Working Party Report, Montenegro committed not to apply or reintroduce any export duty as from the date of accession.[102] It thus accepted an absolute limit on the use of export duties, on whatever products, and irrespective of their rationale. At the time of accession, however, Montenegro did not apply any export duties.[103]

4.4.1.10 Lao People's Democratic Republic

An interesting case is represented by Lao People's Democratic Republic (Lao PDR). Similar to Vietnam and Ukraine, the country enumerated the export duties applied at the time of accession under Table 14 of its Working Party Report,[104] but it clarified under Paragraph 100 of the Report that such '[e]xport duties were collected for revenue purposes contributing, on average, about ... 1 per cent of total Government revenues' and that 'Lao PDR had no plans, at present, to phase [them] out'.[105] Moreover, Lao PDR reclaimed its right to impose export duties on products not listed in Table 14. Paragraph 101 of Lao PDR's Working Party Report states:

> Asked to commit not to introduce export duties on products not included in the previous Table 14 of document WT/ACC/SPEC/LAO/5, the representative of Lao PDR reiterated the importance of export duties to government revenue and the need to maintain this measure for development purposes. Lao PDR applied export duties to a limited number of products, as mentioned in the aforementioned Table 14. He took note of Members' suggestion of applying alternative measures that could foster the

102 Report of the Working Party on the Accession of Montenegro to the World Trade Organization, WT/ACC/CGR/38, 5 December 2011. Paragraph 132 was incorporated into Montenegro's Accession Protocol by means of Paragraph 281 of Montenegro's Working Party Report.

103 Ibid., para. 131.

104 Report of the Working Party on the Accession of Lao People's Democratic Republic to the World Trade Organization, WT/ACC/LAO/45, 1 October 2012.

105 Ibid., para. 100.

> production of those domestic products subject to export duties. However, he stated that Lao PDR had limited possibilities to foster value-added production and/or revenue sources. He added that Lao PDR would comply with the GATT 1994 and WTO provisions with regard to export duties.[106]

Neither Paragraph 100 nor Paragraph 101 of Lao PDR's Working Party Report are incorporated into its Accession Protocol. Yet, the enforceability of such paragraphs would not have compromised LAO PDR's ability to use export duties in accordance with the GATT rules applicable in the case of non-existence of export duty concessions (i.e. Article XI:1 GATT). As matter of fact, according to Paragraph 101, the country remains free to both increase the export duty rates applied to the products indicated in Table 14 of its Working Party Report and introduce export duties on products not listed therein, provided such measures are in accordance with GATT rules. As indicated in Appendix 1, however, Lao PDR does not impose export duties on critical minerals and metals.

4.4.1.11 Tajikistan

Tajikistan has agreed upon a strict regime on export duties. Paragraph 169 of Tajikistan's Working Party Report reads:

> The representative of Tajikistan confirmed that Tajikistan shall not introduce and shall eliminate all duties, taxes, fees and charges applied to exports, unless specifically provided for in Table 9 of this Protocol or applied in conformity with the provisions of Article VIII of the GATT 1994. Tajikistan shall notify to the WTO any changes of its duties, taxes, fees and charges applied to exports on the day of their publication and in any case at least 30 days before any such changes enter into force. The Working Party took note of these commitments.[107]

Table 9 indicates the 'Maximal Applicable Export Duties Rates' for more than 300 tariff lines covering a wide range of products, from live animals to agricultural commodities, from agricultural raw materials to mineral fuels, from raw hides and skins to non-energy minerals and metals. All other products cannot be taxed on exportation as per Paragraph 169, although export fees and charges may be applied in accordance with Article VIII GATT. In this sense, the structure and the

[106] Ibid., para. 101.

[107] Report of the Working Party on the Accession of Tajikistan to the World Trade Organization, WT/ACC/TJK/30, 6 November 2012. Paragraph 169 was incorporated into Tajikistan's Accession Protocol by means of Paragraph 351 of Tajikistan's Working Party Report.

legal technique of Tajikistan's WTO-plus obligations on export duties resemble those of China's specific commitments, inasmuch as they both agreed to eliminate any export duties other than those admitted by virtue of Article VIII GATT and those explicitly indicated under a specific Annex (China) or Table (Tajikistan) included in their accession protocols. The difference between China and Tajikistan is that the latter maintained the export duties unbound for many of the products listed under Table 9 by indicating 'Unbound' in the corresponding 'Export Duty Rate (%)' column. Among such listed products are various forms of minerals and metals such as silver and aluminium. Tajikistan, however, committed to bind the export duties applied on different ferrous waste and scrap forms as well as aluminium and copper waste and scrap.[108]

4.4.2 *Limited flexibility of WTO-plus obligations on export duties*

The analysis of the export duty commitments assumed to date under post-1994 accession protocols shows that these commitments vary greatly in terms of scope and coverage. Countries such as Montenegro, China and Tajikistan agreed to a general elimination obligation, mitigated, in the cases of China and Tajikistan, by the existence of a list of products for which maximum applicable export duty rates are indicated. Other countries such as Latvia, Mongolia and Saudi Arabia promised to eliminate the export duties applied on a limited number of products; Vietnam and Ukraine committed to phase down and bind the export duty rates imposed on specific products, whereas the Russian Federation bound more than 700 tariff lines in its Schedule of Concessions. These countries remain free to impose export taxes on products other than those for which they undertook specific commitments. Finally, Croatia and Lao People's Democratic Republic simply confirmed their commitment to impose export duties in accordance with the GATT obligations.

At the same time, the legal techniques used to assume these country-specific obligations also differ. On a general level, all countries apart from the Russian Federation have undertaken WTO-plus obligations by virtue of individual provisions of their accession protocols (or specific provisions of the related Working Party reports that are expressly incorporated into such protocols). In contrast, the Russian Federation included its export

108 In the case of copper, the maximal applicable export duty rate is 30 per cent, but not less than 300 Euro/1,000 kg. Tajikistan also bound the export duty rates for various energy products (petroleum oils, natural gas, etc.), certain hides and skins, and several forms of cotton and silk. Ibid., Table 9.

duty concessions in a Part of its GATT Schedule specifically devoted to 'Export Duties' (Part V). These differences carry important implications for the 'policy space' left to WTO Members to use export duties.

Inasmuch as newly acceded WTO Members assume export duty commitments under individual accession protocol provisions without including them in their GATT Schedules, such commitments cannot be adjusted in accordance with the GATT-specific procedures in principle available to export duty concessions. Rather, the extent to which such export duty commitments can be renegotiated depends on whether accession protocol provisions can be amended.

The existing WTO accession protocols, however, do not contain any amendment provisions. While the matter is still far from settled, Qin notes that two alternative views were put forward in the attempt to bring clarity in this respect. According to the first, accession terms are 'permanent and immutable, except for the market access commitments incorporated into the schedules of GATT and the General Agreement on Trade in Services (GATS), which can be adjusted according to the GATT and GATS procedures respectively'.[109] Accession terms would, in other words, constitute 'pre-conditions for the WTO Membership of the acceding country and as such c[ould] be renegotiated once the accession is completed'.[110] Under this scenario, therefore, '[t]he only way the acceding country c[ould] escape the terms of its accession is to withdraw from the WTO altogether'.[111] As to the second view, inasmuch as accession protocols are integral parts of the WTO Agreement, they could be amended through the same procedure envisaged for amending other provisions of the WTO Agreement, in accordance with Article X of the WTO Agreement.[112] A majority of two thirds would thus be required. The renegotiation of export duty commitments assumed under specific accession protocol provisions would thus be far more burdensome than the deconsolidation procedure envisaged under Article XXVIII GATT for the modification and/or the withdrawal of concessions incorporated into GATT schedules[113] and with arguably very low chances to be instigated and succeed.[114]

[109] Qin, 'Reforming WTO Discipline', p. 1157-8. See General Agreement on Trade in Services, Annex 1B to the Agreement Establishing the World Trade Organization, 1869 U.N.T.S. 183, entered into force 1 January 1995 [hereinafter GATS].

[110] Qin, 'Reforming WTO Discipline', p. 1157. [111] Ibid. [112] Ibid.

[113] See above, Section 4.3.2.

[114] See J. Y. Qin, 'The Challenge of Interpreting "WTO-plus" Provisions', *Journal of World Trade* 44 (2010), 127, pp. 134–5.

In light of the above, newly acceded WTO Members are practically unable to renegotiate any of the export duty commitments assumed under individual accession protocol provisions, inasmuch as the renegotiation procedure provided for in Article XXVIII GATT is only applicable to export duty concessions included in GATT schedules and, therefore, incorporated into the GATT framework. This element of 'ultra-rigidity' makes them WTO-plus obligations.[115]

The lack of flexibility of WTO-plus obligations on export duties introduces a further element of differentiation among WTO Members, distinguishing newly acceding WTO Members that assumed export duty commitments under individual accession protocol provisions from the Russian Federation and any other WTO Member exclusively bound by GATT obligations and willing to negotiate export duty concessions in its GATT Schedule (e.g. Australia). Moreover, it adds to the existing asymmetry in the legal treatment of import and export duty concessions.[116]

4.5 Transparency and notification requirements applicable to export duties

As export duties proliferate, prompt availability of accurate information concerning the introduction, the extension and/or the elimination of an export tax, as well as any adjustment to applied duty rates, is not only fundamental to ensure that no WTO-inconsistent export duties are in place but is also necessary to provide certainty for markets and trading partners. In the mining sector, in particular, the uncertainty created by lack of transparency in the use of export duties may be particularly detrimental, as mining companies plan investments in advance based on envisaged price and demand patterns.[117]

As discussed above, transparency is the main objective pursued by Article X GATT. WTO Members are required to publish all measures 'pertaining to ... rates of duty, taxes or other charges ... on ... exports' in a prompt and accessible manner (Article X:1) and to administer such measures 'in a uniform, impartial and reasonable manner' (Article X:3). Yet, Article X does not include any specific notification requirements and no other GATT provision specifically addresses the issue of notification of export duties.

115 The expression 'ultra-rigidity' borrows from Qin, 'Reforming WTO Discipline', p. 1186.

116 See above, Section 4.3.2.

117 See above, Chapter 3, Section 3.2.3.

Export duties were, however, explicitly included in the Indicative List of Notifiable Measures annexed to the 1993 Uruguay Round Ministerial Decision on Notification Procedures. The inclusion of export duties under this list does not *per se* create an obligation for WTO Members to specifically notify export taxes inasmuch as a Note to the List of Notifiable Measures posits: '[t[his list does not alter existing notification requirements in the Multilateral Trade Agreements in Annex 1A to the WTO Agreement or, where applicable, the Plurilateral Trade Agreements in Annex 4 of the WTO Agreement'. More generally, WTO Members reaffirmed their commitment to the existing obligations regarding publication and notification under such agreements, while also abiding by a general obligation to notify, to the maximum extent possible, their adoption of trade measures affecting the operation of the GATT 1994 'guided, as appropriate, by the annexed list of measures'.[118] A Central Registry of Notifications, established under the responsibility of the WTO Secretariat in pursuance of Article 2 of the Decision, records the information provided in the WTO Members' notifications (measures, purpose, trade coverage and requirement under which a measure is notified) and cross-references its records of notifications by Member and obligation. It also annually informs every WTO Member of their regular notification obligations for the following year and solicits them when those remain unfulfilled. Finally, it provides information regarding individual notifications to WTO Members, upon request.

Although WTO Members are encouraged to notify export taxes under the Uruguay Round Decision on Notification Procedures, in practice export duties are not consistently and systematically notified to the WTO.[119] Moreover, the quality of information concerning export duties varies greatly across governments. According to the OECD, it is generally 'not up to the standards reflected in the checklist of core transparency requirements derived from rules and recommendations

[118] See Article 1 of the Decision. In this respect, WTO Members reaffirmed the undertaking set out in the 1979 Understanding Regarding Notification, Consultation, Dispute Settlement and Surveillance. In its Paragraph 3, the Understanding stated: 'Contracting parties moreover undertake, to the maximum extent possible, to notify their adoption of trade measures affecting the operation of the General Agreement, it being understood that such notification would of itself be without prejudice to views on the consistency of measures with or their relevance to rights and obligations under the General Agreement'. See Understanding Regarding Notification, Consultation, Dispute Settlement and Surveillance adopted on 28 November 1979 (BISD 26S/210).

[119] This fate is common to export restrictions in general. Fliess *et al.*, 'Recent Developments', p. 31.

developed by GATT/WTO'.[120] Fliess *et al.* note in particular that, if the proliferation of export duties (and restrictions in general) is 'a worrying development, equally disconcerting is their lack of transparency.[121] An advancement in this respect is represented by the WTO Agreement on Trade Facilitation agreed in Bali in 2013, opened for acceptance by WTO Members in November 2014 and not yet entered into force at the time of writing.[122] Section I of the Agreement updates and reinforces disciplines on transparency and notification requirements. Article 1, labelled 'Publication and Availability of Information', contains various paragraphs applicable to export duties. Paragraph 1.1 on 'Publication' prescribes that each WTO Member promptly publish information on, among others, 'applied rates of duties and taxes of any kind imposed on or in connection with . . . exportation' (subparagraph (b)). Members are required to publish such information 'in a non-discriminatory and easily accessible manner in order to enable governments, traders and other interested parties to become acquainted with them'.

Under Paragraph 3.1, moreover, WTO Members shall establish enquiry points to answer reasonable enquiries of governments, traders and other interested parties on matters covered by Paragraph 1.1. The transparency obligations are also complemented with a notification requirement. Under Paragraph 4.1, WTO Members shall notify to the Committee on Trade Facilitation[123] 'the official place(s) where the items in subparagraphs 1.1 (a) to (j) have been published'.

4.6 Conclusions

In contrast to the conventional wisdom on the meaninglessness of WTO disciplines on export duties, this chapter has shown that the GATT

[120] O. R. Agatiello and B. Fliess, 'Export Restrictions: Benefits of Transparency and Good Practices', OECD Trade Policy Papers, No. 146 (OECD Publishing, 2013), p. 32.

[121] Ibid., p. 33.

[122] The Agreement was annexed to the Protocol Amending the Marrakesh Agreement Establishing the World Trade Organization approved and opened for acceptance through the General Council Decision of 27 November 2014 (WTO Doc. WT/L/940, 28 November 2014). Upon the entry into force of the Protocol by acceptance of two thirds of WTO Members in accordance with Article X:3 of the Marrakesh Agreement, the Agreement on Trade Facilitation will be inserted into Annex 1A to the WTO Agreement after the Agreement on Safeguards. See Paragraph 4 of the Protocol Amending the Marrakesh Agreement Establishing the World Trade annexed to the General Council Decision.

[123] See Article 13 of the Agreement on Trade Facilitation.

contains various provisions relevant and applicable to export duties. Based on Article II:1(a) and Article XXVIII(*bis*) GATT, WTO Members can enter into negotiations of export duty concessions to be included in their GATT schedules in the same way as for import tariff concessions. Export duty concessions can also be renegotiated similar to import tariff concessions in accordance with the deconsolidation procedure envisaged under Article XXVIII GATT. Yet, as a result of the historical marginality of export duties at the time of the GATT creation, not all of the GATT adjustment procedures and exceptions envisaged for import tariff concessions seem to be applicable to export duty concessions. Examples include Article XVIII GATT, which allows low-income countries to deviate from import duty concessions in order to promote economic development, and Article XIX GATT, which permits WTO Members to withdraw or modify tariff concessions if, as a result of unforeseen developments, a sudden, sharp and substantial surge of imports of such product significantly impairs the position of a domestic industry. The asymmetrical rigidity of export duty concessions compared with import tariff bindings may be at the origin of the general paucity of commitments in this area, especially considering that many countries purport to use export duties as developmental tools.[124] So far, only one WTO Member (i.e. Australia) has negotiated export duty commitments on a specific set of products in the form of concessions inscribed in its Schedule as admitted by virtue of GATT provisions. All other WTO Members bound exclusively by GATT obligations have refrained from doing so. They can thus apply export duties on any products as per Article XI: GATT irrespective of their rationale, albeit with certain restrictions: they have to be applied on an MFN basis (Article I GATT), not be prohibitive (Article XI:1 GATT), be published in a prompt and accessible manner (Article X:1 GATT) and be administered in a uniform, impartial and reasonable manner (Article X:3 GATT).

Against this backdrop, eleven newly acceded WTO Members have undertaken country-specific export duty commitments in the context of their WTO accession. These commitments are very uneven in scope and coverage, with some countries abiding by general elimination obligations and others committing to phase down and bind the export duty rates applied on a specific list of products. Generally, newly acceded WTO Members have assumed export duty commitments under individual provisions of their accession protocols. The

[124] See above, Part I.

only way for them to renegotiate such commitments would thus be to amend the accession protocol provisions themselves, provided it is possible.[125] This element of 'ultra-rigidity'[126] contrasts with the flexibility of export duty concessions negotiated in the framework of the GATT as per Article XXVIII, and characterizes export duty commitments included in specific accession protocol provisions as WTO-plus obligations. The only exception in this respect is the Russian Federation, which assumed export duty commitments on a far greater number of tariff lines compared with many other newly acceded WTO Members, but did so by inscribing them in the form of export duty concessions in a Part of its GATT Schedule specifically devoted to 'Export Duties', as explicitly allowed under Article II:1(a) and Article XXVIII(*bis*). In this respect, its export duty concessions are incorporated into the GATT framework and do not technically constitute WTO-plus obligations.

In conclusion, when assessing whether the proliferation of export duties on minerals and metals, and critical minerals and metals in particular, was made possible by the irrelevance of WTO disciplines, three factors should be taken into consideration. First, among the leading producers mainly responsible for such proliferation are newly acceded WTO Members that have undertaken country-specific commitments on the use of export duties in their accession protocols. For some of them, such as China, these obligations capture most of the export duties that they currently apply to minerals and metals and, more specifically, to critical minerals and metals. In contrast, in the case of other large exporters responsible for the vertical proliferation of export duties on such materials (e.g. the Russian Federation and, to a limited extent, Vietnam and Ukraine), the export duties in place are not in breach of such obligations on a systematic basis. In other words, of the proliferating number of export taxes applied to critical minerals and metals, and notwithstanding the fact that many of the countries most frequently resorting to export taxes on such products have abided by specific WTO obligations on their use, China seems to be the only country systematically applying WTO-inconsistent export duties. This has mostly to do with the different scope and coverage of the additional commitments agreed by those countries and, in particular, the fact that Vietnam, Ukraine and the Russian Federation have consented to bind

[125] See above, Section 4.4.2.

[126] The expression borrows from Qin, 'Reforming WTO Discipline', p. 1186.

the export duty rate applied to specific products rather than generally undertaking an obligation to eliminate export duties altogether.

Second, other countries significantly contributing to the proliferation of export duties on critical minerals and metals, and, in general, any other original WTO Member apart from Australia, remain free to impose export duties as per Article XI:1 inasmuch as they have not negotiated export duty concessions in their GATT schedules. Among them are countries such as Argentina, India, Indonesia, Malaysia, the Democratic Republic of the Congo, Zambia and Gabon.

Finally, independently of whether export duties are consistent with existing WTO obligations, the general lack of transparency surrounding the use of such instruments, often purported to be temporary but causing uncertainties in mining investments and in world markets, is amplified by the weakness of the WTO notification requirements applicable to export duties.

5

WTO disciplines on export quantitative restrictions

5.1 Introduction

Article XI:1 of the General Agreement on Tariffs and Trade (GATT) differentiates between export duties or taxes, on the one hand, and export 'prohibitions or restrictions, whether made effective through quotas, . . . export licences or other measures', on the other hand. As per a consistent body of World Trade Organization (WTO) jurisprudence, the latter category encompasses all quantitative restrictions (QRs), intended as non-tariff border measures that restrict the volume of exports.

This chapter gives an account of the existing WTO obligations regulating the use of export restrictions other than export duties. Section 5.2 provides an overview of the GATT rules relevant and applicable to export QRs. Section 5.3 discusses the GATT exceptions available to quantitative restrictions on exportation. Section 5.4 illustrates the transparency issues raised by the use of export QRs. Some conclusions on the relevance of the existing WTO disciplines on export quantitative restrictions are finally drawn in light of the specific features of the current wave of QRs that is affecting exports of critical minerals and metals.

5.2 GATT rules on export quantitative restrictions

5.2.1 The general elimination obligation under Article XI:1 GATT

Article XI:1 GATT, headed 'General Elimination of Quantitative Restrictions', outlaws all quantitative restrictions to trade. The structure of this Article is symmetrical for imports and exports:

> *No prohibitions or restrictions* other than duties, taxes or other charges, *whether made effective through quotas,* import or *export licences or other measures,* shall be instituted or maintained by any contracting party on the importation of any product of the territory of any other [Member] or *on the exportation* or sale for export of any product destined for the territory of any other [Member] (emphasis added).

Article XI:1 prescribes the elimination of both prohibitions and restrictions on exports (and imports). The term 'prohibitions' unambiguously applies to measures that totally impede the exportation (or the importation) of a product, namely, complete bans. Existing WTO jurisprudence on Article XI:1 GATT confirms this interpretation. Import bans were deemed incompatible with Article XI:1 in a number of WTO cases, including *Canada–Periodicals, US–Shrimps* and *EC–Asbestos.* On the export side, the Panel condemned a ban imposed by Canada on exports of unprocessed forms of salmon and herrings in the GATT case *Canada–Herring and Salmon.*

Article XI does not expressly mention the whole range of 'restrictions' falling within its remit. The expression '*whether* made effective through quotas, import or export licences *or other measures*' (emphasis added) indicates that the forms of restriction explicitly evoked are only illustrative and that other types of restrictive measures can potentially fall within the scope of Article XI:1 as 'other measures'. In this respect, the interpretation of the term 'restrictions' is critical to determine the reach of Article XI:1 GATT.

A consistent body of WTO case law has broadly interpreted the term 'restrictions' to cover a wide range of measures not solely including measures that may be considered formal quantitative restrictions, such as quotas, but also a whole variety of means that can have a limiting effect on trade, that is, the effect of reducing the volume of imports or exports. In *India–Quantitative Restrictions*, the Panel posited that 'the scope of the term "restriction" is broad, as seen in its ordinary meaning, which is "a limitation on action, a limiting condition or regulation"'.[1] In *India–Autos,* the Panel endorsed this interpretation and further elaborated on the meaning of the term 'restrictions':

> On a plain reading, it is clear that a 'restriction' need not be a blanket prohibition or a precise numerical limit. Indeed, the term 'restriction' cannot mean merely 'prohibition' on importation, since Article XI:1 expressly covers both 'prohibition and restriction'. Furthermore, the Panel considers that the expression '*limiting condition*' used in *India–Quantitative Restrictions* to define the term 'restriction' and which this Panel endorses, is helpful in identifying the scope of the notion in the context of the facts before it. That phrase suggests the need to identify not merely a condition placed on importation, but a condition that is limiting, i.e. that has a limiting effect. In the context of Article XI, that limiting effect must be on importation itself.[2]

[1] Panel Report, para. 5.128. [2] Panel Report, para. 7.270.

It inferred that '*any* form of limitation imposed on, or in relation to importation constitutes a restriction on importation within the meaning of Article XI:1'.[3] In *Colombia–Ports of Entry*, the Panel further clarified that the term 'restrictions' in the sense contemplated by Article XI:1 encompasses 'measures that create uncertainties and affect investment plans, restrict market access for imports, or make importation prohibitively costly'.[4] In *China–Raw Materials*, the Panel further added that any measure having 'the very *potential* to limit trade ... constitute[s] a "restriction" within the meaning of Art. XI:1 of the GATT 1994'.[5]

Consistent with such an interpretation, WTO case law has also interpreted the expression 'other measures' in a broad manner. In *Argentina–Hides and Leather*, the Panel clarified that the formulation 'other measures' is to be interpreted as a 'broad residual category'.[6] Accordingly, in *India–Autos*, the Panel argued that the express terms of Article XI:1 GATT are 'all-encompassing',[7] and confirmed that its formulation 'suggests a broad scope to the types of measures which can be considered to fall within the meaning of Article XI:1'.[8]

Against this backdrop, a network of measures has been considered to constitute a 'restriction' for the purposes of Article XI:1 GATT in the existing WTO case law. Most disputes have so far dealt with Article XI:1-inconsistent 'restrictions' on the import side. Yet, the conclusions reached by WTO dispute settlement bodies as to the types of measures covered by Article XI:1 also potentially apply to measures imposed on the export side, given the symmetrical structure of this Article. Past WTO jurisprudence on Article XI:1-inconsistent import restrictions has indeed been cited with approval by the Panels and the Appellate Body dealing with claims under Article XI:1 with respect to measures applied on exportation.

Until now, four disputes have specifically arisen out of the use of Article XI-inconsistent 'restrictions' on the export side; one case, *Japan–Semiconductors*, was addressed in the GATT era; *Argentina–Hides and Leather, China–Raw Materials* and *China–Rare Earths* were brought before the WTO adjudicators. The last two cases were also appealed.[9] Overall, the WTO dispute settlement bodies have condemned

[3] Ibid., para. 7.265. [4] Panel Report, para. 7.240. [5] Panel Reports, para. 7.1081.

[6] Panel Report, para. 11.17. [7] Panel Report, para. 7.249. [8] Ibid., para. 7.246.

[9] For a complete overview of the case law relating to export-restrictive measures see B. Karapinar, 'China's Export Restriction Policies: Complying with "WTO-plus" or Undermining Multilateralism', *World Trade Review* 10 (2011), 389. Another dispute, *US–Treating Export Restrictions as Subsidies*, dealt with the use of export restrictions on softwood lumber products by the United States. However, Canada challenged these

under Article XI:1 GATT a wide range of measures, irrespective of their legal status or of whether they *de jure* or *de facto*, potentially or actually, limited exports.

In *Japan–Semiconductors*, the GATT Panel clarified that Article XI:1 GATT not only covers legally binding instruments (laws, regulations, etc.) but also applies more broadly to non-mandatory 'measures' such as voluntary guidelines insofar as they result in a *de facto* restriction on exportation.[10] Accordingly, it condemned the system of administrative guidance put in place by the Japanese government to encourage Nipponese companies to increase the price of semiconductors sold abroad. It considered such measures to amount to *de facto* restrictions inasmuch as they could induce Japanese semiconductor producers to reduce their exports to the European market.[11] Mavroidis correctly notes in this respect that the Panel 'satisfied itself that Japan was in violation of [Article XI:1 GATT] by the mere fact that it provided incentives that could (but not necessarily) affect the behaviour of private parties'.[12]

This broad interpretation was somewhat narrowed in *Argentina–Hides and Leather*. In this case, the Panel reiterated the applicability of Article XI:1 GATT to restrictions having the same effect as formal quantitative restrictions setting an explicit numerical ceiling (i.e. the effect of reducing the volume of exports), but clarified that, in such cases, a causal link between the contested measure and the reduced level of exports (or imports) had to be demonstrated in order to establish a violation of Article XI:1 GATT.[13] In other words, the Panel affirmed that a different evidentiary standard applied to *de jure* and *de facto* restrictions for the purposes of Article XI:1 GATT, even if starting from the assumption that 'Article XI:1, like Articles I, II and III of the GATT 1994, protects competitive opportunities of imported products, not

measures under the WTO Agreement on Subsidies and Countervailing Measures (ASCM) by claiming that they amounted to a subsidy to US downstream industries. The Panel thus did not make any findings with respect to whether such measures were inconsistent with Article XI:1 GATT. See below, Chapter 7, Section 7.4.2.

10 GATT Panel Report, paras. 104–17.

11 The series of measures adopted by the Japanese government were aimed at monitoring the costs and prices of semiconductor exports to the United States in accordance with the US–Japan Semiconductor Pact. The United States aimed to protect its domestic industries from the inflow of cheaper semiconductors coming from Japan. Yet, as the European Union completely relied on Japanese imports at that time, European industries suffered from the price increase of Nipponese semiconductors. Ibid., para. 30.

12 P. C. Mavroidis, *Trade in Goods: The GATT and the Other Agreements Regulating Trade in Goods*, 2nd edn (Cambridge University Press, 2012), p. 67.

13 Panel Report, paras. 11.12–22.

trade flows'.[14] In the Panel's view, greater weight was to be attached to the actual trade effects of *de facto* restrictions so that 'a demonstration of causation must consist of a persuasive explanation of precisely how the measure at issue causes or contributes to the low level of exports'.[15] Accordingly, the Panel was not persuaded that the measure in question, an Argentinean regulation that allowed the presence of domestic tanner delegates (i.e. representatives from a domestic downstream industry) at the customs inspection for hides, would discourage exports by delaying customs procedures and thus amount to a 'restriction' within the meaning of Article XI:1 GATT.[16]

Within these broad coordinates, many types of export 'restrictions' as identified in the first part of this study have already been found to be in breach of Article XI:1 GATT by WTO adjudicators. Evidently, export quotas unambiguously fall within the scope of Article XI:1 GATT. In *China–Raw Materials*, the Panel cited with approval a whole body of WTO jurisprudence on Article XI:1 related to import quotas, and clarified that, in the same way, '[e]xport quotas are inconsistent with Members' obligation by virtue of Article XI:1 because they have a restrictive or limiting effect on exportation'.[17] Similarly, Chinese export quotas on rare earth elements, molybdenum and tungsten were also condemned as per Article XI:1 GATT.[18]

Based on the existing WTO case law, export licensing schemes may also qualify as 'restrictions' within the purview of Article XI:1 GATT under certain circumstances. In *Japan–Semiconductors*, the Panel considered Article XI:1 GATT to apply to non-automatic export licensing causing 'undue delays' in the issue of export licences.[19] In *China–Raw Materials*, the Panel elaborated on the applicability of Article XI:1 GATT to export licensing schemes. It clarified that the categorization of a licensing system as automatic or non-automatic serves no purpose for determining whether or not a measure is permissible under Article XI:1; instead, what is useful is 'the design and the structure of the licence to determine if it has a "limiting" or

[14] Ibid., para. 11.20, citing *Japan–Taxes on Alcoholic Beverages II*, AB Report, p. 16 and *Korea–Taxes on Alcoholic Beverages*, AB Report, paras. 119–20 and 127.

[15] *Argentina–Hides and Leather*, para. 11.19.

[16] Ibid., para. 11.35.

[17] *China–Raw Materials*, Panel Reports, paras. 7.206–7, citing GATT Panel Report, *France–Import Restrictions*; GATT Panel Report, *Japan–Agricultural Products I*; GATT Panel Report *US–Sugar*; and GATT Panel Report, *US–Sugar Waiver*.

[18] *China–Rare Earths*, Panel Reports, para. 7.200.

[19] *Japan–Semiconductors*, Panel Report, para. 118.

"restrictive" effect'.[20] The Panel noted that a licensing system that operates by granting an import or export licence to each and every applicant (i.e. an automatic export licensing scheme) would not breach Article XI:1 GATT in that it would not imply 'any restriction or limiting effect on importation or exportation in connection with the application and granting of the licence'.[21] At the same time, a system requiring an applicant to satisfy certain prerequisites before being granted an import or export licence would not necessarily run afoul of Article XI:1. In this respect, the Panel distinguished between discretionary and non-automatic licensing schemes. In its view, the former would always run counter to Article XI:1 GATT because a scheme where 'a licensing agency has discretion to grant or deny a licence based on unspecified criteria ... by its very nature would always have a restrictive or limiting effect'.[22] In contrast, the latter would not *per se* be impermissible under Article XI:1 GATT unless the prerequisites imposed created a restriction or limiting effect on exportation.[23]

Minimum export prices have also been considered in violation of Article XI:1 GATT. In *Japan–Semiconductors*, the GATT Panel opined that 'the complex of measures [adopted by the Japanese government] amounts to a coherent system restricting the sale for export of monitored semiconductors at prices below company-specific costs to markets other than the United States'.[24] Accordingly, it cited the GATT Panel Report on *EEC–Minimum Import Price* and applied the same reasoning on the export side. It concluded that a regulation preventing exportation below a minimum price level was a restriction on exportation inconsistent with Article XI:1 GATT.[25] In the same vein, the Panel in *China–Raw Materials* further elaborated:

> the authority to determine and require exporters to follow a particular export price level and not deviate below it without facing what amounts to a strict penalty, including revocation of the right to export altogether, has the potential to restrict trade. The restriction or limitation on exportation arises from the possibility that a price is set at such level that exporters

[20] Panel Reports, *China–Raw Materials*, para. 7.915. Indeed, the Panel noted: 'the obligation set forth in Article XI:1 does not distinguish between types of import or export licences that would be prohibited, be they automatic, non-automatic or discretionary. Rather, it concerns "prohibitions and restrictions" including those "made effective through ... import or export licences"'.

[21] Ibid., para. 7.916. [22] Ibid., para. 7.921. [23] Ibid., paras. 7.917–8.

[24] *Japan–Semiconductors*, GATT Panel Report, para. 117. [25] Ibid.

> cannot find a potential buyer in order to sell their product ... The Panel considers the very *potential* to limit trade to constitute a 'restriction' within the meaning of Article XI:1 of the GATT 1994.[26]

It thus concluded that China's system of coordinated prices amounted to a restriction within the meaning of Article XI:1 GATT.[27]

Finally, in *Colombia–Ports of Entry*, ports of entry restrictions were also condemned under Article XI:1 GATT.[28] By the same token, ports of exit restrictions may run afoul of Article XI:1 GATT. Other measures falling within the remit of Article XI:1 GATT include export restrictions made effective through state trading enterprises (STEs).[29] Although no relevant case law exists yet, the applicability of Article XI:1 GATT to these types of measures is explicitly mentioned in the Note *Ad* Articles XI, XII, XIII, XIV and XVIII:

> Throughout Articles XI, XII, XIII, XIV and XVIII, the terms 'import restrictions' or 'export restrictions' include restrictions made effective through state-trading operations.

In conclusion, Article XI:1 GATT has consistently been interpreted by WTO dispute settlement bodies in a broad manner, so as to encompass a wide range of measures having, *de jure* or *de facto*, restricting or limiting effects on exports (and imports), either actual or potential. Accordingly, any types of export restriction identified in the first part of this study, including those measures not yet specifically addressed by WTO adjudicators, may fall within the scope of Article XI:1 GATT as long as the complainants are able to demonstrate that they have limiting or restricting effects on exports.

5.2.2 *Other GATT provisions relevant to quantitative restrictions on exports*

Although the use of quantitative restrictions is prohibited under Article XI:1 GATT, a number of exceptions are available to WTO Members under the GATT to justify the use of Article XI:1-inconsistent measures under certain conditions.[30] Any export QRs imposed lawfully under the GATT needs however to be applied in a non-discriminatory manner. As

[26] Panel Reports, *China–Raw Materials*, para. 7.1081. [27] Ibid., para. 7.1082.

[28] *Colombia–Ports of Entry*, Panel Report, para. 7.275.

[29] L. Ehring and G. F. Chianale, 'Export Restrictions in the Field of Energy', in Y. Selivanova (ed.), *Regulation of Energy in International Trade Law: WTO, NAFTA and Energy Charter* (Kluwer Law International, 2012), p. 130.

[30] See below, Section 5.3.

seen above, the most favoured nation (MFN) obligation embodied in Article I:1 GATT applies 'with respect to all rules and formalities in connection with importation and exportation'. The scope of the expression 'rule in connection with exportation' has not yet been interpreted by the WTO dispute settlement bodies. In *US–Poultry (China)*, however, the Panel condemned a discriminatory import ban as inconsistent with Article I:1 GATT.[31] It could thus be argued that an export ban applied selectively on certain WTO Members only may run afoul of Article I:1 GATT. However, it is difficult to assess whether other types of export restrictions may be covered by Article I:1 GATT, as the existing WTO case law has generally been hesitant to make findings under Article I:1 after condemning a measure under Article XI:1 GATT for judicial economy purposes.[32]

The MFN principle is, however, instilled in Article XIII GATT on 'Non-discriminatory Administration of Quantitative Restrictions'. Article XIII:1 states:

> No prohibition or restriction shall be applied by any contracting party on the importation of any product of the territory of any other contracting party or on the exportation of any product destined for the territory of any other contracting party, unless the importation of the like product of all third countries or the exportation of the like product to all third countries is similarly prohibited or restricted.

The language of Article XIII:1 requires that import or export restrictions lawfully applied under the GATT be administered in a non-discriminatory manner, thus imposing an MFN-like obligation.[33] In this sense, any export restrictions in breach of Article I:1 GATT will also violate Article XIII:1 GATT.

Article XIII:2 GATT, moreover, sets out a number of rules related to the administration of restrictions with respect to the distribution of allowed trade. The chapeau of Article XIII:2 states:

> In applying import restrictions to any product, contracting parties shall aim at a distribution of trade in such product approaching as closely as possible the shares which the various contracting parties might be expected to obtain in the absence of such restrictions . . .

[31] Panel Report, para. 7.410.

[32] See Ehring and Chianale, 'Export Restrictions', p. 127, citing *US–Wool Shirts and Blouses*, AB Report, p. 19 and *US–Shrimp*, Panel Report, paras. 7.22–3.

[33] P. Van den Bossche and W. Zdouc, *The Law and Policy of the World Trade Organization*, 3rd edn (Cambridge University Press, 2013), p. 492.

Although it refers to import restrictions only, the applicability of Article XIII:2 to restrictions on exports is confirmed by Article XIII:5, according to which ' ... in so far as applicable, the principles of this Article shall also extend to export restrictions'. Various sub-paragraphs of Article XIII GATT contain specific requirements to be met by WTO Members when imposing a lawful quantitative restriction. Under Article XIII:2(a)–(b), a quota limiting the total amount of imports (or exports), whether allocated among supplying (or, in the case of export restrictions, importing) countries or global, is to be preferred over export licences or permits without a quota. Moreover, when allocating a quota among supplying/importing countries, WTO Members 'may seek agreement with respect to the allocation of shares in the quota with all other contracting parties having a substantial interest in supplying [or importing] the product concerned' pursuant to Article XIII:2(d) GATT. If this method is 'not reasonably practicable', however, the WTO Member shall allocate the quota shares among supplying/importing countries in accordance with historical market shares:

> the contracting party shall allot to contracting parties having a substantial interest in supplying [or importing] the product shares based upon the proportions, supplied [or imported] by such contracting parties during a previous representative period, of the total quantity or value of imports [or exports] of the product, due account being taken of any special factors which may have affected or may be affecting the trade in the product.

The three years preceding the imposition of the quota have generally been selected as the 'representative period'.[34] WTO Members are required to allocate a part of the quota to all Members having an interest in supplying/importing but do not have the obligation to allot quota shares to Members without a substantial interest. Yet, where a Member decides to allot quota shares to Members with minor market shares, it has to do so with respect to all such Members.[35] Article XIII:2 GATT was criticized by Mavroidis:

> it is now non-discriminatory administration of quotas that is privileged through Article XIII GATT; it is respect for historical market shares – a

[34] See GATT Panel Report, *EEC–Apples (Chile I)*, para. 4.16 and GATT Panel Report, *EEC–Dessert Apples*, para. 12.22, cited in Van den Bossche and Zdouc, *The Law and Policy*, p. 494. As per Article XIII:4 GATT, the Member imposing the quota selects the period but any Member having a substantial interest in supplying/importing the restricted product may request to be consulted regarding 'the need for an adjustment of the proportion determined or of the base period selected, or for the reappraisal of the special factors involved ... '.

[35] Panel Report, *EC–Bananas III*, para. 7.73.

> rather discriminatory policy . . . Art. XIII thus, by preserving historic shares, does not account for changes in supply and demand.[36]

Other authors recognize that, by allowing discrimination between Members having a substantial interest in supplying/importing and Members with minor market shares, Article XIII:2 is inconsistent with Article XIII:1 GATT. Yet, Van den Bossche and Zdouc interpret Article XIII:2 as a *lex specialis* to Article XIII:1 GATT.[37]

Article XIII:3, moreover, requires WTO Members imposing export (and import) restrictions to provide promptly all the relevant information concerning the application and the administration of the restrictions.

Article XIII GATT applies to GATT-consistent export (and import) restrictions. In other words, otherwise GATT-consistent export restrictions become unlawful if they do not comply with Article XIII GATT. However, an export restriction that cannot be justified under any of the GATT exceptions is then automatically in breach of Article XI:1 GATT, whether applied in a non-discriminatory manner or not.[38]

Finally, Article XVII GATT is also relevant to export restrictions to the extent that it covers the conduct of STEs. As shown above, in certain countries (e.g. China) exports of specific products are subject to state trading.[39] According to Article XVII:1(a), STEs are defined as either 'state enterprises' established or maintained by a WTO Member or 'any enterprises' to which governments 'grant . . ., formally or in effect, exclusive or special privileges'. As per the same provision, these enterprises shall act 'in accordance with the general principles of non-discriminatory treatment'.[40] Pursuant to Article XVII:1(b) GATT, moreover, the requirements set out in Article XVII:1(a) 'shall be understood to require that such enterprises shall, having due regard to the other provisions of this

36 Mavroidis, *Trade in Goods*, p. 85.

37 Van den Bossche and Zdouc, *The Law and Policy*, p. 494, citing Panel Report, *EC–Bananas III*, para. 7.74.

38 See Mavroidis, *Trade in Goods*, p. 86.

39 See above, Chapter 2, Section 2.5.4.

40 See GATT Panel Report, *Canada–Administration of the Foreign Investment Review Act (FIRA)*; AB Report, *Canada–Wheat*. However, it is not clear from the text of the provision whether Article XVII:1 GATT is meant to cover both standards of the non-discrimination principle (i.e. the most favoured nation and the national treatment requirement). In *Korea–Various Measures on Beef* and *Canada–Wheat*, the Panel seemed to suggest that this might be the case (para. 753 and paras. 6.44–50, respectively). See V. Pogoretskyy, 'Energy Dual Pricing in International Trade: Subsidies and Anti-dumping Perspectives', in Y. Selivanova (ed.), *Regulation of Energy in International Trade Law: WTO, NAFTA and the Energy Charter* (Kluwer Law International, 2012), pp. 197–8.

Agreement, make any such purchases or sales *solely in accordance with commercial considerations*, including price, quality, availability, marketability, transportation and other conditions of purchase or sale ... ' (emphasis added). In *Canada–Wheat*, the Panel shed light on the relationship between Article XVII:1(a) and Article XVII:1(b) GATT. In its view, the former establishes a non-discrimination obligation, whereas the latter clarifies the scope of such obligation. Accordingly, any analysis on the consistency of a STE's conduct with Article XVII:1(b) GATT cannot be pursued before having first found a breach of Article XVII:1(a).[41] The Note *Ad* Article XVII, paragraph 1, furthermore, specifies that:

> The charging by a state enterprise of different prices for its sales of a product in different markets is not precluded by the provisions of this Article, provided that such different prices are charged for commercial reasons, to meet conditions of supply and demand in export markets.

Inasmuch as STEs responsible for exporting raw materials act in accordance with commercial considerations, their conduct cannot be challenged as discriminatory within the meaning of Article XVII GATT. Yet, WTO Members cannot implement export restrictions through state trading enterprises. This possibility is explicitly precluded by virtue of the Note *Ad* Articles XI, XII, XIII, XIV and XVIII:

> Throughout Articles XI, XII, XIII, XIV and XVIII, the terms 'import restrictions' or 'export restrictions' include restrictions made effective through state-trading operations.

5.3 GATT exceptions applicable to quantitative restrictions on exports

The GATT contains a number of exceptions that allow WTO Members to lawfully resort to quantitative restrictions. Some of them are Article XI:1-specific exceptions, whereas other provisions were originally drafted as exceptions to Article XI but then evolved into exceptions encouraging WTO Members to adopt price-based measures (e.g. the balance of payments exceptions). Some others are general GATT exceptions available to justify any GATT-inconsistent measure.[42] Importantly, GATT exceptions specific to quantitative types of restrictions often treat imports and exports asymmetrically.

[41] Panel Report, paras. 77–125.

[42] This type of exceptions will be dealt with in Chapter 6.

5.3.1 *The critical shortage of supply clause under Article XI:2(a) GATT*

Article XI:2 GATT provides for a number of Article XI:1-specific exceptions. WTO Members may thus lawfully adopt Article XI:1-inconsistent measures under the circumstances laid out in Article XI:2 GATT. According to Article XI:2(a), the general prohibition contained in Article XI:1 shall not extend to:

> [e]xport prohibitions or restrictions temporarily applied to prevent or relieve critical shortages of foodstuffs or other products essential to the exporting contracting party.

Article XI:2(a) specifically applies to *export* prohibitions or restrictions, and allows WTO Members to apply such restrictions to cope with situations of supply shortages. This provision appears particularly important for dealing with export restrictions on mineral resources, and particularly critical minerals and metals, as they are often applied with the aim of ensuring adequate domestic supply.[43]

The shortage of supply clause under Article XI:2(a) GATT was interpreted for the first time only recently in the *China–Raw Materials* dispute. As a preliminary issue, the Panel clarified that the burden to prove that the requirements under Article XI:2(a) are met, and therefore that no inconsistency arises under Article XI:1, lies on the defendant party.[44] Accordingly, the Panel clarified that a measure cannot be justified under Article XI:2 by arguing that the complainants failed to demonstrate that a measure did not fall within the terms of Article XI:2(a).[45]

The Panel and the Appellate Body then interpreted Article XI:2(a) GATT by shedding light on the meaning of each of the requirements contained therein, considered individually and in concert:[46] (i) the 'temporarily applied' requirement; (ii) the 'critical shortage' requirement; and, (iii) the 'essentialness' requirement. The first requirement was interpreted to demand that the period of implementation of an export restriction or prohibition be finite, that is, limited in time.[47] According to the Appellate Body, this would not necessarily imply that the duration of the measure must be fixed in advance.[48]

43 See above, Chapter 3, Section 3.3.1.
44 Panel Reports, *China–Raw Materials*, para. 7.210. 45 Ibid., para. 7.211.
46 AB Reports, *China–Raw Materials*, para. 328.
47 Panel Reports, *China–Raw Materials*, paras. 7.255 and 7.260; AB Reports, *China–Raw Materials*, para. 323.
48 AB Reports, *China–Raw Materials*, para. 331, rejecting the Panel's view that the temporariness of a measure should be clearly indicated at the outset. For a critique of the Appellate Body's interpretation see Karapinar, 'Defining the Legal Boundaries', p. 451.

Yet, any measure maintained in the long-run 'with no indication that [it] would be withdrawn and every indication that it will remain in place until the reserves have been depleted' would run afoul of Article XI:2(a),[49] irrespective of whether it is subject to annual review.[50] In other words, situations of *permanent* shortages cannot be addressed within the scope of Article XI:2(a).[51]

In the view of the Appellate Body, this interpretation finds supports when considering the second requirement. According to the WTO adjudicator, the 'temporarily applied' requirement contextually informs the notion of 'critical shortage' in that the restrictions should be applied in the interim, to provide relief in extraordinary conditions in order to bridge a passing need'.[52] Thus, the finite availability of a product is insufficient to constitute a 'critical shortage' because 'if there is no possibility for the shortage ever to cease to exist, it would not be possible to "relieve or prevent" it through an export restriction applied only for a limited period of time'.[53] Accordingly, the 'critical shortage' requirement refers to 'those deficiencies in quantity that are crucial, that amount to a situation of decisive importance or that reach a vitally important or decisive stage, or a turning point'.[54]

As to the essentialness requirement, the Appellate Body agreed with the Panel that Article XI:2(a) GATT does not limit *a priori* the types of 'other products' that may be subject to export restrictions in derogation from Article XI:1 but requires that the product in question be '"important" or "necessary" or "indispensable"' to a particular Member'.[55] The determination of whether a particular product is 'essential' within the meaning of Article XI:2(a) should take into account 'the particular circumstances faced by that Member at the time when a Member applies a restriction or prohibition'.[56] At the same time, however, 'the mere designation of a product as essential or the imposition of conservation-related restrictions imposed on extraction or processing should not be relevant to the assessment of whether a product is "essential" to a Member' for it would in effect 'allow a Member to manufacture

49 Panel Reports, *China–Raw Materials*, para. 7.350.
50 AB Reports, *China–Raw Materials*, paras. 339–44.
51 Panel Reports, *China–Raw Materials*, para. 7.351.
52 AB Reports, *China–Raw Materials*, para. 330. See also Panel Reports, *China–Raw Materials*, para. 7.297.
53 Ibid.
54 AB Reports, para. 324. See also Panel Reports, *China–Raw Materials*, paras. 7.284–5.
55 AB Reports, *China–Raw Materials*, para. 326.
56 Panel Reports, *China–Raw Materials*, para. 7.277.

"essentialness" when none exists'.[57] Neither the Panel nor the Appellate Body attempted to list exhaustively the guiding factors that could inform the essentialness requirement. Yet, they agreed in considering criteria such as the product's importance in use, the contribution to national economic development, and the complexity in substitutability as indicative of the essentialness of a product.[58]

The WTO dispute settlement bodies found support for a narrow interpretation of the shortage of supply exception not only when considering the ordinary meaning of the words constituting the provision, as imparting meaning to each other, but also when considering their relevant context. To this end, they examined other GATT exceptions related to the issue of exhaustibility, namely Article XX(g) and Article XX(j). According to Article XX(g), WTO Members may justify a measure in breach of a GATT obligation when it is 'related to the conservation of exhaustible natural resources' and provided it is 'made effective in conjunction with restrictions on domestic productions or consumption'.[59] Based on the principle of effective interpretation,[60] the Appellate Body clarified that the two types of exceptions 'are intended to address different situations and must mean different things'.[61] Article XX(g) is a general exception available to WTO Members to seek justification for GATT-inconsistent measures aimed at protecting exhaustible natural resources. Accordingly, it 'incorporates additional protections in its chapeau to ensure that the application of a measure does not result in arbitrary or unjustifiable discrimination or amount to a disguised restriction'.[62] On the other hand, Article XI:2(a) is an Article XI:1-specific obligation, which limits the scope of the obligation not to impose quantitative restrictions. In the Appellate Body's view, this implies that the reach of the two provisions is different: 'where the

[57] Ibid., para. 7.345. In reaching this conclusion, the Panel drew a contrast with the language of Article XXI(b) GATT, underscoring that the latter reads: '[no]thing in this Agreement shall be construed … to prevent any contracting party from taking any action *which it considers necessary* for the protection of its essential security interests'. Ibid., para. 7.276.

[58] Ibid., paras. 7.340–5. [59] See below, Chapter 6, Section 6.3.2.

[60] Panel Reports, *China–Raw Materials*, para. 7.256; AB Reports, *China–Raw Materials*, para. 333. According to the principle of effective interpretation, all applicable provisions of a treaty should be interpreted in a way so as to give effect to all of them without rendering them useless, redundant or irrational. See I. Van Damme, *Treaty Interpretation by the WTO Appellate Body* (Oxford University Press, 2009).

[61] AB Reports, *China–Raw Materials*, para. 337, citing Panel Reports, *China–Raw Materials*, para. 7.300.

[62] AB Reports, *China–Raw Materials*, para. 333, citing Panel Reports, para. 7.258.

requirements of Article XI:2(a) are met, there would be no scope for the application of Article XX, because no obligation exists'.[63]

It follows that, while WTO Members cannot resort indistinguishably to either Article XI:2(a) or Article XX(g),[64] nothing *a priori* precludes an export restriction applied within the purview of Article XI:1(a) from operating simultaneously with a conservation measure adopted under Article XX(g):

> We do not exclude that a measure falling within the ambit of Article XI:2(a) could relate to the same product as a measure relating to the conservation of an exhaustible natural resource. It would seem that Article XI:2(a) measures could be imposed, for example, if a natural disaster caused a 'critical shortage' of an exhaustible natural resource, which, at the same time, constituted a foodstuff or other essential product.[65]

Article XX(j) GATT allows WTO Members to take measures that are 'essential to the acquisition or distribution of products in general or local supply' in derogation of GATT provisions. It requires that 'any such measures shall be consistent with the principle that all contracting parties are entitled to an equitable share of the international supply of such products, and that any such measures, which are inconsistent with the other provisions of the Agreement shall be discontinued as soon as the conditions giving rise to them have ceased to exist'. The Appellate Body found support for its strict interpretation of Article XI:2(a) by drawing a contrast between the wording of the two provisions:

> Contrary to Article XI:2(a) . . . Article XX(j) does not include the word 'critical' or another adjective further qualifying the short supply. We must give meaning to this difference in the wording of these provisions. To us, it suggests that the kinds of shortages that fall within Article XI:2(a) are more narrowly circumscribed than those falling within the scope of Article XX(j).[66]

In light of the above, Article XI:2(a) GATT cannot be invoked to justify export restrictions responding to situations of physical scarcity and/or exhaustibility of essential mineral resources. At the same time, the fact that an essential product is exhaustible does not necessarily preclude a WTO Member from legitimately imposing an export restriction under Article XI:2(a). WTO Members must however demonstrate that such a restriction aims at alleviating or preventing an imminent situation of

[63] AB Reports, *China–Raw Materials*, para. 334.

[64] Ibid., para. 333, citing Panel Reports, *China–Raw Materials*, para. 7.257.

[65] AB Reports, *China–Raw Materials*, para. 337.

[66] Ibid., para. 325.

decisive importance or extraordinary crisis. This assessment requires a case-by-case analysis on the part of WTO dispute settlement bodies taking into account all the requirements of Article XI:2(a) GATT. Yet, as correctly noted by Howse and Josling, the narrow approach adopted in *China–Raw Materials* suggests that Article XI:2(a) cannot be invoked to justify 'intransigently beggar-thy-neighbour response[s] to the scarcity or finitude of natural resources'.[67] Interestingly, China did not seek to justify the export restrictions challenged in *China–Rare Earths* under Article XI:2(a) GATT.

5.3.2 *Limited applicability of other Article XI-specific exceptions*

In addition to the critical shortage of supply clause, another two exceptions are provided for in Article XI:2 GATT. As per Article XI:2(b), the general elimination of quantitative restrictions shall not extend to 'import and export prohibitions or restrictions necessary to the application of standards or regulations for the classification, grading or marketing of commodities in international trade'. This provision was interpreted by the GATT Panel in *Canada–Herring and Salmon*. In the case at issue, Canada sought justification for an export ban imposed on certain fisheries products claiming that it was necessary to maintain its quality standards. The Panel rejected this claim after noticing that 'Canada prohibited export of certain unprocessed salmon and unprocessed herring even if they could meet the standards generally applied to fish exported from Canada'.[68] The Panel also provided a narrow interpretation of Article XI:2(b) GATT. In its view, this provision only covered those regulations that specifically apply to marketing and not just any regulation that facilitates foreign sales. In the case of export restrictions, these measures could be justified under Article XI:2(b) only when 'designed to promote the marketing of a commodity by spreading supplies of the restricted product over a longer period of time'.[69] In this respect, Article XI:2(b) GATT appears of limited relevance to justify export restrictions applied to mineral products.

Finally, according to Article XI:2(c) the provisions of Article XI:1 GATT do not apply to '[i]mport restrictions on any agricultural or fisheries product ... necessary to the enforcement of governmental measures'. Article XI:2(c) GATT only applies to 'restrictions' applied on the

[67] R. Howse and T. Josling, 'Agricultural Export Restrictions and International Trade: A Way Forward', International Food and Agricultural Trade Policy Council (IPC) Position Paper, 2012, http://www.agritrade.org/Publications/documents/Howse_Josling_Export_Restriction_final.pdf (accessed November 2014), p. 14.

[68] GATT Panel Report, para. 4.2. [69] Ibid., para. 4.3.

importation of agricultural or fisheries products and thus it is not relevant to regulate the use of export restrictions.

5.3.3 *Non-availability of other GATT exceptions applicable to quantitative restrictions*

A number of GATT provisions allow WTO Members to apply otherwise Article XI:1-inconsistent quantitative restrictions. Yet, these exceptions are not always symmetrical on the import and export side. In most cases, they are import specific. This is the case of the balance of payments (BoP) exception contained in Article XII and Article XVIII GATT. This exception allows WTO Members to adopt measures in derogation of Article XI:1 GATT in order to address balance of payments difficulties. Article XII:1 allows WTO Members to 'restrict the quantity or value of merchandise permitted to be *imported*' (emphasis added), provided the requirements laid out in the following paragraphs are met. These requirements are somewhat relaxed as per Article XVIII(b) GATT (including paragraphs 8 to 12), which contains the balance of payments exceptions for developing countries.[70] Article XVIII:8 reads:

> In order to safeguard its external financial position and to ensure a level of reserves adequate for the implementation of its programme of economic development, a [WTO Member with low standards of living and in the early stages of development] may, subject to the provisions of paragraphs 10 to 12, control the general level of its imports by restricting the quantity or value of merchandise permitted to be imported; *Provided* that the import restrictions instituted, maintained or intensified shall not exceed those necessary:
> (*a*) to forestall the threat of, or to stop, a serious decline in its monetary reserves, or
> (*b*) in the case of a contracting party with inadequate monetary reserves, to achieve a reasonable rate of increase in its reserves . . .

The balance of payments exception has proved to be of very limited relevance for developed countries in the era of floating exchange rates.[71]

[70] For a detailed account see Van den Bossche and Zdouc, *The Law and Policy*, pp. 636–45; Mavroidis, *Trade in Goods*, pp. 75–80.

[71] The exception was drafted at the time of fixed parities under the assumption that, given certain conditions, an import restriction would correspond to a nominal devaluation. Mavroidis, *Trade in Goods*, pp. 75–6. According to the author, 'this provision, although still present in the GATT, is largely of historical interest'.

By contrast, developing countries have resorted to this exception quite frequently, particularly in the 1980s and 1990s, often adopting, however, price-based measures instead of quantitative restrictions.[72] While the existing exception contemplates that these countries adopt import restrictions meant to alleviate BoP imbalances that are caused by excessive imports and/or insufficient exports, recent experience shows that resource-based economies that are vulnerable to boom-bust price cycles may also be exposed to severe terms of trade shocks caused by surges of exports. Externally induced balance of payments disequilibria on the export side, in particular, may also hamper the efforts of low-income commodity-dependent countries to achieve economic development through economic diversification and lead to the Dutch disease.[73] This circumstance is indirectly acknowledged in Article XVIII:11, which states:

> In carrying out its domestic policies, the contracting party concerned shall pay due regard to the need for restoring equilibrium in its balance of payments on a sound and lasting basis and to the desirability of assuring an economic employment of productive resources.

The asymmetrical structure of the balance of payments exception, by neglecting to discipline the use of temporary BoP measures on the export side, fails to adequately take into account that terms of trade shocks may intervene at either end of the boom-bust price cycle with just as harmful consequences for the economic development prospects of commodity-dependent countries.

Another exception that only applies to imports is Article XVIII(c) GATT (including paragraphs 13 to 21), which allows countries with low standards of living and in the early stages of development to take measures inconsistent with other GATT provisions when 'governmental

[72] Although not originally contemplated in the explicit wording of Article XII and Article XVIII(b) GATT, the use of tariff-like measures was first endorsed in the 1979 Declaration on Trade Measures Taken for Balance-of-Payments Purposes and then formally recognized as the preferred balance of payments tool in the Understanding on Balance-of-Payments Provisions of the GATT 1994. Pursuant to Paragraph 2 of the Understanding, in particular, WTO Members commit to give preference to price-based measures such as import surcharges, import deposit requirements and other equivalent trade measures with an impact on the price of imported goods. These measures can be 'applied by a Member in excess of the duties inscribed in the Schedule of that Member'. Accordingly, Paragraph 3 of the Understanding requires WTO Members to seek to avoid as much as possible the imposition of quantitative restrictions and in any case to explain why price-based measures would not be adequate to arrest the deterioration of the balance of payments.

[73] See above, Chapter 3, Section 3.3.1.

assistance is required to promote the establishment of a particular industry',[74] provided certain substantive and procedural requirements are met. Yet, Article XVIII:14 specifies that the WTO Members facing such circumstances 'shall indicate the specific measure affecting *imports* which it proposes to introduce in order to remedy these difficulties'. The same holds true for countries whose economy is in the process of development as per Article XVIII:4(b), and 'desiring, in the interest of the development of its economy, to introduce a measure of the type described in paragraph 13 in respect of the establishment of a particular industry'.

WTO Members are also prevented from using export quantitative restrictions in situations where a surge of exports is seriously injuring a domestic industry. Under Article XIX GATT, the use of safeguard measures taking the form of quantitative restrictions is only envisaged to 'give "breathing space" to a domestic industry to adapt itself to the new market situation by temporarily restricting *imports*'.[75] Similar to the use of safeguards in the form of tariff increases, therefore, the use of quantitative restrictions is solely contemplated to provide a safety valve in cases of increased imports. Article XIX GATT reads:

> If, as a result of unforeseen developments and of the effect of the obligations incurred by a contracting party under this Agreement, ... any product is being imported into the territory of that contracting party in such increased quantities and under such conditions as to cause or threaten serious injury to domestic producers in that territory of like or directly competitive products, the contracting party shall be free, in respect of such product, and to the extent and for such time as may be necessary to prevent or remedy such injury, to suspend the obligation in whole or in part ...

The same concept informs the WTO Agreement on Safeguards.[76] WTO Members are thus left with no safeguard tool to cope with situations of a sudden, sharp and significant increase of exports. As seen above, however, these situations are quite common in international commodities markets and, in the case of minerals and metals and critical minerals and metals in particular, they may be induced by the expansion of world demand as well as by speculative behaviours.[77]

[74] See Article XVIII:13.

[75] Van den Bossche and Zdouc, *The Law and Policy*, p. 624 (emphasis added).

[76] See Article 2.1 of the WTO Agreement on Safeguards.

[77] For a full account see below, Chapter 7.

Moreover, export restrictions cannot be imposed in derogation of Article XI:1 GATT to avoid dumping practices that are inconsistent with Article VI GATT. This was the argument invoked by the Japanese government in *Japan–Semiconductors*.[78] The GATT Panel, however, affirmed that a WTO Member cannot invoke Article VI to justify a restriction on exports because Article VI only refers to actions adopted by importing countries:

> The Panel noted that Article VI:1 declared that dumping was to be condemned if it caused or threatened material injury to an established industry or materially retarded the establishment of an industry and that Article VI:2 allowed contracting parties to levy a duty on dumped products, subject to certain specified conditions. The provision was silent on actions by exporting countries. The Panel therefore found that Article VI did not provide a justification for measures restricting the exportation or sale for export of a product inconsistently with Article XI:1.[79]

In conclusion, the GATT does provide for a number of exceptions to Article XI:1 GATT. Yet, in some notable cases, these exceptions are import specific. They cannot therefore be invoked by WTO Members, especially low-income commodity-dependent countries, to justify restrictions imposed on exports even in cases where such measures are inspired and/or respond to the same rationale evoked under these provisions.

5.4 Transparency and notification requirements applicable to export quantitative restrictions

In light of the general obligation to eliminate quantitative restrictions on (imports and) exports enshrined in the GATT, sound transparency and notification requirements applicable to export restrictions represent a fundamental tool to ensure timely availability of information and ultimately ensure that GATT rules on quantitative restrictions are respected.

Article X GATT applies to quantitative restrictions on exports in the same way that it applies to tariff measures on exports. WTO Members are required to publish all measures 'pertaining to ... requirements, restrictions or prohibitions on imports or exports ... ' in a prompt and accessible manner as per Article X:1, and to administer such measures 'in a uniform, impartial and reasonable manner' as per Article X:3.

Moreover, many GATT exceptions to Article XI:1 GATT, such as Article XI:2(c), Article XII and Article XVIII(b) and (c), prescribe

[78] GATT Panel Report, *Japan–Semiconductors*, paras. 45–8. [79] Ibid., para. 120.

specific consultation and notification requirements to be followed in order to lawfully implement a quantitative restriction. Article XIII:3 (a)–(c) GATT also require that WTO Members promptly provide all relevant information concerning any quantitative restrictions imposed in accordance with GATT rules. Under Article XVII:4 GATT, moreover, WTO Members must notify to other Members about the products that are (imported into or) exported from their territories by STEs. As seen above, however, only a few of these GATT provisions target quantitative restrictions on exports as well.

Similar to export taxes, 'export restrictions' are explicitly included in the Indicative List of Notifiable Measures annexed to the 1993 Uruguay Round Ministerial Decision on Notification Procedures. Moreover, 'export restrictions' are listed among the categories of quantitative restrictions annexed to the Decision on Notification Procedures for Quantitative Restrictions adopted by the Council for Trade in Goods on 1 December 1995 and revised in 2012.[80] Up to 2012, the implementation of the Decision had been quite modest. The Decision committed WTO Members to make complete notifications of the quantitative restrictions that they maintained biennially and allowed them to make reverse notifications, but did not specify whether WTO Members had to notify only WTO-inconsistent quantitative restrictions or all of them. Moreover, WTO Members could consult the database compiled by the WTO Secretariat on the basis of the notifications only upon request. In the 2012 revision, the Decision clarifies that:

> Members shall make complete notifications of *all* quantitative restrictions in force by September 2012 and at two yearly intervals thereafter. They should also notify changes to those quantitative restrictions as soon as possible, but not later than six months from their entry into force. Members will also retain the right to notify, at any time, corrections to their notifications, as well as to reverse notifications to which they are subject. Notifications shall be circulated in a new document series and will automatically be included in the agenda of the meeting of the Committee on Market Access.[81]

WTO Members are furthermore required to make notifications pursuant to detailed requirements. Notifications have to include information on the type of restriction, the tariff line code and the product description, the WTO justification and the national legal basis for imposing the

[80] WTO Doc. G/L/59, 10 January 1996; and WTO Doc. G/L/59/Rev.1, 22 June 2012.
[81] Ibid., para. 1 (emphasis added).

restriction.[82] The WTO Secretariat is required to create a database inscribing all the information provided by Members, which is now available to the public,[83] and to circulate a report based on such information every year.[84] WTO Members remain free to make reverse notifications.[85] In this respect, the existing WTO notification requirements on export restrictions appear stricter and more effective than those imposed on Members on the use of export duties. Moreover, a further improvement in transparency disciplines for export restrictions will be realized with the entry into force of the Agreement on Trade Facilitation.[86] Like export taxes, export 'restrictions or prohibitions' are in fact included in the list of measures to which Article 1 of the new WTO Agreement, labelled 'Publication and Availability of Information', applies. Members will thus be required to promptly publish all relevant information concerning such measures 'in a non-discriminatory and easily accessible manner in order to enable governments, traders and other interested parties to become acquainted with them' (Paragraph 1.1 on 'Publication', subparagraph (f)). Accordingly, WTO Members are also to establish enquiry points as per Paragraph 3.1 of the new WTO Agreement, and to notify to the Committee on Trade Facilitation the official place(s) where such information has been published in accordance with Paragraph 4.1.

5.5 Conclusions

The analysis of WTO disciplines relevant and applicable to export restrictions has shown that the existing GATT rules capture the full range of restrictions applied to exports of critical minerals and metals identified in the first part of this study. Article XI:1 GATT contains a general obligation to eliminate any types of quantitative restriction on exports, from prohibitions to quotas, from licensing schemes to minimum export prices, from ports of exit restrictions to any other measures that have a limiting effect on the volume of exports. In this sense, Article XI:1 has consistently been interpreted in a broad manner by WTO dispute settlement bodies on the export side as much as on the import side. It is important to recall at this stage that the stringency of Article XI:1 GATT is a reflection of the traditional preference for tariff measures over quantitative restrictions inspiring the overall GATT architecture. As explained by the Panel in *Turkey–Textiles*:

[82] Ibid., para. 3. [83] Ibid., para. 4. [84] Ibid., para. 7. [85] Ibid., para. 5.
[86] See above, Chapter 4, Section 4.5.

> The prohibition on the use of quantitative restrictions forms one of the cornerstones of the GATT system. A basic principle of the GATT system is that tariffs are the preferred and acceptable form of protection ... The prohibition against quantitative restrictions is a reflection that tariffs are GATT's border protection 'of choice'.[87]

Inasmuch as non-tariff border restrictions are symmetrically prohibited under Article XI:1 GATT, whether applied to imports or exports, the case for a WTO regulatory deficiency on the export side finds no foundation. Transparency and notification requirements are also much more developed in the case of export quantitative restrictions. This chapter has shown that, in the case of quantitative restrictions within the meaning of Article XI:1 GATT, the asymmetry in the treatment of imports and exports arises in the case of exceptions. The GATT does provide a set of exceptions to Article XI:1 GATT aimed at addressing specific temporary needs, such as critical shortages of essential products (Article XI:2(a)), balance of payments disequilibria (Article XII and Article XVIII(b)), and the protection of infant industries (Article XVIII(c)). However, the critical shortage of supply clause was interpreted by the Appellate Body in a way that cannot address the concerns over availability of domestic supply that have inspired the current use of export restrictions on mineral raw materials in developing countries. These countries cannot lawfully resort to export restrictions in order to achieve economic development goals such as those recognized under Article XVIII(b) and Article XVIII(c) GATT on the import side. As per Article XIX GATT, moreover, WTO Members may resort to safeguard measures in derogation of Article XI GATT only when a surge of imports is causing or threatening to cause a serious injury to a domestic industry.

The stronger 'rigidity' of GATT disciplines on export restrictions, combined with the recognition that non-tariff border restrictions have more disruptive effects on trade, may explain why WTO Members have generally made use of quantitative restrictions more sparingly than export duties, including in the minerals and metals sector. Although export licensing schemes are the most frequently reported type of export restriction, not all countries are using them to affect export volumes, that is, as a 'restriction' within the meaning of Article XI:1 GATT. As shown in the first part of this study, moreover, the other types of export restrictions are much less diffused. China is the only country making extensive

[87] Panel Report, *Turkey–Textiles*, para. 9.63. The arguments in favour of tariffs over quantitative restrictions are explained above, Chapter 3, Section 3.2.

use of export quotas on mineral resources.[88] Relatively few large suppliers have introduced and/or maintained export prohibitions, in many cases temporarily, and other measures such as minimum export prices, domestic market regulation or captive mining are also residually applied by very few countries on a handful of products.

Finally, inasmuch as Article XI:1-inconsistent export restrictions are not subject to the whole range of GATT exceptions specifically available to import QRs, WTO Members may lack the incentive to engage in the negotiation of export duty concessions to be inscribed in their GATT schedules, as explicitly admitted by virtue of Article II:2(a) and Article XVIII(*bis*) GATT. The reduced policy space left to WTO Members to use export QRs could in other words have restrained them from raising the level of their commitments in the area of export duties, that is, the only area where they still maintain ample margins of flexibility on the export side. This is particularly true for those low-income commodity-dependent countries that cannot avail themselves of those exceptions provided for in the GATT to pursue economic development goals that, *mutatis mutandis*, could well be relevant to export restrictions, especially when applied in the minerals and metals sector.[89]

In conclusion, for all the reasons stated above, the asymmetry in the treatment of import and export restrictions may have a bearing on the paucity of commitments in the area of export duties on the part of those WTO Members exclusively bound by GATT obligations as to their use of these instruments. As such, it adds to the asymmetry in the treatment of import and export duties in determining the vertical proliferation of export duties on critical minerals and metals.

[88] A reason for the proliferation of Chinese export quotas is given below, Chapter 6, Section 6.2.2.

[89] See above, Chapter 3, Section 3.3.

6

GATT general exceptions relevant and applicable to WTO-inconsistent export restrictions

6.1 Introduction

In addition to the specific flexibilities available in the General Agreement on Tariffs and Trade (GATT) to World Trade Organization (WTO) Members willing to deviate from the provisions regulating the use of export duties and quantitative restrictions, the so-called general exceptions listed under Article XX GATT allow WTO Members to introduce and/or maintain measures in derogation of any GATT obligation. WTO Members may thus in principle invoke one of the grounds listed under Article XX GATT to seek justification for any GATT-inconsistent export duties and/or quantitative restrictions. Among the public policy goals enumerated under Article XX GATT are a number of rationales frequently evoked by WTO Members when resorting to export restrictions in the minerals and metals sector. These include non-economic grounds such as the protection of human, animal and plant life and health, the conservation of exhaustible natural resources, and economic public policy goals such as the assurance of adequate domestic supply in particular conditions. Although the relevance of some of these exceptions for export restrictions on minerals and materials and, in particular, critical minerals and metals, is unquestionable, Article XX serves as an exception to GATT obligations only. Whether it could also serve as an exception to obligations contained in accession protocols is a matter of much debate. The issue has so far arisen with respect to the applicability of Article XX to China's WTO obligations on the use of export duties in *China–Raw Materials* and *China–Rare Earths*.

This chapter gives an account of the GATT general exceptions relevant and applicable to export restrictions on minerals and materials and, in particular, critical minerals and metals. Section 6.2 addresses the preliminary question of the availability of Article XX GATT to export duties in breach of WTO-plus obligations assumed under post-1994 accession protocols in light of recent WTO case law. Section 6.3 examines the room left to WTO Members to impose export restrictions to address environmental concerns based on the Appellate Body's findings in *China–Raw Materials* and *China–Rare Earths*. Section 6.4 discusses the relevance of the 'industrial policy' clauses listed under Article XX GATT to justify export restrictions used as developmental tools. Finally, the chapter provides some conclusions on the policy space left to WTO Members to justify WTO-inconsistent export restrictions on minerals and metals under Article XX exceptions in light of recent WTO jurisprudence.

6.2 Availability of Article XX GATT to WTO-plus obligations on export duties assumed under post-1994 accession protocols

As discussed in Chapter 4, WTO Members bound exclusively by GATT obligations as to their use of export duties may negotiate export duty concessions under Article II:1(a) and Article XXVIII(*bis*) GATT but remain free to introduce and/or maintain export duties on any products for which no commitment was listed in GATT schedules by virtue of Article XI:1 GATT. However, a number of newly acceded WTO Members have assumed country-specific export duty commitments under individual provisions of their accession protocols. As these provisions are not contained in the GATT, the question arises as to whether Article XX GATT exceptions are applicable to such obligations. The chapeau of Article XX GATT explicitly states that 'nothing *in this Agreement* shall be construed to prevent the adoption or enforcement by any contracting party of' measures coming under one of the particular grounds listed under paragraphs (a) to (j) (emphasis added). Accordingly, as per the language of Article XX GATT itself, GATT general exceptions are available to excuse violations of GATT obligations only. Based on existing WTO case law, however, Article XX GATT may serve as an exception to obligations assumed under post-1994 accession protocols depending on the specific language of the individual provision that is being violated. The Appellate Body clarified the preconditions for the applicability of Article XX GATT to accession protocol provisions in *China–Raw Materials* and *China–Rare Earths*.

6.2.1 The existence of an 'objective link' as a precondition for the applicability of Article XX GATT to post-1994 accession protocol provisions

The issue of the availability of Article XX GATT to obligations assumed under accession protocols has arisen in three WTO disputes so far, namely *China–Audiovisuals, China–Raw Materials* and *China–Rare Earths*. In the latter two cases, the WTO dispute settlement bodies specifically dealt with the question of whether violations of China's WTO obligations on export duties, as set forth in Paragraph 11.3 of its Accession Protocol, could be justified under Article XX GATT. In both cases, WTO adjudicators found that Paragraph 11.3 of China's Accession Protocol was not subject to the general exceptions provided for in Article XX GATT. In *China–Rare Earths*, the Appellate Body also elaborated on the 'systemic' relationship between China's Accession Protocol provisions, on the one hand, and the Marrakesh Agreement and the Multilateral Trade Agreements annexed thereto (thus, including the GATT), on the other hand. The overall approach of the Appellate Body is in principle applicable to other obligations assumed under specific provisions of post-1994 accession protocols, not necessarily confined to export duty commitments.

6.2.1.1 Unavailability of Article XX GATT to violations of Paragraph 11.3 of China's Accession Protocol

In *China–Raw Materials*, the Appellate Body examined for the first time whether Article XX GATT could be available to justify the use of export duties in breach of China's WTO-plus obligations assumed under Paragraph 11.3 of its Accession Protocol. The Panel had already found that Paragraph 11.3-inconsistent export duties could not be subject to Article XX GATT defences.[1] The Appellate Body upheld this finding after a thorough analysis of Paragraph 11.3 based on the customary rules of treaty interpretation codified by Articles 31 and 32 of the Vienna Convention on the Law of Treaties[2], as required under Article 3.2 of the WTO Understanding on Dispute Settlement.[3]

1 Panel Reports, *China–Raw Materials*, para. 7.160.

2 See Vienna Convention on the Law of Treaties, Vienna, 23 May 1969, 1155 U.N.T.S. 33, entered into force 27 January 1980 [hereinafter VCLT].

3 See Understanding on Rules and Procedures Governing the Settlement of Disputes, Annex 2 to the Agreement Establishing the World Trade Organization, 1869 U.N.T.S. 401, entered into force 1 January 1995 [hereinafter DSU]. According to Article 3.2 DSU, the dispute settlement system serves 'to clarify the existing provisions of those agreements in accordance with customary rules of interpretation of public international law'. WTO adjudicators have thus traditionally applied the general rules of treaty interpretation laid out in Articles 31 and 32 VCLT. Accordingly, in the words of the Appellate Body, '[a] treaty interpreter must begin with, and focus upon, the text of the particular provision to be interpreted. It is in the words constituting that provision, read in their context, that the object and purpose of the States parties to the treaty must first be sought. Where the meaning imparted by the text itself is equivocal or inconclusive, or where confirmation of

Starting with the analysis of the text of Paragraph 11.3, the Appellate Body noted that it referred to Article VIII GATT alone while leaving out any express reference to other specific GATT provisions,[4] as well as any general references to the GATT 1994 or to the WTO Agreement that could be interpreted as indicating that Paragraph 11.3 incorporates the flexibilities of Article XX GATT.[5] The Appellate Body noted that this was in contrast to other provisions of China's Accession Protocol such as Paragraph 5.1 on state trading obligations, which contained the phrase 'without prejudice to China's right to regulate trade in a manner consistent with the WTO Agreement' in its introductory clause, and was accordingly interpreted in *China–Audiovisuals* so as to allow breaches thereof to be subject to Article XX GATT.[6]

Turning to the relevant context, the Appellate Body observed that the narrow and specific language of Paragraph 11.3 contrasted with the wording of the other sub-paragraphs of Paragraph 11 – which both include the phrase 'in conformity with the GATT 1994'.[7] The Appellate Body thus found there was no basis for China to invoke Article XX exceptions to justify violations of Paragraph 11.3 of its Accession Protocol:

> as China's obligation to eliminate export duties arises exclusively from China's Accession Protocol, and not from the GATT 1994, we consider it reasonable to assume that, had there been a common intention to provide access to Article XX of the GATT 1994 in this respect, language to that effect would have been included in Paragraph 11.3 or elsewhere in China's Accession Protocol.[8]

In the view of the Appellate Body, this conclusion was confirmed by the language of the relevant provisions of the Working Party Report, which prohibited the use of export duties providing for the same set of specific exceptions referred to under Paragraph 11.3, namely those covered in Annex 6 to China's Accession Protocol and in Article VIII GATT, without incorporating any general GATT 1994 flexibilities.[9] Moreover, the

the correctness of the reading of the text itself is desired, light from the object and purpose of the treaty as a whole may usefully be sought'. AB Report, *US–Shrimp*, para. 114.

4 AB Reports, *China–Raw Materials*, para. 291. Moreover, the Appellate Body clarified that the reference to Article VIII GATT in Paragraph 11.3 of China's Accession Protocol could not be interpreted, as China had claimed, to confirm the availability of Article XX GATT, as the two provisions have a different reach. Ibid., para. 290. This conclusion was cited by the Appellate Body in *China–Rare Earths*, para. 5.64.

5 AB Reports, *China–Raw Materials*, paras. 304–7.

6 AB Report, *China–Audiovisuals*, paras. 299–333.

7 AB Reports, *China–Raw Materials*, para. 293.

8 Ibid.

9 AB Reports, *China–Raw Materials*, paras. 294–9.

Appellate Body found that such a conclusion was not inconsistent with the object and purpose of the WTO Agreement considered as a whole, as China had argued by stating that 'the Panel distorted the balance of rights and obligations established in [its] Accession Protocol by assuming that China has "abandon[ed]" its right to impose export duties "to promote fundamental non-trade related interests, such as conservation and public health"'.[10] The Appellate Body did note that such fundamental interests were explicitly recognized among the various objectives listed in the Preamble to the WTO Agreement,[11] but interpreted the language of the Preamble to comport with its understanding that:

> the WTO Agreement, as a whole, . . . reflect[s] the balance struck by WTO Members between trade and non-trade-related concerns. However, none of the objectives listed above, nor the balance struck between them, provides specific guidance on the question of whether Article XX of the GATT 1994 is applicable to Paragraph 11.3 of China's Protocol of Accession.[12]

China's argument based on the right to regulate trade 'in a manner that promotes conservation and public health' was then reiterated before the Panel in *China–Rare Earths*. According to China, the Appellate Body in *China–Raw Materials* had summarily dismissed the interpretative value of the WTO's fundamental objectives,[13] so that its interpretation of Paragraph 11.3 would be tantamount to ruling that 'trade liberalization must be promoted at whatever cost – including forcing Members to endure environmental degradation and the exhaustion of their scarce natural resources' in contrast to the object and purpose of the WTO Agreement.[14] The Panel importantly recognized that:

> an interpretation of the covered agreements that resulted in sovereign States being legally prevented from taking measures that are necessary to protect the environment or human, animal or plant life or health would likely be inconsistent with the object and purpose of the WTO Agreement.

[10] AB Reports, *China–Raw Materials*, para. 305.

[11] The first recital of the Preamble to the Marrakesh Agreement recognizes that trade relations between Members 'should be conducted with a view to raising standards of living, ensuring full employment and a large and steadily growing volume of real income and effective demand, and expanding the production of and trade in goods and services, *while allowing for the optimal use of the world's resources in accordance with the objective of sustainable development, seeking both to protect and preserve the environment* and to enhance the means for doing so in a manner consistent with their respective needs and concerns at different levels of economic development' (emphasis added).

[12] AB Reports, *China–Raw Materials*, para. 306.

[13] Panel Reports, *China–Rare Earths*, para. 7.109.

[14] Ibid., paras. 7.108–10.

> In the Panel's view, such a result could even rise to the level of being 'manifestly absurd or unreasonable'.[15]

Yet, the Panel considered that the result of the Appellate Body's ruling in *China–Raw Materials* would not hamper China's ability to address fundamental environmental concerns inasmuch as other instruments and means – by the admission of China itself – could be used to pursue those interests, and because all that Paragraph 11.3 implied was to require China to use other such instruments instead of export duties for those products which are non-listed under Annex 6 to its Accession Protocol.[16]

China's argument, related to the object and purpose of the WTO Agreement, was one of four allegedly novel legal arguments raised by China to ask the Panel to reverse the Appellate Body's finding on the unavailability of Article XX GATT to violations of Paragraph 11.3 of its Accession Protocol in *China–Raw Materials*. The other claims related to: (i) the meaning of textual silence in Paragraph 11.3; (ii) the existence of an 'intrinsic relationship' between Paragraph 11.3 and the GATT 1994 and, (iii) the interpretation of the term 'nothing in this Agreement' in the chapeau of Article XX GATT.[17] In examining each argument, the Panel did not conduct a *de novo* determination but merely aimed at discerning whether they presented '"cogent reasons" for departing from the prior adopted finding, by the Appellate Body [in *China–Raw Materials*], on the same question of law presented to this Panel'.[18] However, it found that China did not present any cogent reason to reverse the Appellate Body's finding.[19] This finding was not appealed, but the Appellate Body confirmed the soundness of the approach endorsed in *China–Raw Materials*:

15 Ibid., para. 7.111. 16 Ibid., paras. 7.112–4. 17 Ibid., paras. 7.62–104.

18 Ibid., para. 7.61. In doing so, the Panel referred to *US–Stainless Steel (Mexico)* and to the body of relevant international law jurisprudence indicating a high threshold for departing from a previous decision. See AB Report, *US–Stainless Steel (Mexico)*, para. 160 and footnote 313.

19 Panel Reports, *China–Rare Earths*, para. 7.115. One panelist, however, dissented by stating: 'I believe that Paragraph 11.3 of China's Accession Protocol is an integral part of China's obligations on trade in goods in the sense that it adds to the basic obligations of the GATT 1994 relating to border tariffs. A proper interpretation on the availability of Article XX of the GATT 1994 to Paragraph 11.3 of China's Accession Protocol should take into account the fact that Paragraph 11.3 must be read cumulatively and simultaneously with related GATT Articles II and XI and as an integral part of the GATT system of rights and obligations. Therefore, in my view, unless China explicitly gave up its right to invoke Article XX of the GATT 1994, which it did not, the general exception provisions of the GATT 1994 are available to China to justify a violation of Paragraph 11.3 of its Accession Protocol'. Ibid., para. 7.138.

'[w]e also see no reason to revisit the ruling of the Appellate Body in *China–Raw Materials*'.[20]

While the approach adopted by the Appellate Body in *China–Raw Materials* was intended to specifically determine whether Paragraph 11.3 of China's Accession Protocol was subject to Article XX GATT, the methodology used is highly informative for eventual future disputes. Based on this methodology, it is possible to identify a number of guiding criteria. Firstly, whether Article XX GATT may be invoked to justify a breach of a protocol provision must be assessed on a case-by-case basis, having due regard to the language of the specific provision, which must be interpreted in accordance with VCLT rules on treaty interpretation. Secondly, the specific accession protocol provision does not necessarily have to contain an explicit textual reference to Article XX flexibilities. General references to the GATT or to the WTO Agreement contained therein or in other relevant provisions of the accession protocol or the related Working Party report may also be interpreted to grant access to Article XX GATT.

6.2.1.2 The relationship between the provisions of post-1994 accession protocols, the Marrakesh Agreement and WTO Multilateral Trade Agreements

In *China–Rare Earths*, the Appellate Body clarified the systemic relationship between individual provisions of accession protocols, on the one hand, and the WTO Agreement and the Multilateral Trade Agreements annexed thereto, on the other hand. China had argued without success before the Panel that Paragraph 11.3 of its Accession Protocol was subject to Article XX GATT because it 'intrinsically related' to the GATT, and in particular to Article II and Article XI regulating the use of export duties. It thus contended that it had to be treated as an 'integral part' of the GATT based on a proper interpretation of Article XII:1 of the Marrakesh Agreement, read in combination with Paragraph 1.2 of China's Accession Protocol. According to China's argument, any provision of its Accession Protocol had to be considered an integral part of the Marrakesh Agreement or one of the Multilateral Trade Agreements to which it intrinsically related, unless language to the contrary was explicitly incorporated therein.[21]

Following China's appeal, the Appellate Body first clarified the meaning of Article XII:1 of the Marrakesh Agreement on 'Accession'. Article

[20] AB Reports, *China–Rare Earths*, para. 5.65. [21] Ibid., paras. 5.13–5.

XII:1 provides: 'Any State . . . may accede to this Agreement, on terms to be agreed between it and the WTO. Such accession shall apply to this Agreement and the Multilateral Trade Agreements annexed thereto'. In the view of the Appellate Body, the second sentence of Article XII:1 indicates that the terms of accession negotiated during country-specific accession processes 'taken as a whole, must reflect, namely, that an acceding Member is subject to the rights and obligations in *all* of the Multilateral Trade Agreements'.[22] In other words, it extends the fundamental principle of the single undertaking, established for existing WTO Members under Article II:2 of the Marrakesh Agreement, to newly acceded WTO Members.[23] Accordingly, the Appellate Body made clear that, contrary to China's view:

> Article XII:1 itself does not speak to the question of the specific relationship between individual provisions of the Marrakesh Agreement and the Multilateral Trade Agreement. In particular, Article XII:1, alone, does not create a substantive relationship, 'intrinsic' or otherwise, between provisions of China's Accession Protocol (such as Paragraph 11.3) and provisions of the covered agreements (such as Article II and XI of the GATT 1994).[24]

The Appellate Body then turned to the analysis of Paragraph 1.2 of China's Accession Protocol. Paragraph 1.2, contained in Part I of China's Accession Protocol on 'General Provisions', reads:

> The WTO Agreement to which China accedes shall be the WTO Agreement as rectified, amended or otherwise modified by such legal instruments as may have entered into force before the date of entry into force of this Protocol. This Protocol, which shall comprise the commitments referred to in paragraph 342 of the Working Party Report, shall be an integral part of the WTO Agreement.

China had argued that the term 'the WTO Agreement' in the first sentence encompassed the Marrakesh Agreement and the Multilateral Trade Agreements annexed thereto, and accordingly claimed that the legal effect of the second sentence of Paragraph 1.2 was to make the individual provisions of China's Accession Protocol integral parts of the Multilateral Trade Agreements to which they 'intrinsically' related.[25] In the Appellate Body's view, however, the relationship between a specific provision of China's Accession Protocol and a relevant covered agreement does not hinge on whether the scope of the term 'the WTO Agreement' refers solely to the Marrakesh Agreement or to the WTO Agreement as a

[22] Ibid., para. 5.28. [23] Ibid., paras. 5.29–33. [24] Ibid., para. 5.34.
[25] Ibid., paras. 5.35–41.

whole (i.e. including the Multilateral Trade Agreements annexed thereto). The Appellate Body reasoned as follows:

> Paragraph 1.2 of China's Accession Protocol, and in particular its stipulation that the Protocol is to be an 'integral part' of 'the WTO Agreement', essentially serves to build a bridge between the package of protocol provisions and the existing package of WTO rights and obligations under the Marrakesh Agreement and the Multilateral Trade Agreements. The bridge created by Paragraph 1.2 between the protocol provisions and the existing package of rights and obligations under the WTO legal framework, however, is of a general nature. The fact that such a bridge exists does not in itself answer the question as to *how individual provisions* in China's Accession Protocol are related or linked to individual provisions of the other WTO agreements. More specifically, this bridge does not dispense with the need to analyse, on a case-by-case basis, the specific relationship between an individual provision in the Protocol, on the one hand, and provisions of the Marrakesh Agreement and the Multilateral Trade Agreements, on the other hand.[26]

The Appellate Body also clarified that the case-by-case analysis of whether a specific accession protocol provision 'has an objective link to specific obligations under the Marrakesh Agreement and the Multilateral Trade Agreements, and whether the exceptions under those agreements may be invoked to justify a breach of such protocol provision, must be answered ... on the basis of the customary rules of treaty interpretation and the circumstances of the disputes'.[27] Moreover, it outlined the methodology to conduct such analysis:

> The analysis must start with the text of the relevant provision in China's Accession Protocol and take into account its context, including that provided by the Protocol itself and by relevant provisions of the Accession Working Party Report, and by the agreements in the WTO legal framework. The analysis must also take into account the overall architecture of the WTO system as a single package of rights and obligations and any other relevant interpretative elements, and must be applied to the circumstances of each dispute, including the measure at issue and the nature of the alleged violation.[28]

In this respect, it confirmed the validity of the analysis conducted in *China–Raw Materials* to assess the existence of a link between Paragraph 11.3 and Article XX GATT.[29]

Inasmuch as the operative term of Paragraph 1.2 of China's Accession Protocol (i.e. 'This Protocol ... shall be an integral part of the WTO

[26] Ibid., para. 5.50. [27] Ibid., para. 5.62. [28] Ibid., para. 5.74. [29] Ibid., paras. 5.63–5.

Agreement') is a standard formula used in all post-1994 accession protocols, the Appellate Body's approach applies in principle to any accession protocol provisions of newly acceded Members. Accordingly, whether a WTO Member that assumed export duty commitments by virtue of a specific provision of its accession protocol may seek to defend a breach of such provision under GATT general exceptions depends on whether there is a discernible 'objective link' between the individual provision at issue and Article XX GATT. In this respect, the Appellate Body opined that express textual references, or the lack thereof, to a covered agreement (such as the GATT), a provision thereof (such as Article XX GATT) or to 'the WTO Agreement' are *not* dispositive in and of themselves.[30] At the same time, it also made clear that 'neither obligations, nor rights may be automatically transposed from one part of [the WTO] legal framework into another'.[31]

6.2.2 Implications for WTO Members assuming specific obligations on export duties under post-1994 accession protocols

Based on the approach of the Appellate Body in *China–Raw Materials* and *China–Rare Earths*, the analysis of whether a particular protocol provision has an objective link to the GATT and/or to Article XX GATT exceptions requires a thorough analysis of the individual provision, which should not be limited to its text but should also take into account its context and the overall architecture of the WTO system. Such a case-by-case analysis is beyond the scope of this study, as it would also need to be applied to the specific circumstances of each case, including – as the Appellate Body in *China–Rare Earths* clarified – the measure at issue and the nature of the alleged violation. Yet, it is possible to infer some general considerations on the implications of the different language of the specific accession protocol provisions on export duties agreed upon by newly acceded WTO Members.

As shown in Chapter 4 above, the universe of WTO-plus commitments on export duties is quite varied. The recourse to different formulas, linked to the existence of varying export duty commitments, also affects the relationship of country-specific accession protocol provisions with the GATT. Only in certain cases have explicit textual references to either the GATT 1994 and/or its exceptions, or to the WTO Agreement, been incorporated into the individual accession protocol provisions,

[30] Ibid., para. 5.61. [31] Ibid., para. 5.74.

whereas a number of newly acceded WTO Members have abided by WTO-plus obligations on the use of export duties more or less similar to those of China.

Hence, Montenegro's absolute obligation not to introduce any export duty is not accompanied by the incorporation of any explicit or implicit reference to Article XX flexibilities.[32] In a similar vein, Latvia committed to abolish all the export duties that it maintained at the time of accession, with the exception of antiques, without envisaging any direct or indirect link to Article XX GATT exceptions.[33] The same holds true for Mongolia and Saudi Arabia, although these countries promised to eliminate export duties on specific products – raw cashmere and iron and steel scrap, respectively.[34] Moreover, similar to China, Tajikistan consented to eliminate all export duties except those admitted by Table 9 of its Working Party Report or applied in conformity with Article VIII GATT, while omitting reference to Article XX GATT or any other general reference to the GATT 1994 or the WTO Agreement.[35]

In contrast, Vietnam agreed to reduce export duties on ferrous and non-ferrous scrap metals in accordance with Table 16 of its Working Party Report while confirming that it would apply such duties 'in conformity with the GATT 1994'.[36] Similarly, Ukraine committed to respect the export duty bindings indicated in Table 20(b) of its Working Party Report 'unless justified under the exceptions of the GATT 1994'.[37] Croatia and Lao People's Democratic Republic, on the other hand, promised to apply export duties in compliance with, respectively, 'the provisions of the WTO Agreement'[38] and 'the GATT 1994 and WTO provisions with regard to export duties'.[39]

No definitive conclusions can be drawn from the language of such accession protocols provisions alone on the basis of the approach of the Appellate Body in *China–Rare Earths*. However, the analysis conducted above shows that, at the very least, WTO-plus obligations on export duties are not only quite uneven as to scope and coverage; they also differ with respect to their relationship with the GATT and, in particular, with

[32] See WTO Doc. WT/ACC/CGR/38, 5 December 2011, para. 133.
[33] See WTO Doc. WT/ACC/LVA/32, 19 February 1999, para. 69.
[34] See WTO Doc. WT/ACC/MNG/9, 27 June 1996, para. 24 and WTO Doc. WT/ACC/SAU/61, 11 December 2005, para. 184.
[35] See WTO Doc. WT/ACC/TJK/30, 6 November 2012, para. 169.
[36] See WTO Doc. WT/ACC/VNM/48, 11 January 2007, para. 260.
[37] See WTO Doc. WT/ACC/UKR/152, 16 May 2008, para. 240.
[38] See WTO Doc. WT/ACC/HRV/59, 29 June 2000, para. 100.
[39] See WTO Doc. WT/ACC/LAO/45, 1 October 2012, para. 101.

the general exceptions under Article XX. In this respect, the Russian Federation appears to be the only newly acceded WTO Member whose right to invoke Article XX GATT follows *automatically* from having assumed its export duty commitments in the form of export duty concessions inscribed in its GATT schedule. Such concessions are in fact incorporated into the GATT framework by virtue of Article II:7 GATT, and can thus be derogated from in accordance with GATT general exceptions. In all other cases, the applicability of Article XX GATT lies in the verdict of the WTO dispute settlement bodies. Yet, as *China–Raw Materials* shows, the more ambiguous the link to the GATT or to the exceptions provided for in the GATT in the individual accession protocol provision at issue, the weaker the chances that such provision would be found to be subject to Article XX flexibilities.

6.2.3 *Systemic imbalances produced by the Appellate Body's approach in light of the fragmentation of WTO disciplines on export duties*

The conditional applicability of Article XX flexibilities to export duty commitments assumed under post-1994 accession protocols adds another level of asymmetry in the already fragmented WTO disciplines on the use of export duties. The existence of country-specific accession protocol provisions regulating the use of export duties on the part of selected newly acceded WTO Members, and the substantial absence of any significant commitments on the use of these instruments on all other WTO Members exclusively bound by GATT obligations, with the exception of the Russian Federation and Australia, creates unequal rights and obligations among WTO Members.

Qin observes that '[a]s a result of the varying arrangements, there are now effectively four-tiers of WTO Members in terms of the rights and obligations concerning export restraints'.[40] According to the author, the first tier includes the majority of WTO Members, which remain free to resort to export duties under Article XI:1 GATT.[41] The second tier

[40] J. Y. Qin, 'Reforming WTO Discipline on Export Duties: Sovereignty over Natural Resources, Economic Development and Environmental Protection', *Journal of World Trade* 46 (2012), 1147, p. 1161.

[41] This tier includes both original WTO Members and newly acceded WTO Members that have not agreed upon any specific obligations on the use of export duties by virtue of their accession protocols. See above, Chapter 4, Section 4.3.1. Moreover, newly acceded WTO Members such as Croatia and Lao PDR also belong to this tier, in that they simply

includes Australia and the Russian Federation, which have negotiated export duty concessions on specific products in their GATT schedules, thus reserving the right to invoke all applicable GATT provisions. The third tier consists of Ukraine and Vietnam. These countries cannot modify or withdraw their export duty commitments under Article XXVIII GATT, but may invoke GATT exceptions to justify a violation of their WTO-plus obligations. Finally, the fourth tier comprises Montenegro, China, Tajikistan, Latvia, Saudi Arabia and Mongolia, namely those newly acceded WTO Members that undertook the obligation to eliminate export taxes without the possibility to invoke any GATT flexibilities.[42]

The approach espoused by the Appellate Body in *China–Raw Materials* and *China–Rare Earths*, and its implications for what has been described as the 'ad hoc rule-making in the WTO accession regime',[43] have raised strong controversy among scholars. A number of authors have focused on the disproportionate burden that such an approach specifically places on China, already subject to one of the most severe WTO-plus regimes on the use of export duties, to the detriment of the cardinal WTO principles of non-discrimination and reciprocity.[44] Many authors, especially in the aftermath of the *China–Raw Materials* ruling, have openly criticized the approach espoused by the Appellate Body and suggested alternatives to revisit its decision on the non-availability of Article XX to violations of Paragraph 11.3 of China's Accession Protocol in the *China–Rare Earths* dispute, either based on the principle of systemic integration[45] or on the inherent right to regulate trade to address fundamental environmental

confirmed their commitment to impose export duties in accordance with GATT obligations in their accession protocols. See above, Chapter 4, Section 4.4.1.

42 Qin, 'Reforming WTO Discipline', pp. 1161–2. 43 Ibid., p. 1162.

44 See ibid., pp. 1161–70; J. F. De Medeiros, 'Global Trade Law – China's Export Restraints Found to be Inconsistent with its Obligations as a Member of the World Trade Organization', *Suffolk Transnational Law Review* 35 (2012), 203; X. Wu, 'No Longer Outside, Not Yet Equal: Rethinking China's Membership in the World Trade Organization', *Chinese Journal of International Law* 10 (2011), 227; L. Ying, 'The Applicability of Environmental Protection Exceptions to WTO-Plus Obligations: In View of the *China–Materials* and *China–Rare Earths* cases', *Leiden Journal of International Law* 27 (2014), 113.

45 See, e.g. E. Baroncini, 'The Applicability of GATT Article XX to China's WTO Accession Protocol in the Appellate Body reports of the *China–Raw Materials* case: Suggestions for a Different Interpretative Approach', *China-EU Law Journal* 1 (2012), 1; E. Baroncini, 'The *China-Rare Earths* WTO Dispute: A Precious Chance to Revise the *China–Raw Materials* Conclusions on the Applicability of GATT Article XX to China's WTO Accession Protocol', *Cuadernos de Derecho Internacionales* 4 (2012), 49; B. Gu, 'Applicability of GATT Article XX in *China–Raw Materials*: A Clash within the WTO Agreement', *Journal of International Economic Law* 15 (2012), 1007.

concerns.[46] Qin noted, in particular, that the Appellate Body treated accession protocols as 'self-contained agreement[s], independent from the rest of the WTO Agreement', thus transforming WTO-plus obligations on export duties into *de facto* permanent obligations.[47] Following the *China–Rare Earths* ruling, she also stressed the inconsistency of the approach of the Appellate Body with Article 30(3) VCLT, according to which an earlier treaty (i.e. the GATT 1994) applies only to the extent that its provisions are compatible with those of the later treaty (i.e. the post-1994 accession protocol) related to the same subject matter. In Qin's view, the Appellate Body erred in dismissing China's claim that Article XI:1 of the GATT 1994 had been modified by Paragraph 11.3 of China's Accession Protocol,[48] so that Paragraph 11.3 and, more generally, any accession protocol provisions on export duties 'modifying' Article XI:1 GATT would have an objective link to the GATT 1994 sufficient to establish the availability of Article XX defences. In particular, she argued that the Appellate Body reasoning for rebutting such argument, namely that Article 30(3) was not 'apposite for understanding the relationship between the different components of the single package of rights and obligations, all of which form part of the "same treaty" to which China acceded in 2001',[49] failed to consider that only after resolving the conflict between an accession protocol provision and a covered agreement can the former be integrated into the single package of WTO rights and obligations in a coherent manner.[50]

Exploring the systemic implications of the Appellate Body's approach in *China– Raw Materials*, Liu contended that, although the Appellate Body might skilfully interpret the individual provisions of accession protocols if needed in the context of a WTO dispute, until the WTO addresses once and for all the issue of the legal status of accession protocols in the WTO legal system, the wrongful interpretation given with respect to the relationship between accessions protocols and other WTO legal instruments could be corrected. He accordingly identified a series of guiding principles that the WTO Ministerial Conference and/or the General Council should endorse

[46] I. Espa, 'The Appellate Body Approach to the Applicability of Article XX GATT in the Light of *China–Raw Materials*: A Missed Opportunity?', *Journal of World Trade* 46 (2012), 1399; Ying, 'The Applicability of Environmental Protection Exceptions', pp. 113–39.

[47] Qin, 'Reforming WTO Discipline', p. 1170.

[48] AB Reports, *China–Rare Earths*, para. 2.26.

[49] Ibid., para. 5.70.

[50] J. Y. Qin, Speech given at the International Centre for Trade and Sustainable Development (ICTSD), Talking Disputes No. 10: The *China–Rare Earths* Disputes, Geneva, World Trade Organization, 8 October 2014.

when seeking to clarify the relationship between accession protocols and WTO covered agreements in order to 'maintain the integrity of the WTO multilateral trading system, to accelerate the realization of the WTO's goals, to implement the international law principle of sovereign equality and the WTO's "non-discrimination" principle and to avoid the severe damage caused by the fragmentation of WTO's rules to the WTO legal system'.[51]

These criticisms notwithstanding, it should be noted that, in *China–Rare Earths*, the Appellate Body did arguably attempt to lay down a guiding framework for deciding on the availability of Article XX GATT to violations of Paragraph 11.3 of China's Accession Protocol, which could also be suitable for determining the existence of an 'objective link' between other accession protocol provisions and the WTO Agreement and/or the Multilateral Trade Agreements annexed thereto. In this sense, it seemed inspired by the broader goal to properly situate accession protocols within the overall WTO legal framework, as a way to ensure predictability and legal security failing specific points of reference in this respect.

Yet, inasmuch as the interpretative outcome reached on the basis of this approach excludes *a priori* the applicability of Article XX GATT to China's Paragraph 11.3-inconsistent export duties, it may create the incentive for China to substitute export duties with more trade-distortive, but GATT-inconsistent, export quantitative restrictions when seeking to address legitimate concerns coming under one of the public policy goals recognized under Article XX GATT. Notably, indeed, China is one of the very few countries applying export quotas, and in any case the only one making extensive use of such instruments in the minerals and metals sector, sometimes in substitution for previously maintained export duties.[52] Given the broader implications of the approach designed by the Appellate Body in *China–Rare Earths*, moreover, other countries belonging to the fourth 'tier' of WTO Members may be tempted to engage in this substitution process in an attempt to reserve their right to invoke GATT general exceptions. The indirect incentive to apply export quantitative restrictions, however, runs counter to the cornerstone principle informing the overall GATT architecture, that is, the choice of tariffs over more trade-distortive measures such as quantitative restrictions as the lawful means of restricting exports, and introduces a further element of inconsistency in

[51] J. Liu, 'Accession Protocols: Legal Status in the WTO Legal System', *Journal of World Trade* 48 (2014), 751, p. 772.

[52] See above, Chapter 2, Section 2.5.2.

the already fragmented and imbalanced WTO regulatory framework on export restrictions.

6.3 The relevance of Article XX GATT to justify export restrictions in light of recent WTO case law

Export duties or other types of export restrictions in breach of GATT obligations or accession protocol provisions having an objective link to the GATT may still be lawful insofar as they can be justified under Article XX GATT. Labelled 'General Exceptions', Article XX lists a number of legitimate public policy goals under paragraphs (a) to (j), which may be invoked to use a measure in derogation of GATT obligations, provided that such measure is not applied, as per the chapeau of Article XX, 'in a manner which would constitute a means of arbitrary or unjustifiable discrimination between countries where the same conditions prevail, or a disguised restriction on international trade'.

The Appellate Body explained in *US–Shrimp* that Article XX GATT sets forth '*limited and conditional* exceptions from the obligations of the substantive provisions of the GATT',[53] in the sense that the grounds listed under Article XX GATT are exhaustive and that justification is granted only to the extent that a measure fulfils the conditions imposed by Article XX GATT. In this respect, a consolidated body of WTO jurisprudence has established that Article XX GATT imposes a so-called two-tiered test. First, the measure at issue must fall within the scope of one of the particular exceptions listed under Article XX GATT and meet the requirements specified therein. Second, a measure provisionally justified under one of the listed exceptions must satisfy the conditions imposed by the chapeau of Article XX.[54] The burden of proof in both steps falls on the defending party invoking an exception under Article XX.[55]

Among the exceptions listed under paragraphs (a) to (j) of Article XX GATT are a number of economic and non-economic public policy goals that may be of particular relevance for WTO Members seeking to justify export restrictions on minerals and metals, and critical minerals and metals in particular. The non-economic grounds available under Article XX aim at providing an avenue for justifying measures

[53] AB Report, *US–Shrimp*, para. 157.

[54] See AB Report, *US–Gasoline*, pp. 20–1; AB Report, *US–Shrimp*, paras. 119–20.

[55] See AB Report, *US–Wool Shirts and Blouses*, p. 335; AB Report, *US–Gasoline*, p. 21.

addressing, *lato sensu*, environmental concerns: under Paragraph (b), WTO Members may justify measures 'necessary to protect human, animal or plant life or health'; under Paragraph (g), measures 'related to the conservation of exhaustible natural resources if such measures are made effective in conjunction with restrictions on domestic production or consumption' may be adopted in derogation of GATT obligations. Other Article XX exceptions protect economic interests more or less related to industrial policy concerns. WTO Members may seek to justify measures 'involving restrictions on exports of domestic materials necessary to ensure essential quantities of such materials to a domestic processing industry during periods when the domestic price of such materials is held below the world price as part of a governmental stabilization plan ... ' under Article XX(i), whereas they may adopt measures 'essential to the acquisition or distribution of products in general or local short supply ... ' as per Article XX(j).

The 'industrial' exceptions under Article XX GATT have not yet been tested in a WTO dispute concerning the use of export restrictions. The 'environmental' exceptions provided for in Article XX GATT, however, have been invoked in a number of WTO cases on export restrictions, including *China–Raw Materials* and *China–Rare Earths*. These two disputes are of particular interest for this study because the measures at issue were applied to mineral products, including several critical minerals and metals. By discussing the conditions upon which China could have successfully fulfilled the requirements imposed by Article XX(b) and (g) GATT, the WTO dispute settlement bodies shed light on the policy space left to WTO Members to use export restrictions when seeking to address environmental concerns.

6.3.1 *Article XX(b) GATT*

Under Article XX(b) GATT, WTO Members may seek to justify measures that are 'necessary to protect human, animal or plant life or health'. This exception was invoked in both *China–Raw Materials* and *China–Rare Earths* with respect to a number of export restrictions imposed by China on various mineral raw materials. In assessing China's arguments, the Panels built on the consolidated body of WTO jurisprudence interpreting Article XX(b) GATT to clarify the meaning and scope of the requirements therein imposed for addressing export restrictions.

The Panels explained that in order to be justified under Article XX(b), a measure must first fall within the range of policies designed to protect human, animal or plant life or health.[56] To this end, the 'design' and 'structure' of a challenged measure have conventionally informed the analysis of whether its objective is the protection of life and health,[57] while WTO adjudicators have generally showed deference to the WTO Members' chosen level of protection.[58]

In this respect, the Panel clarified that a 'mere assertion' that export restrictions form part of a 'comprehensive policy that has the goal to reduce pollution and to protect the health of China's population, its animals and plants' is not sufficient to demonstrate that they are *designed* to achieve such objectives,[59] even if it is proved that the mining and production of the restricted products do cause environmental damage and thus harm the life and health of humans, animals and plants.[60] Having this in mind, the Panels examined the broad collection of documents composing China's comprehensive environmental policy and found no 'persuasive evidence of a connection between environmental protection standards and export restrictions',[61] nor any mention of how export restrictions would operate to reduce pollution caused by their production.[62] Moreover, they observed that some of the evidence submitted seemed to 'indicate that, contrary to China's assertions, the export duties at issue are designed and structured to promote increased domestic production of high value-added downstream products that use the raw materials at issue in this dispute as inputs'.[63] Hence, the Panels concluded that, to the extent that a defending party 'simply produce[s] a list of measures referring, *inter alia*, to environmental protection and polluting products [without showing] how these instruments fulfil the

[56] Panel Reports, *China–Raw Materials*, para. 7.479, citing Panel Report, *EC–Tariff Preferences*, para. 7.199. See also Panel Reports, *China–Rare Earths*, para. 7.145.

[57] Panel Reports, *China–Raw Materials*, para. 7.479, citing Panel Report, *US–Gasoline*, para. 6.21; Panel Report, *Brazil–Retreaded Tyres*, paras. 7.108 and 7.115. See also Panel Reports, *China–Rare Earths*, para. 7.145.

[58] Panel Reports, *China–Raw Materials*, para. 7.479, citing AB Report, *US–Gasoline*, p. 30; AB Report, *EC–Asbestos*, para. 168; AB Report, *Brazil–Retreaded Tyres*, paras. 140 and 210. See also Panel Reports, *China–Rare Earths*, para. 7.145.

[59] Panel Reports, *China–Raw Materials*, paras. 7.507–8 and Panel Reports, *China–Rare Earths*, paras. 7.159–60.

[60] Panel Reports, *China–Rare Earths*, paras. 7.150–6.

[61] Panel Reports, *China–Raw Materials*, para. 7.507 and Panel Reports, *China–Rare Earths*, para. 7.160.

[62] Panel Reports, *China–Raw Materials*, para. 7.508 and Panel Reports, *China–Rare Earths*, para. 7.166.

[63] Panel Reports, *China–Rare Earths*, para. 7.169. See also Panel Reports, *China–Raw Materials*, para. 7.499.

objective it claims to address', the first test imposed by Article XX(b) cannot be met.[64]

Assuming that the first requirement is fulfilled, the Panels explained that the next step is to prove that a measure is 'necessary' within the meaning of Article XX(b). Citing *Brazil–Retreaded Tyres*, the Panels confirmed that '[i]n order to determine whether a measure is "necessary" within the meaning of Article XX(b) of the GATT 1994, a panel must consider the relevant factors, particularly the importance of the interests or values at stake, the extent of the contribution to the achievement of the measure's objective and its trade restrictiveness'.[65] They also reiterated that the necessity test imposes two sub-requirements: first, the measure must 'bring about' or be 'apt to make' a 'material contribution' to the achievement of its objective;[66] second, there needs to be a 'genuine relationship of ends and means between the objective pursued and the measures at issue'.[67] When complex public health or environmental problems require a 'comprehensive policy comprising a multiplicity of interacting measures', moreover, a necessary measure could contribute to one of the objectives recognized under Article XX(b) as part of such a comprehensive policy by means of operating together with other measures forming part of the same policy framework.[68] The demonstration of the material contribution 'could consist of quantitative projections in the future, or qualitative reasoning based on a set of hypotheses that are *tested and supported by sufficient evidence*'.[69] If such analysis leads to the preliminary conclusion that the measure is necessary, the second step is to assess whether other least trade-restrictive measures could be reasonably available to achieve the same level of protection sought by the defendant party.[70]

64 Panel Reports, *China–Raw Materials*, para. 7.511. See also Panel Reports, *China–Rare Earths*, para. 7.171.

65 Panel Reports, *China–Raw Materials*, para. 7.481 and Panel Reports, *China–Rare Earths*, para. 7.146, citing AB Report, *Brazil–Retreaded Tyres*, para. 178.

66 Panel Reports, *China–Rare Earths*, para. 7.146, citing AB Report, *Brazil–Retreaded Tyres*, para. 151.

67 Panel Reports, *China–Rare Earths*, para. 7.146, citing AB Report, *Brazil–Retreaded Tyres*, para. 145.

68 AB Report, *Brazil–Retreaded Tyres*, paras. 151 and 172, cited in Panel Reports, *China–Raw Materials*, para. 7.484 and Panel Reports, *China–Rare Earths*, para. 7.146.

69 Panel Reports, *China–Rare Earths*, para. 7.173, citing AB Report, *Brazil–Retreaded Tyres*, para. 151 (emphasis added in *China–Rare Earths*).

70 Panel Reports, *China–Raw Materials*, paras. 7.489–91, citing AB Report, *US–Gambling*, paras. 307–11; AB Report, *Brazil–Retreaded Tyres*, para. 156; and AB Report, *Korea–Various Measures on Beef*, para. 166.

Based on these guiding criteria, the Panels seemed to suggest that export restrictions would struggle to qualify as necessary measures within the meaning of Article XX(b) GATT as long as the environmental and public health concerns they claim to address depend on domestic *production* and not on foreign consumption, unless the imposing country is able to demonstrate that the decrease in foreign demand is associated with a decrease in domestic production. China used several arguments in attempting to substantiate its proposition that, by reducing the demand for exported mineral supplies, export restrictions would help reduce production domestically and thus be apt to materially contribute to the achievement of health and environmental protection. In *China–Raw Materials*, it provided a series of quantitative estimations aimed at proving the effects of the challenged export restrictions on domestic production, but the Panel considered them 'highly speculative' and methodologically flawed inasmuch as they were conducted 'in isolation', failing to take into account the 'pollution that may be generated by additional production in the downstream sectors' likely to accompany the imposition of export duties and quotas on mineral products given the vertical structure of the metals industries.[71] China also claimed that the export restrictions at issue were apt to make a material contribution to the protection of human, animal and plant life and health by triggering a 'virtuous circle of development and environmental protection',[72] as higher levels of growth are associated with environmental gains.[73] Yet, according to

[71] Panel Reports, *China–Raw Materials*, para. 7.533. For the same reasons, the Panel also rejected China's contention that export restrictions on scrap metals would foster the consumption of less energy-intensive and polluting products by reducing the price of secondary production relative to primary production of metals. Ibid., paras. 7.597–609.

[72] Ibid., para. 7.521. China referred to the theory of so-called export sophistication externalities. See R. Hausmann and D. Rodrik, 'What You Export Matters', *Journal of Economic Growth* 12 (2007), 1.

[73] The reference is to the so-called environmental Kuznets curve. As explained by the Panel, the empirical correlation may be caused by a number of factors including: '(i) the composition of production and/or consumption that moves away from natural resources goods; (ii) the preference for environmental quality; (iii) the development of institutions introducing the proper regulatory measures to address environmental problems; and/or (iv) the arising economies of scale associated with pollution abatement technologies'. Panel Reports, *China–Raw Materials*, para. 7.551. For a full account see A. Levinson, 'Environmental Kuznets Curve', in *The New Palgrave Dictionary of Economics*, 2nd edn (London: Palgrave Macmillan, 2008).

the Panel, even assuming that export restrictions would increase a country's aggregate growth,[74] Article XX(b) could not be interpreted to allow such 'restrictions on any raw materials simply because they help increase growth, and, in turn, eventually reduce pollution'[75] when there is no persuasive evidence that they are necessary for environmental gains.[76] In *China–Rare Earths*, China contended that export duties were apt to make a contribution to health and environmental protection by increasing the price of the restricted raw materials as a way of internalizing the environmental costs of producing such materials. The Panel noted, however, that export taxes would only increase the price at which a product is sold abroad and China failed to identify 'any corresponding measure that it applies to increase the price of the products at issue when destined for consumption *inside* China'.[77]

Finally, the Panels also seemed to suggest that WTO-consistent or, in any case, least trade-restrictive but equally effective measures would be reasonably available as alternatives to export restrictions when addressing environmental concerns. In both *China–Raw Materials* and *China–Rare Earths*, the complainants indicated several of such alternative measures, including investment in environmentally friendly technologies and recycling infrastructures, requirements for strict compliance with environmental standards, requirements for mines to make deposits for ecological recovery, increases in resource taxes on ores, increases in volume restrictions on mining and production and the introduction of a pollution tax.[78] In the Panels' view, the contention that all such measures were already in place, on the one hand, implicitly qualified them as reasonably available alternative measures within the meaning of Article XX(b)[79] and, on the other hand, failed to discharge a complainant from demonstrating why such environmental measures could not be strengthened to serve as an alternative to export restrictions.[80]

[74] In the Panel's view, the establishment of such link would have corroborated China's contention that export restrictions were used as environmental tools. Panel Reports, *China–Raw Materials*, paras. 7.547–9.

[75] Panel Reports, *China–Raw Materials*, para. 7.515. [76] Ibid., paras. 7.553–4.

[77] Panel Reports, *China–Rare Earths*, paras. 7.173–6.

[78] Panel Reports, *China–Raw Materials*, paras. 7.575–83 and 7.610; Panel Reports, *China–Rare Earths*, paras. 7.182–5.

[79] Panel Reports, *China–Raw Materials*, para. 7.568.

[80] Ibid., para. 7.590; Panel Reports, *China–Rare Earths*, para. 7.186.

6.3.2 *Article XX(g) GATT*

Article XX(g) GATT allows WTO Members to deviate from GATT obligations when using measures 'relating to the conservation of exhaustible natural resources if such measures are made effective in conjunction with restrictions on domestic production or consumption'. Article XX(g) was invoked in *China–Raw Materials* and *China–Rare Earths* to justify a number of export quantitative restrictions imposed on various (mostly critical) minerals and metals.[81] In *China–Rare Earths*, the Panel and the Appellate Body set out the interpretation of Article XX(g), as a whole, in order to determine 'the legal test to be applied in considering [specific] Article XX(g) defence[s]'.[82] The Appellate Body clarified that 'Article XX(g) always calls for a holistic assessment of all its constituents elements' to be carried out 'on a case-by-case basis, through careful scrutiny of the factual and legal context in a given dispute'.[83] In this respect, the 'unitary test' construed under Article XX(g) requires the fulfilment of all the parameters imposed by the first and second prongs. Such requirements were first thoroughly examined and then applied in considering China's arguments in defence of its export restrictions.

Under the first prong, Article XX(g) requires that a measure 'relates to' the 'conservation' of an 'exhaustible natural resource'. As to the term 'exhaustible natural resources', the Panel noted that 'the precise point at which processed raw materials cease to be considered "exhaustible natural resources" has never been addressed in WTO dispute settlement bodies'.[84] Yet, in the Panel's view, the Appellate Body made clear in *US–Gasoline* that the measures for which Article XX(g) is invoked do not need to be imposed on 'raw natural resources, so long as the[ir] object ... is to conserve, directly or indirectly, such raw natural resources'.[85] Accordingly, the Panel considered it unnecessary to define the precise scope of the term inasmuch as export restrictions applied on semi-processed or processed

[81] In the GATT era, Article XX(g) GATT was invoked by Canada to justify its export prohibition on certain unprocessed salmon and herring products. As the measure was targeted at foreign processors and consumers only, excluding from its scope domestic processors and consumers, the Panel concluded that it could not be deemed to be 'primarily aimed at the conservation' of salmon and herring stocks and at 'rendering effective' the restrictions on the harvesting of these fish species. See GATT Panel Report, *Canada–Herring and Salmon*, paras. 4.6–7.

[82] See Panel Reports, *China–Rare Earths*, paras. 7.244–337 and AB Reports, *China–Rare Earths*, paras. 5.85–101.

[83] Ibid., paras. 5.94–5. See also Panel Reports, *China–Rare Earths*, para. 7.240.

[84] Panel Reports, *China–Rare Earths*, para. 7.248. [85] Ibid., para. 7.250.

products could also in principle fall within the scope of the concept of 'exhaustible natural resources'.[86]

The term 'conservation' was interpreted in a broad manner by the Panel in both *China–Raw Materials* and *China–Rare Earths*. According to the Panels, a proper reading of the term 'conservation' should take into account the objective of sustainable development as recognized in the Preamble to the Marrakesh Agreement[87] and the international law principle of sovereignty over natural resources as a 'relevant rule of international law applicable in the relations between the parties' in accordance with Article 31(3) VCLT.[88] WTO Members are thus entitled to determine their own conservation policies on the basis of a full range of policy considerations and goals, including their own economic and sustainable development needs,[89] also by means of a 'comprehensive policy comprising a multiplicity of interacting measures'.[90] At the same time, however, the right to adopt conservation policies cannot be relied upon to 'excuse export restrictions adopted in aid of economic development if they operate to increase protection of [a] domestic industry' in contradiction of other Article XX exceptions, such as Article XX(i)'.[91] Based on such a broad interpretation of the term 'conservation', China was able to persuade the Panel that it had in place a comprehensive conservation policy consisting of 'a series of interconnected measures and programmes, including extraction and production caps and enforcement actions, which are designed to manage the extraction and supply of [the challenged] resources'.[92]

86 AB Reports, *China–Rare Earths*, para. 5.78. For a full account of the interpretation of the term 'exhaustible natural resources' see M. Chi, '"Exhaustible Natural Resource" in WTO Law: GATT Article XX(g) Disputes and Their Implications', *Journal of World Trade* 48 (2014), 939.

87 Pursuant to Article 31(2) VCLT, the Preamble to the WTO Agreement forms part of the relevant context for Article XX(g) GATT. Panel Reports, *China–Rare Earths*, para. 7.261. Panel Reports, *China–Raw Materials*, para. 7.373, citing AB Report, *US–Shrimp*, para. 153.

88 Panel Reports, *China–Raw Materials*, paras. 7.377–81; Panel Reports, *China–Rare Earths*, para. 7.262. The principle of sovereignty over natural resources allows states to 'freely use and exploit their natural wealth and resources wherever deemed desirable by them for their own development'. UN General Assembly, 'Resolution 1803 (XVII) Permanent Sovereignty over Natural Resources', 14 December 1962. For a full account of the principle of sovereignty over natural resources see below, Chapter 7, Section 7.3.1.

89 Panel Reports, *China–Raw Materials*, paras. 7.375 and 7.381; Panel Reports, *China–Rare Earths*, paras. 7.265–7.

90 Panel Reports, *China–Raw Materials*, para. 7.375, citing AB Report, *Brazil–Retreaded Tyres*, para.151.

91 Panel Reports, *China–Raw Materials*, para. 7.386; Panel Reports, *China–Rare Earths*, para. 7.270.

92 Panel Reports, *China–Rare Earths*, paras. 7.378–83, paras.7.690–9, and paras. 7.855–61.

As to the 'relating to' requirement, the Appellate Body confirmed in *China–Rare Earths* the validity of the rational connection test applied in *US–Shrimp* and endorsed in *China–Raw Materials* according to which, for a measure to 'relate to' conservation within the meaning of Article XX(g) GATT, there must be a 'close and genuine relationship of ends and means'[93] to be established on a case-by-case basis.[94] It also posited that the assessment of whether a measure relates to conservation must focus on 'the design and structure' of the measure concerned,[95] as an 'objective methodology … diminishing the uncertainty that would arise in basing such assessment on actual effects or the occurrence of subsequent events'.[96] The analysis of the design and structure of the challenged measures, however, must not be taken 'in isolation' but needs to consider such measures 'in their policy and regulatory context'.[97] In this respect, the mere existence of a comprehensive conservation policy incorporating the challenged measures does not *per se* establish the fulfilment of the 'relating to' requirement. At the same time, a measure adopted as part of a policy framework comprising a multiplicity of interacting measures may satisfy the 'relating to' criterion even if it reflects different policy objectives.[98] Evidence relating to 'the actual operation or the impact of a measure at issue' may moreover be considered in the assessment under Article XX(g), even if this paragraph does not prescribe an empirical or actual effects test.[99]

In *China–Rare Earths*, the Panel and the Appellate Body clarified the implications of these guiding criteria in addressing a number of claims advanced by China to demonstrate that its export quotas related to conservation. In response to the argument that a 'substantial, close and real' relationship to the conservation goal could be discerned from the textual references in the measures constituting China's export quotas, the Panel posited that 'references to "conservation" are not sufficient on their

93 AB Reports, *China–Raw Materials*, para. 355, citing AB Report, *US–Shrimp*, para. 136; AB Reports, *China–Rare Earths*, para. 5.90.

94 AB Reports, *China–Rare Earths*, para. 5.113. See also Panel Reports, *China–Rare Earths*, para. 7.292.

95 AB Reports, *China–Rare Earths*, para. 5.96 and paras. 5.111–4, citing AB Report, *US–Shrimp*, para. 137; AB Report, *US–Gasoline*, pp. 17, 20; AB Reports, *China–Raw Materials*, para. 355.

96 AB Reports, *China–Rare Earths*, para. 5.112.

97 AB Reports, *China–Rare Earths*, para. 5.108, citing Panel Reports, *China–Rare Earths*, para. 7.289.

98 Panel Reports, *China–Rare Earths*, paras. 7.288–9, citing Panel Reports, *China–Raw Materials*, paras. 7.375–6.

99 AB Reports, *China–Rare Earths*, paras. 5.113–4.

own to prove that the challenged measures are "related to" conservation [and] cannot insulate measures from challenge on the basis that their design and architecture do not "relate to the conservation of exhaustible natural resources"'.[100] In this regard, the mere mention of the goal of conservation, especially when combined with references to other policy objectives, can be considered towards establishing the required connection, but it 'cannot substitute for a full and proper explanation of how the [challenged measures] are designed to promote conservation'.[101]

Turning to an examination of the design and structure of China's export quotas, the Panel and the Appellate Body shed light on the policy space left to WTO Members to use such instruments under the conservation exception. First, the Panel considered that the imposition of quantitative restrictions 'on the amount of legally produced goods that can be exported' is 'overbroad', and falls under the *US–Shrimp* definition of a measure which is 'disproportionately wide in its scope and reach in relation to the policy objective of protection and conservation'.[102] At the same time, the Panel opined that the use of a measure restricting exports, and thus burdening foreign consumers only, could not be considered to have a rational connection to the goal of reducing consumption of illegally produced goods (and thus achieving conservation) *domestically*.[103]

The Appellate Body also elaborated on the so-called signalling function of export quotas. It accepted that export quotas 'do or at least can send conservation-related signals to foreign users' inasmuch as they stimulate foreign consumers and investors to explore alternative sources of supply.[104] However, it considered that such signals were undermined by the 'perverse signals' that export quotas are liable to send to domestic consumers:

> Whereas export quotas may reduce foreign demand for [the restricted goods], they will also *stimulate* domestic consumption by effectively reserving a supply of low-price raw materials for use by domestic downstream industries. They may also encourage relocation of rare earth-consuming industries to China.[105]

100 Panel Reports, *China–Rare Earths*, para. 7.390.

101 Ibid., para. 7.406. See also Panel Reports, *China–Raw Materials*, paras. 7.422–6.

102 Panel Reports, *China–Rare Earths*, paras. 7.425–30, citing AB Report, *US–Shrimp*, para. 141.

103 Panel Reports, *China–Rare Earths*, paras. 7.436–9.

104 AB Reports, *China–Rare Earths*, para. 5.156, citing Panel Reports, *China–Rare Earths*, para. 7.443.

105 AB Reports, *China–Rare Earths*, para. 5.156, citing Panel Reports, *China–Rare Earths*, para. 7.444.

In this regard, the Appellate Body recognized that while extraction and/or production caps could in principle counteract such perverse signals, their actual capacity to do so would depend on the stringency of the production quotas and on how they interact with the export quotas. Hence, the mere existence of a conservation policy framework incorporating domestic production caps cannot be considered sufficient *per se* to mitigate any perverse signals associated with export quotas, unless, in the design of the conservation programme, there was a 'mechanism to ensure that the export quota is set at such a level that, in combination with the extraction and/or production caps, no perverse incentives will be sent to domestic users'.[106]

Moreover, the Panel clarified in *China–Rare Earths* that Article XX(g) could not be invoked to justify the use of export quotas as 'safeguard tools' aimed at 'moderating speculative demand surges that could upset market balance and certainty sought by China [and] thus contribut[ing] to sustainable development'.[107] In the Panel's view, the protection of a domestic industry from speculative surges cannot come under the conservation exception but should be met through other WTO provisions addressing industrial policy needs. In concluding so, the Panel specified to what extent sustainable development considerations may fit into the conservation exception:

> Although . . . a Member's sustainable economic development needs are a legitimate consideration that may be taken into account when deciding whether and how to design and administer a conservation policy, they are not a standalone justification for the imposition of measures otherwise WTO-inconsistent . . . the panel in *China–Raw Materials* did not suggest that 'sustainable economic development' is itself a goal that can be pursued under Article XX(g), but simply indicated that resource-endowed Members may take their sustainable economic development needs into account in designing a conservation policy that 'manages the supply and use' of exhaustible resources in a way that 'take[s] into account the challenge of using and managing resources in a sustainable manner that ensures the protection and conservation of the environment while promoting economic development'.[108]

The Panel also clarified the relationship between 'conservation' within the meaning of Article XX(g), sustainable development and permanent

[106] AB Reports, *China–Rare Earths*, paras. 5.159–60, citing Panel Reports, *China–Rare Earths*, paras. 7.446–8.

[107] Panel Reports, *China–Rare Earths*, paras. 7.449–50.

[108] Panel Reports, *China–Rare Earths*, para. 7.451, citing Panel Reports, *China–Raw Materials*, para. 7.375.

sovereignty over natural resources. It stated that the exercise of sovereignty over national resources cannot be intended to enable Article XX(g) to allow a WTO Member to allocate the available stock of a product between foreign and domestic consumers because, once extracted and in commerce, natural resources are subject to WTO law.[109] In the Panel's view, WTO Members cannot defend measures which have a sustainable economic development objective, such as supply management, as conservation-related measures.[110]

Finally, the Panel made clear that the way in which export quotas are set may be a factor to take into account into establishing whether they fulfil the 'relating to' criterion, provided their design demonstrates a 'substantial' connection to the goal of conservation 'regardless of the subjective thoughts or intentions of those who set it'.[111] In this respect, the domestic availability of unused export quota volumes does not in itself exclude a system of export quotas from being conservation-related within the meaning of Article XX(g).[112]

Under the second legal benchmark of Article XX(g), a measure which relates to conservation must be 'made effective in conjunction with restrictions on domestic production or consumption'. In a first step, a defending party must thus be able to demonstrate that it imposes 'restrictions' on domestic production or consumption. In *China–Rare Earths*, the Appellate Body agreed with the Panel that the interpretation of the term 'restrictions' given in *China–Raw Materials* in the context of its analysis of Article XI:1 GATT could be applied to describe a 'restriction' within the meaning of Article XX(g) GATT. Accordingly, any measure 'which restricts someone or something, a limitation on action, a limiting condition or regulation' can be considered to fall within the ambit of Article XX(g).[113]

Based on this interpretation, the Panel reviewed several alleged categories of domestic restrictions, clarifying what types of measure may fall within the remit of this requirement. It excluded access conditions and environmental regulatory requirements from qualifying as 'restrictions' inasmuch as the former cannot control the amounts of materials

[109] Panel Reports, *China–Rare Earths*, para. 7.462.

[110] Ibid., para. 7.460. The Panel reiterated that 'measures adopted for the purpose of economic development . . . are not "measures relating to conservation" but measures relating to industrial policy'.

[111] Ibid., paras. 7.465–85.

[112] Ibid., paras. 7.486–8.

[113] AB Reports, *China–Rare Earths*, para. 5.91, citing AB Reports, *China–Raw Materials*, para. 319. See also Panel Reports, *China–Rare Earths*, para. 7.307.

extracted or produced by enterprises that are already in the market,[114] and the latter impose compliance costs ordinarily imposed on enterprises to internalize the environmental externalities caused by mining activities.[115] In contrast, the imposition of a resource tax aimed at increasing costs borne by mining companies operating domestically could be considered a restriction, provided that a defendant party is able to submit persuasive evidence that this tax was designed in such a way as to have a limiting effect on domestic consumption and/or production (e.g. by proving that the introduction and/or increase of a resource tax led to a reduction in the sales of a taxed product).[116] Extraction and production quotas, moreover, comply with the 'restriction' criterion to the extent that they are set at a level which is below the level of expected demand and that they are designed so as to counteract the perverse incentives associated with export quotas, including appropriate enforcement measures.[117] The same holds true, *mutatis mutandis*, for volume restrictions on consumption.[118]

Finally, the Appellate Body elaborated on the requirement that conservation-related restrictions are 'made effective in conjunction with' restrictions on domestic production or consumption. In its view, this benchmark requires that restrictions imposed with respect to international trade must operate jointly and concurrently with restrictions on domestic production or consumption towards a conservation objective.[119] Citing *US–Gasoline*, the Appellate Body also confirmed that this requirement embodies the notion of 'even-handedness' in the imposition of restrictions.[120] Accordingly, the balancing required under Article XX(g) cannot be understood as meaning that the burden of conservation must be 'evenly distributed between foreign producers, on the one hand, and domestic producers or consumers, on the other hand'.[121] Rather, the notion of even-handedness as reflected in the second clause of Article XX(g) demands that a Member 'impose "real" restrictions on domestic production or consumption that reinforce and complement the restrictions on international trade, and particularly so in circumstances where domestic

114 Panel Reports, *China–Rare Earths*, para. 7.498. 115 Ibid., paras. 7.556–66.
116 Ibid., para. 7.554. 117 Ibid., para. 7.521–44. 118 Ibid., para. 7.545–50.
119 AB Reports, *China–Raw Materials*, para. 356 and AB Reports, *China–Rare Earths*, para. 5.93.
120 The Appellate Body thus clarified that such a notion does not impose a separate requirement in addition to being 'made effective in conjunction with'. Ibid., paras. 5.123–7, citing AB Report, *US–Gasoline*, p. 20.
121 Ibid., paras. 5.133–6.

consumption accounts for a major part of the exhaustible natural resource to be conserved'.[122] In this regard, the design and structure of domestic and export restrictions should be given primacy, while the market effects of such restrictions may also be taken into account.[123] Factors considered by the Panel as indicating the fulfilment of the even-handedness requirement in *China–Rare Earths* include coordination in the level and timing of the announcements of domestic and export restrictions, correspondence in the product coverage of such measures and the absence of a temporal disconnect between domestic and export restrictions.[124]

6.3.3 *Industrial policy exceptions under Article XX GATT*

Along with non-economic concerns, Article XX GATT also protects industrial policy needs. Articles XX(i) and (j), in particular, may be invoked to justify restrictions imposed on industrial materials, either when such restrictions form part of a governmental stabilization plan, or when they are essential to the acquisition or distribution of products in general or local short supply, under certain conditions. These exceptions, however, have never been interpreted by WTO dispute settlement bodies apart from general references to either one or other paragraph of Article XX as providing context for interpreting Article XX(g) GATT.

Article XX(i) GATT essentially allows WTO Members to restrict exports of domestic materials to assist their downstream industry. Such export restrictions must be part of a governmental stabilization plan aimed at keeping the domestic price of such inputs below the world price. The rationale for the exception lies in the need to prevent the effects of such a plan leading to the dismantling of a domestic processing industry.[125] Yet, governmental stabilization plans such as those described in Article XX(i) have fallen into disuse.[126]

Even if this condition could be met, moreover, Article XX(i) pays due regard to the interests of foreign producers, providing that the restrictions 'shall not operate to increase the exports of or the protection afforded to such domestic [downstream] industry, and shall not depart from the provisions of this Agreement relating to non-discrimination'. In

122 AB Reports, *China–Rare Earths*, paras. 5.132. 123 Ibid., para. 5.140.

124 Panel Reports, *China–Rare Earths*, paras. 7.572–99.

125 See EPCT/A/PV/36, cited in P. C. Mavroidis, *Trade in Goods: The GATT and the Other Agreements Regulating Trade in Goods*, 2nd edn (Cambridge University Press, 2012), p. 354.

126 Ibid.

other words, the restrictions cannot be justified if they operate in a way that increases protection of the domestic industry in breach of the core GATT principle of non-discrimination. In *China–Raw Materials* and *China–Rare Earths*, the Panel referred to this specific condition to support its interpretation of Article XX(g) and, in particular, to put a limit on the extent to which Members could 'stretch' the conservation exception in the name of sustainable development:

> the right to adopt 'conservation programmes' does not permit the exercise of boundless discretion such that WTO Members may adopt GATT-inconsistent measures as they see fit. Members cannot, for instance, regulate under Article XX(g) in a way that achieves indirectly what other subparagraphs of Article XX, including Article XX(i), prohibit directly.[127]

Article XX(j) was originally conceived as a temporary provision instrumental in addressing the severe shortages caused by the Second World War.[128] It was then maintained in view of handling emergency situations such as natural disasters.[129] According to the WTO Secretariat, this exception would provide WTO Members with 'some flexibility to take trade-restrictive action when a particular resource becomes temporarily scarce.[130] The negotiating history of the term 'general or local short supply' seems to indicate that this requirement was 'intended to apply to "cases where a product, although in international short supply, was not necessarily in short supply in . . . every country importing a commodity"'.[131]

In *China–Raw Materials*, the Appellate Body discerned from the absence of the word 'critical' that the kind of shortages coming under Article XX(j) are less narrowly circumscribed than those falling within the ambit of Article XI:2(a) GATT.[132] Yet, if a product is in short supply within the meaning of Article XX(j), the restrictions essential to the acquisition and distribution of such product shall ensure that all Members have access to an 'equitable share' of the product. The 1950 Working Party on The Use of Quantitative Restrictions for Protective

[127] Panel Reports, *China–Rare Earths*, para. 7.269, citing Panel Reports, *China–Raw Materials*, para. 7.386.

[128] In the intentions of the drafters, the provision was to be eliminated by 1 January 1951. See EPCT/A/PV/30, cited in Mavroidis, *Trade in Goods*, p. 355.

[129] GATT Doc. BISD 3S/249, para. 42, cited in Mavroidis, *Trade in Goods*, p. 355. Article XX(j) became a permanent provision by virtue of GATT Decision L/3361, 17S/18 adopted on 20 February 1970.

[130] WTO, 'World Trade Report 2010: Trade in Natural Resources', www.wto.org/english/res_e/booksp_e/anrep_e/world_trade_report10_e.pdf (accessed December 2014), p. 168.

[131] Ibid. [132] AB Reports, *China–Raw Materials*, para. 325.

and Other Commercial Purposes clarified that this principle differs from the principle of non-discrimination and noted that equitableness should be assessed on a case-by-case basis.[133] The Panel in *China–Rare Earths* opined that 'this Article . . . seems to us to reflect a general principle that, in the case of restrictions on trade, Members' access to goods and materials should reflect as closely as possible the situation that would prevail in the absence of these restrictions'.[134] Finally, measures protected under Article XX(j) must be discontinued 'as soon as the conditions giving rise to them have ceased to exist'. Any Article XX(j)-consistent measure is therefore meant to be applied for the time period strictly necessary to recover from the situation of emergency that generated the shortage.

6.3.4 *The chapeau of Article XX GATT*

China–Rare Earths was the first WTO dispute on export restrictions where the Panel completed the analysis under the chapeau of Article XX GATT on an *arguendo* basis. The Panel first explained the legal test imposed by the chapeau in light of the existing WTO case law and then applied its interpretation to China's export restrictions. It thus contributed to clarifying what factors could have a bearing on establishing whether export restrictions are applied in a manner that constitutes 'arbitrary or unjustifiable discrimination between countries where the same conditions prevail, or a disguised restriction on international trade'.

Recalling *US–Gasoline*, the Panel stated that the chapeau aims at addressing the manner in which a measure is applied and preventing abuses of Article XX exceptions.[135] In this respect, the requirements imposed by the chapeau 'impart meaning to one another [so that] the kind of considerations pertinent in deciding whether the application of a particular measure amounts to "arbitrary or unjustifiable discrimination" may also be taken into account in determining the presence of a "disguised restriction" on international trade'.[136]

As to the 'arbitrary or unjustifiable discrimination' clause, the Panel cited *US–Shrimp* to recall that it imposes three conditions: (i) the application of a measure must result in a discrimination; (ii) the discrimination must be arbitrary or unjustifiable; and (iii) it must occur in

[133] GATT/CP.4/33 (Sales No. GATT/1950–3), para. 12.
[134] Panel Reports, *China–Rare Earths*, para. 7.359.
[135] Ibid., para. 7.348, citing AB Report, *US–Gasoline*, p. 22.
[136] Panel Reports, *China–Rare Earths*, para. 7.349, citing AB Report, *US–Gasoline*, p. 22.

countries where the same conditions prevail.[137] The first requirement refers to both most favoured nation (MFN) and national treatment types of discrimination.[138] Accordingly, the Panel rejected China's claim that its export duties were applied consistently with the 'arbitrary or unjustifiable discrimination' clause because they did not make a distinction 'according to the destination of the products being produced'.[139]

The Panel also endorsed the Appellate Body's statement in *Brazil–Retreaded Tyres* that the analysis of whether discrimination is justifiable under the chapeau of Article XX should be based on whether the 'reasons given for this discrimination bear no *rational connection* to the objective falling within the purview of a paragraph of Article XX, or would go against that objective', and also take into account, as a relevant factor, the effects of the discrimination.[140] Accordingly, the Panel found that the mere fact that an export quota remains unfilled does not necessarily imply that it has no discriminatory impact as long as its persistence over time induces international markets to adjust to long-term distortions.[141] Furthermore, the Panel cast doubts on the contention that an unfilled export quota could not cause a price gap between foreign and domestic prices, and posited that the decreasing of the price gap cannot be persuasively established when the effects of other factors, such as the price effects of export duties applied concurrently with export quotas, are difficult to isolate.[142] Other aspects that were considered relevant by the Panel in its assessment include: (i) differences in the denomination of export and production quotas, which could lead to 'inflating the amount that appears to be exported';[143] (ii) differences in the scope of the production and export quotas reflecting a 'structural imbalance' in favour of domestic users to the detriment of foreign consumers;[144] and, (iii) differences in the criteria used for the allocation of the quotas between domestic and foreign users.[145]

The Panel also noted that, according to *US–Gasoline,* discrimination may be arbitrary or unjustifiable 'where *alternative measures* exist which

[137] Panel Reports, *China–Rare Earths*, para. 7.350, citing AB Report, *US–Gasoline*, para. 150.
[138] Ibid. [139] Panel Reports, *China–Rare Earths*, paras. 7.189–90.
[140] Ibid., para. 7.352, citing AB Report, *Brazil–Retreaded Tyres*, paras. 227–30.
[141] Panel Reports, *China–Rare Earths*, paras. 7.630–5.
[142] Ibid., paras. 7.636–49. In the words of the Panel: 'the presence of the export duty may at times lead to the quota to have no apparent effect on foreign price, in the case in which the duty reduces the quantity demanded to below the export quota level. However, it would be erroneous to conclude that in this circumstance the export quota has no effect at all'. Ibid., para. 7.644.
[143] Ibid., para. 7.650. [144] Ibid., para. 7.651. [145] Ibid., paras. 7.652–7.

would have avoided or at least diminished the discriminatory treatment'.[146] In this regard, the Panel noted that, when reasonably available WTO-consistent alternative measures exist, which could achieve the same desired level of protection as export quotas while avoiding the discrimination effects entailed by such quotas, the 'arbitrary or unjustifiable discrimination' clause under the chapeau of Article XX cannot be met.[147] Moreover, the alternative means must be available to fulfil the same goals invoked in justification of a measure under Article XX.[148]

Finally, the Panel recalled that in *US–Gasoline*, the Appellate Body confirmed that '"disguised restriction" ... may properly be read as embracing restrictions amounting to arbitrary or unjustifiable discrimination in international trade taken under the guise of a measure formally within the terms of an exception listed in Article XX'.[149] Accordingly, it stated that a mere assertion that an export duty is an intrinsic part of a WTO Member's environmental policy, without any explanation of the criteria used to determine the duty rates, cannot be considered as 'tailored to' protect the environment.[150]

6.4 Conclusions

This chapter has given an overview of GATT general exceptions relevant to export restrictions on minerals and metals. The analysis showed that there is little policy space available to WTO Members for justifying the use of export restrictions under Article XX GATT. A first preliminary issue concerns the availability of Article XX GATT to violations of export duty commitments assumed under individual accession protocol provisions, as Article XX exceptions are available *per se* for GATT-inconsistent measures only. The Appellate Body made clear in *China–Rare Earths* that measures imposed in breach of an accession protocol provision may be subject to Article XX GATT defences insofar as such a provision has an objective link to the GATT or Article XX itself, to be established on a case-by-case basis. Based on this approach, those newly acceded WTO Members that have undertaken export duty commitments by virtue of individual provisions of their accession protocols may not have the option to invoke Article XX GATT, depending on the specific language of such commitments interpreted in their relevant context and in light of

[146] Ibid., para. 7.354, citing AB Report, *US–Gasoline*, pp. 26–7.

[147] Panel Reports, *China–Rare Earths*, paras. 7.675–7.

[148] Ibid., para. 7.676.

[149] Ibid., para. 7.349, citing AB Report, *US–Gasoline*, p. 16.

[150] Panel Reports, *China–Rare Earths*, paras. 7.191–2.

the overall architecture of the WTO Agreement and the specific circumstances of each case. Such conditionality in the applicability of Article XX introduces an additional element of asymmetry in the already varied network of rights and obligations of WTO Members with regard to the use of export duties. The fragmentation of WTO disciplines on export duties, and its implications, as emerge from the approach by the Appellate Body in *China–Raw Materials* and *China–Rare Earths* do not just create, as Qin calls it, a 'four-tiered' WTO Membership undermining the core WTO principles of non-discrimination and reciprocity.[151] As the case of China shows, they also create systemic imbalances ultimately running counter to the cornerstones of the GATT architecture. One example is the incentive to substitute least trade-distortive export duties with quantitative types of Article XI:1 GATT-inconsistent export restrictions for newly acceded WTO Members willing to reserve the right to invoke Article XX defences.[152]

The analysis of the meaning and scope of Article XX exceptions relevant to export restrictions on minerals and metals, and critical minerals and metals in particular, reveals that WTO Members have very limited margin of manoeuvre to use export restrictions for sustainable development purposes under the 'environmental' exceptions provided for in Article XX GATT.[153] In the case of Article XX(b), the legal test applied under the necessity clause makes it very unlikely that an export restriction could be considered as bringing *additional* environmental benefits compared with reasonably available environmental measures:

> export restrictions are not an efficient policy to address environmental externalities when these derive from domestic *production* rather than exports or imports. This is because generally the pollution generated by the production of the goods consumed domestically is not less than that of the goods consumed abroad. So the issue is the production itself and not the fact that it is traded.[154]

[151] See Qin, 'Reforming WTO Discipline', pp. 1161–2.

[152] See also E. Baroncini, 'Obblighi WTO-Plus, tutela dell'ambiente, della salute e preservazione delle risorse naturali: il Report del Panel nel caso *China–Raw Materials*', *Diritto Comunitario e degli Scambi Internazionali* 50 (2011), 627.

[153] For a critique of the 'narrow' interpretation of Article XX GATT in *China–Raw Materials* see H. W. Liu and J. Maughan, 'China's Rare Earths Export Quotas: Out of the *China–Raw Materials* Gate But Past the WTO's Finish Line?', *Journal of International Economic Law* 15 (2012), 971.

[154] Panel Reports, *China–Raw Materials*, para. 7.586.

The very nature of export restrictions as 'measures that increase the cost of [a raw material] to foreign consumers but decrease their costs to domestic users' was also considered by the Panels in *China–Raw Materials* and *China–Rare Earths* as 'difficult to reconcile with the goal of conservation' recognized under Article XX(g) GATT.[155] This notwithstanding, it should be noted that the Appellate Body clarified that 'Article XX(g) of the GATT 1994 does not exclude, *a priori*, export quotas or any other type of measure from being justified by a WTO Member pursuing the conservation of an exhaustible natural resource'.[156] The concrete implementation options, however, must satisfy all the requirements imposed by Article XX(g). In this respect, the Panels clarified that the conservation exception should be interpreted harmoniously with the principle of sovereignty over natural resources to allow WTO Members to 'manage the supply and use of non-renewable resources through conservation-related measures that foster sustainable development of their domestic economies',[157] even if such measures are part of a comprehensive policy framework comprising a multiplicity of objectives. Yet, a Member's sovereign rights must be exercised within the parameters of Article XX(g) and cannot be invoked to transform it into an exception meant to assist downstream processing industries.

Finally, the general exceptions more directly addressing industrial policy needs, such as Articles XX(i) and (j) GATT, target very specific situations of emergency. This makes them hardly suitable for justifying export restrictions for sustainable economic development purposes.

155 Panel Reports, *China–Raw Materials*, para. 7.434; Panel Reports, *China–Rare Earths*, para. 7.541.

156 AB Reports, *China–Rare Earths*, para. 5.162.

157 Panel Reports, *China–Raw Materials*, para. 7.404.

PART III

Regulatory prospects

Part II has identified the World Trade Organization (WTO) rules relevant and applicable to all types of export restrictions imposed on minerals and metals and, in particular, on critical minerals and metals. The analysis of the existing WTO disciplines on the export side served as the second step of a two-tiered analysis aimed at investigating whether the proliferation of such export restrictions, as qualified in Part I, resulted from an overall deficiency of relevant WTO disciplines. Part II showed that, while the proliferation of export restrictions on minerals and metals cannot be directly attributed to a lack of substantive WTO obligations *per se*, its peculiar features have been shaped by the lack of consistency in such obligations, which intervenes at several levels.

The most evident shortcoming of the WTO legal framework on export restrictions is the absence of a system-wide discipline on export duties. The fragmentation of WTO rules on export duties carries implications of systemic relevance for the WTO legal system, ranging from the existence of substantial disparities in the scope and coverage of obligations between the original and newly acceded WTO Members as well as among the different newly acceded WTO Members themselves, to the varying degrees of flexibility enjoyed by such Members in the use of export duties in revocation and/or derogation of such obligations, including when they are aimed at pursuing legitimate public policy objectives. As shown in the case of China, this could lead to incentives for the use of more trade-restrictive types of export restrictions. Another element of inconsistency is the asymmetrical treatment of import and export quantitative restrictions (QRs) in the General Agreement on Tariffs and Trade (GATT), particularly with respect to the unavailability of economic development and emergency exceptions to Article XI:1 GATT-inconsistent export QRs. These factors, matched with the limited policy space left to WTO Members to use export restrictions for sustainable development purposes under Article XX GATT, may have restrained WTO Members from engaging in the negotiation of export duty concessions in their GATT schedules.[1] The overall paucity of commitments in this area is indeed the main element of weakness in the WTO treatment of export restrictions in general, together with the lack of transparency affecting the use of export restrictions due to the loose formulation of applicable notification requirements.

Part III explores avenues for reforming the existing WTO disciplines on export restrictions, which could allow for the identified inconsistencies to be resolved. It first gives an account of the long-standing frictions that continue to ensue between net importing WTO Members and resource-endowed developing country Members over the use of export restrictions on raw materials in the WTO (Chapter 7). Chapter 8 critically examines the proposals for reform of WTO rules on the export side put forward in the course of the Doha negotiations within the Non-agricultural Market Access (NAMA) Negotiating Group. Chapter 9 concludes the study by presenting the ideal constitutive elements and potentially available implementing options for designing more coherent and equitable WTO disciplines on export restrictions.

[1] The ambiguity formally surrounding the applicability of Article XXVIII GATT to export duty concessions may have also contributed to the lack of commitment. See above, Chapter 4, Section 4.3.2.

7

Understanding the multifaceted discourse on export restrictions in the WTO: a brief conceptual framework

7.1 Introduction

This chapter attempts to disentangle the network of conflicting interests at the basis of the divide between import-dependent and resource-endowed countries over the function of export restrictions and the policy space to be left under WTO rules for legitimately resorting to such instruments. These tensions have consistently frustrated any constructive discussion on export restrictions in the WTO since the 1970s, and are as strong as ever now that the proliferation of export restraints on raw materials, and particularly on critical minerals and metals, has triggered momentum on the issue in the multilateral trading system. The objective of this mapping exercise is to identify the scope and targets that a proposal for reform of WTO rules on export restrictions would need to embrace in order to respond to the legitimate concerns of both sides and thus have a realistic chance to find consensus.

The chapter starts with an overview of the long-standing issues that have characterized the debate over export restrictions since the GATT era (Section 7.2). It then looks more specifically into the arguments in favour of export restrictions as legitimate sustainable economic development tools, addressing the question of whether and on what legal basis the developmental function of export restrictions should be acknowledged in the WTO legal framework in light of existing international law principles relevant for global natural resources management (Section 7.3). This is followed by an analysis of the arguments in support of a narrow interpretation of existing WTO rules on export restrictions and/or the introduction of stricter disciplines on the use of beggar-thy-neighbour export restrictions (Section 7.4). The arguments advanced on both sides are applicable to the broader category of industrial raw materials more generally, but remain of particular relevance for the case of export restrictions applied to critical minerals and metals. Finally, Section 7.5 provides some preliminary conclusions on the way that the interplay between such different vectors should inform discussions on the reform of WTO rules on export restrictions.

7.2 Access to supplies vs. development needs in the debate over export restrictions in the GATT era

The issue of export restrictions on raw materials was first addressed during the GATT era in the 1970s. Renewed attention towards the topic arose during the Tokyo Round, triggered by the first wave of export restrictions that followed the 1973 oil crisis.[1] Since their inception, however, discussions over the need to improve access to supplies have been marked by strong resistance on the part of commodity exporting countries[2]. These countries have claimed that any reassessment of GATT provisions relating to export restrictions would be unbalanced if it did not concurrently address the question of access to markets, and that the principle of permanent sovereignty over natural resources had to be preserved in any GATT effort in this area.[3] Accordingly, they trongly advocated that the elimination and/or reduction of export barriers be matched with a parallel reduction of restrictions imposed by trading partners on imports of finished goods produced using the raw materials affected by export restrictions.[4]

The question of whether and how to ensure greater comparability between GATT rules on import and export restrictions emerged in the Framework Group of the Tokyo Round and soon ignited a vivid debate over concrete implementing options for reform. Net importing WTO Members supported the adoption of a code of conduct containing a blanket prohibition of export restrictions, whereas resource-based developing country Members insisted on the need to differentiate concessions on a product-by-product basis ‘provided that major importing countries made substantial moves in the direction of dismantling trade barriers which they maintained against imports of processed resource products’.[5]

[1] See above, Introduction, Section I.1.

[2] J. Y. Qin, ‘Reforming WTO Discipline on Export Duties: Sovereignty over Natural Resources, Economic Development and Environmental Protection’, *Journal of World Trade* 46 (2012), 1147, p. 1180.

[3] GATT Doc. MTN.GNG/NG2/W/40, 8 August 1989.

[4] Ibid., para. 13.

[5] Ibid., para. 15.

Inevitably, the stalemate resulted in a *de minimis* outcome. The Framework Group did draft an Understanding Regarding Export Restrictions and Charges,[6] which was adopted by the Trade Negotiating Committee at the end of the Tokyo Round.[7] However, no concrete progress was made on the matter on that occasion, as the participants requested the contracting parties 'to reassess, as one of their priority tasks after the conclusion of the Tokyo Round, the provisions relating to export restrictions and charges, in the context of the international system as a whole, *taking into account the development, financial and trade needs of the developing countries*'.[8] The developing countries' delegations reiterated that any further action in this regard would have to be inspired by the principle of sovereignty over natural resources and take into account 'the need for developing countries to utilize their resources for their development in the most optimal manner as considered appropriate for them, including processing of their raw materials, setting up industries to diversify their economies and ensuring supplies to domestic industries'.[9] Some discussions were also held over the issue of export restrictions applied to mineral resources within the Group on the Sector Approach of the Tokyo Round negotiations. Within this group, the contracting parties agreed on the need to elaborate provisions establishing the 'purposes and conditions under which export restrictions might or might not be used as well as relating to the binding of export taxes and the inconsistencies surrounding their use'.[10] Yet, the failure to provide concrete answers to the most immediate concerns of both sides left the issue of export restrictions unresolved; the item was accordingly excluded from the Declaration adopted at the end of the Ministerial meeting in 1982.[11]

Interestingly, the same fundamental tensions continue to frustrate the attempts to find consensus on how to reform WTO disciplines on the export side now that the second wave of export restrictions on raw materials, and particularly the proliferation of such measures applied to critical minerals and metals, has triggered again momentum on the issue. Some authors have even suggested that the WTO accession regime on export duties and all its inconsistencies are themselves a by-product of the failure of the WTO multilateral negotiations to date.[12] Accordingly,

[6] GATT Doc. MTN/FR/6, 16 July 1979. [7] GATT Doc. MTN/28, 12 April 1979.

[8] GATT Doc. MTN.GNG/NG2/W/40, para. 15 (emphasis added). [9] Ibid.

[10] Ibid., para. 14, citing GATT Doc. MTN/SEC/W/21, 24 January 1978.

[11] GATT Doc. BISD 29/S/9, 29 November 1982.

[12] In the words of Julia Ya Qin: 'proposals to strictly limit the use of export duties have met continued resistance from developing countries. Unable to effect a systemic change, the

as the state of the Doha negotiations also shows,[13] today, as in the 1970s, the success of any attempt to reassess WTO disciplines on export restrictions lies in achieving a balance between the need for net importing WTO Members to secure fair access to supplies against beggar-thy-neighbour export restrictions, and the legitimate development concerns of resource-endowed WTO Members.[14]

7.3 Arguments in favour of export restrictions as developmental tools

As discussed above, net exporting WTO Members, including both emerging economies and low-income commodity-dependent developing countries (CDDs), have consistently advocated for the right to use export restrictions as developmental tools. The development function of export restrictions has been defended by developing countries irrespective of their Membership 'status' within the WTO. The differences in terms of the rights and obligations of different 'tiers' of WTO Members,[15] however, have influenced the choice of arguments invoked as the basis for the entitlement to introduce and/or maintain export restrictions. WTO Members bound exclusively by GATT obligations have repeatedly insisted on the legitimacy of export duties, which they refrained to schedule and bind as it is permitted under the GATT.[16] The same holds true for newly acceded WTO Members undertaking narrowly tailored WTO-plus obligations on the use of export duties in their accession protocols, such as Vietnam.[17] In contrast, China, which cannot use export duties due to the general elimination obligation assumed under its Accession Protocol, has sought to defend the right to implement quantitative export restrictions for sustainable economic development purposes under Article XX(g) GATT after being unsuccessful in arguing that it could not be deprived of the right to impose export duties when addressing fundamental environmental concerns.

Against this backdrop, the efforts to give legal expression to the need of resource-rich developing countries not to permanently lose their right to

WTO has nonetheless required a few acceding countries to make sweeping export-duty commitments'. Qin, 'Reforming WTO Discipline', pp. 1185–6.

[13] See below, Chapter 8.

[14] For an account of the body of economic literature finding a negative correlation between resource abundance and economic growth see above, Chapter 3, Section 3.3.1.

[15] The idea of a 'four-tiered' WTO Membership borrows from Qin, 'Reforming WTO Discipline', pp. 1161–2. For a full account see above, Chapter 6, Section 6.2.3.

[16] See above, Chapter 3, Section 3.3.1. [17] See above, Chapter 4, Section 4.4.1.

use export duties and other types of restrictions for sustainable economic development purposes has progressively rested on the general international law principle of permanent sovereignty over natural resources. Export restrictions have also in some cases been defended as 'safeguard' measures aimed at protecting domestic industries from the impact of export surges. Finally, commodity-dependent developing countries (CDDCs) have mainly advocated the right to use such instruments to relieve commodity dependence and vulnerability to price volatility.

7.3.1 *The regained centrality of the principle of sovereignty over natural resources*

The principle of sovereignty over natural resources has taken center stage in the debate over the alleged discriminatory implications inherent to the fragmentation of the existing WTO rules on export restrictions, due to its strong developmental and environmental connotations. This principle developed in the 1950s and progressively evolved into a principle of general international law as a corollary to the right to self-determination of colonial peoples.[18] Gradually, the emphasis shifted from peoples to states with the aim of enhancing prospects for economic development.[19] The first document proclaiming the notion of

[18] For a full account of the origin of the principle, its evolution and status under international law, and the rights and duties emanating from it, see P. J. I. M. Waart, 'Permanent Sovereignty over Natural Resources as a Cornerstone for International Economic Rights and Duties', *Netherlands International Law Review* 24 (1977), 304; M. S. Rajan, *Sovereignty over Natural Resources* (Radiant Publishers, 1978); H. S. Zakariya, 'Sovereignty over Natural Resources and the Search for a New International Economic Order', in K. Hossain (ed.), *Legal Aspects of a New International Economic Order* (Frances Pinter, 1980), pp. 208–19; N. J. Schrijver, *Sovereignty over Natural Resources: Balancing Rights and Duties*, 2nd edn (Cambridge University Press, 2008); V. Zambrano, *Il principio di sovranità permanente sulle risorse naturali tra vecchie e nuove violazioni* (Giuffrè, 2009).

[19] See, e.g. United Nations (UN) General Assembly Resolution 837 (IX), Recommendations Concerning International Respect for the Right of Peoples and Nations to Self-Determination, 14 December 1954; UN General Assembly Resolution 1314 (XIII), Recommendations Concerning International Respect for the Right of Peoples and Nations to Self-Determination, 12 December 1958, and UN General Assembly Resolution 1803 (XVII), Permanent Sovereignty over Natural Resources, 14 December 1962, cited in Schrijver, *Sovereignty over Natural Resources*, pp. 387–9. From the mid-1980s, however, the emphasis on peoples was re-established. In the Declaration on the Right to Development, for instance, Article 1, paragraph 2, affirms that the 'inalienable' right to development (see Article 1, paragraph 1) also implies 'the full realization of the right of peoples to self-determination, which includes, subject to the relevant provisions

sovereignty over natural resources was the United Nations (UN) General Assembly Resolution 626 (VII), according to which states are entitled to 'freely use and exploit their natural wealth and resources wherever deemed desirable by them for their own progress and economic development'.[20] In the following decades, as newly independent developing countries struggled to achieve economic self-determination, the concept of sovereignty over natural resources served as a core principle to set the foundations for a new international economic order (NIEO).[21] Based on this principle, 'all states' had the right to 'permanently' exercise sovereign rights over natural resources in the interests of national economic development and the well-being of their population.[22]

The failure of developing countries to impose their own vision of the principles governing international economic relations and establish the NIEO also seemed to affect the fortune of the principle of sovereignty over natural resources for some time.[23] Yet, various factors, ranging from the emergence of global value chains through the rapid industrialization of developing countries to the increasing competition over access to natural resources, have recently triggered renewed attention towards a principle which has been 'extensively recognized as giving rise to a series of resource-related and foreign-investment-related rights or as re-emphasizing rights emanating from other principles such as territorial

of both International Covenants on Human Rights, the exercise of their inalienable right to full sovereignty over all their natural wealth and resources'. See UN General Assembly Resolution 41/28, Declaration on the Right to Development, 4 December 1986. The idea that peoples are the ultimate beneficiary of permanent sovereignty over natural resources has gradually gained prominence along with the erosion of the concept of sovereignty. Accordingly, the principle has been recognized as giving rise to both rights *and* duties. Ibid., pp. 306–67. In the same vein, Qin acknowledges that the state is 'merely a representative of its citizens in exercising this right and ... has the duty to exercise such right diligently and in the best interest of its population'. Qin, 'Reforming WTO Discipline', p. 1165.

[20] UN General Assembly Resolution 626 (VII), Right to Freely Exploit Natural Wealth and Resources, 21 December 1952, para. 1.

[21] For a full account see G. Sacerdoti, 'New International Economic Order (NIEO)', in R. Wolfrum (ed.), *Max Planck Encyclopedia of Public International Law* (Oxford University Press, 2013), Vol. VII, pp. 659–668.

[22] UN General Assembly Resolution 3201 (S-VI), Declaration on the Establishment of a New International Economic Order, 1 May 1974; UN General Assembly Resolution 3202 (S-VI), Programme of Action on the Establishment of a New International Economic Order, 1 May 1974; UN General Assembly Resolution 3281 (XXIX), Charter of Economic Rights and Duties of States, 12 December 1974.

[23] Schrijver, *Sovereignty over Natural Resources*, pp. 171–2.

sovereignty and economic jurisdiction'.[24] Among these rights are: the right to 'possess, use and freely dispose of natural resources within its own territories'; the right to 'determine freely and control the exploration, development, exploitation, use and marketing of natural resources' and the right to 'manage and conserve natural resources pursuant to national developmental and environmental policies'.[25] Importantly, however, such rights also correspond to parallel duties incumbent on states, such as the duty to exercise sovereignty over natural resources in the interests of national development, having 'due care for the environment' and managing natural resources 'so as to ensure sustainable production and consumption'.[26] In this respect, the link between the principle of sovereignty over natural resources and sustainable development has progressively gained prominence, so that the former is now recognized as 'an important cornerstone of . . . international sustainable development law'.[27]

7.3.1.1 Implications of the principle of sovereignty over natural resources for the WTO accession regime on export duties

The principle of sovereignty over natural resources has been invoked to question the legitimacy of the WTO accession regime on export duties. Qin has in particular argued that the current configuration of WTO-plus obligations on export duties, as obligations contained in individual provisions of selected newly acceded WTO Members' accession protocols, not only denies the legitimate developmental function of export duties but permanently deprives such Members from 'the only legitimate means available under WTO law . . . to exercise their sovereign right over natural resources for the purpose of developing domestic industries'.[28]

The point of departure of Qin's approach is that export duties have proved to be fundamental tools for the promotion of downstream

[24] Ibid., p. 391. [25] Ibid.

[26] Ibid., pp. 391–2. See also W. Shan, P. Simons and D. Singh (eds.), *Redefining Sovereignty in International Economic Law* (Hart Publishing, 2008); J. P. Puissochet, *L'Etat souverain dans le monde d'aujourd'hui* (Pedone, 2008).

[27] Schrijver, *Sovereignty over Natural Resources*, p. 394. See also N. J. Schrijver, *The Evolution of Sustainable Development in International Law: Inception, Meaning and Status* (Martinus Nijhoff Publishers, 2008), pp. 162 ff; G. Sacerdoti, 'La disciplina del commercio internazionale e la protezione dell'ambiente', in P. Fois (ed.), *Il principio dello sviluppo sostenibile nel diritto internazionale ed europeo dell'ambiente*, (Editoriale Scientifica, 2007), p. 63 ff.

[28] Qin, 'Reforming WTO Discipline', p. 1169.

processing industries and economic diversification, thus being instrumental to the achievement of economic development. Accordingly, the developmental function of export duties should be preserved, as it is preserved on the import side.[29] In her view, 'the legitimacy of export duties for economic development stems ultimately from the principle of sovereignty over natural resources'.[30] In other words, export duties applied on national (mineral) natural resources cannot be assimilated to beggar-thy-neighbour policies condemned under WTO law, even when used as economic development tools, because they are aimed at exploiting a country's own resources through the exercise of its sovereign rights.[31] Qin has moreover insisted on the 'permanent' qualification of sovereignty over natural resources, arguing that it is recognized as implying that 'sovereignty is the rule and can be exercised at any time [while] limitations are the exceptions and cannot be permanent, but limited in scope and time'.[32]

Following this premise, Qin argues that the WTO accession regime on export duties, as it currently stands, cannot be reconciled with the principle of sovereignty over natural resources. In particular, she posits that the approach espoused by the Appellate Body in *China–Raw Materials* has the effect of making each accession protocol a 'self-contained agreement' and the WTO-plus commitments on export duties assumed therein *de facto* permanent stand-alone obligations, thus 'effectively turn[ing] them into more "sacred" obligations than the most fundamental principles of the WTO'.[33] The ultra-rigidity of WTO-plus obligations assumed under access protocol provisions would thus leave the newly acceded WTO Members with no available means to modify and/or withdraw their export duty commitments as is allowed for tariff concessions included in GATT schedules.[34] Such 'virtual immutableness' amounts, in Qin's view, to a 'permanent alienation of a Member's ownership right to claim a larger share of its natural resources for domestic use [. . .] arguably inconsistent with the concept of permanent sovereignty over natural resources'.[35]

Qin's argument lies in the premise that sovereignty over natural resources cannot be alienated for good, as it 'can always be regained,

[29] For a full account of the GATT-specific flexibilities available to modify and/or withdraw import duty concessions, including for economic purposes, see above, Chapter 4.

[30] Qin, 'Reforming WTO Discipline', p. 1163. [31] Ibid., p. 1164.

[32] Schrijver, *Sovereignty over Natural Resources*, p. 263, quoting G. Abi-Saab, 'Progressive Development of the Principles and Norms of International Law Relating to the New International Economic Order', UN Doc. A/39/504/Add.1, 23 October 1984.

[33] Qin, 'Reforming WTO Discipline', p. 1156.

[34] For a full account see above, Chapter 4, Section 4.3.2.

[35] Qin, 'Reforming WTO Discipline', p. 1186.

notwithstanding contractual obligations to the contrary'.[36] In this regard, it should be noted that there is still much controversy as to the extent to which sovereignty over natural resources can be exercised in breach of other international obligations.[37] *China–Raw Materials* and *China–Rare Earths* show that this is also the case when it comes to WTO rules.[38] Yet, Qin has given voice and a legal expression to the arguments in favour of the legitimacy of export duties as economic development tools on the part of developing countries and, in particular, of those newly acceded WTO Members acquiring awareness of the implications of the legal technique used to assume export duty commitments in their accession protocols. Moreover, and importantly, she has advocated a system-wide reform of WTO disciplines on export duties as the only way to ensure that export duties remain the lawful means for achieving economic development objectives.[39]

7.3.1.2 The principle of sovereignty over natural resources as a basis to defend export restrictions addressing sustainable development concerns

The principle of sovereignty over natural resources has been invoked by China to defend its Article XI:1 GATT-inconsistent export quotas under Article XX(g) GATT in *China–Raw Materials* and *China–Rare Earths*, due to the strong developmental and environmental connotations of the principle and its link to the principle of sustainable development. China argued that the conservation exception provided for in Article XX(g) had to be interpreted in the light of the principle of permanent sovereignty over natural resources, as a relevant rule of international law applicable to the parties as per Article 31(3) of the Vienna Convention on the Law of Treaties (VCLT), and the principle of sustainable development recognized in the Preamble to the Marrakesh Agreement as relevant context.[40] The Panels accepted that the term 'conservation' in Article XX(g) incorporated the notion of exercising rights over natural resources in the interests of a Member's economic and sustainable development, and accordingly

[36] Ibid., p. 1165.

[37] Although some scholars have argued that the principle acquired the status of *jus cogens*, this thesis has not been widely accepted. Schrijver, *Sovereignty over Natural Resources*, pp. 374–7. Moreover, the duty to respect international agreements seems now to be recognized as forming part of the corpus of duties qualifying the principle of sovereignty over natural resources. Ibid., para. 392.

[38] See above, Chapter 6, Sections 6.3.2 and 6.4 and below, Section 7.3.1.2.

[39] Qin, 'Reforming WTO Discipline', pp. 1180–1. For a full account see below, Chapter 9, Section 9.2.2.

[40] For a full account see above, Chapter 6, Section 6.3.2.

recognized the right of WTO Members to design their conservation programmes based on 'their own assessment of various, sometimes competing, policy considerations and in a way that responds to their own concerns and priorities'.[41] However, they clarified that, while 'conservation' policies may take sustainable economic development into account, measures adopted for the purpose of economic development cannot be pursued under the rubric of 'conservation' within the meaning of Article XX(g) GATT. In other words, Article XX(g) cannot be 'stretched' into an exception protecting measures that pursue industrial policy goals.[42]

The Panels' conclusion lies in the premise that the permanent sovereignty over natural resources has to be exercised within the limits imposed by the requirements of Article XX(g) GATT:

> As noted by the panel in *China–Raw Materials*, a State's sovereignty is also expressed in its decision to ratify an international treaty and accept the benefits and obligations that such ratification entails. In becoming a WTO Member, China has of course not forfeited permanent sovereignty over its natural resources, which it enjoys as a natural corollary of its statehood. Nor . . . has China or any other WTO Member 'given up' its right to adopt export quotas or any other measure in pursuit of conservation. China has, however, agreed to exercise its rights in conformity with WTO rules, and to respect WTO provisions when developing and implementing policies to conserve exhaustible natural resources.[43]

In this regard, the Panel specified that the fulfilment of Article XX(g) parameters should be assessed on a case-by-case basis in light of the factual circumstances of each dispute. It also posited that the rejection of China's arguments did not stem from an absolute prohibition on using export quotas as conservation policy instruments.[44] The Appellate Body confirmed that Article XX(g) GATT 'should not be understood as suggesting that export quotas can never relate to conservation'.[45] In China's view, however, the interpretation of Article XX(g) given in *China–Rare Earths* by the WTO dispute settlement bodies 'suggest[s] that export quotas are *per se* incompatible with Article XX of the GATT 1994, and . . . prevent[s] . . . any "realistic options" from falling within the scope of the conservation exception'.[46]

[41] Panel Reports, *China–Rare Earths*, para. 7.459. [42] Ibid., paras. 7.451–2 and 7.459–60.
[43] Ibid., para. 7.270. [44] Ibid., paras. 7.291–2.
[45] AB Reports, *China–Rare Earths*, para. 5.161, citing Panel Reports, *China–Raw Materials*, para. 7.293.
[46] Panel Reports, *China–Rare Earths*, para. 5.161.

7.3.2 *Export restrictions as safeguard tools*

Export restrictions have also been defended as 'safeguard' measures aimed at protecting domestic industries from the impact of export surges. Export surges of raw materials may intervene for various reasons, including actual supply shortages and panic buying, aggressive stockpiling strategies[47] and speculative behaviour in commodity exchanges.[48] The importance in use of and the difficulty of substituting specific industrial inputs play a major role in determining potential exposure to a sudden and sharp expansion of demand, while uneven distribution of global production may also induce countries vulnerable to supply disruptions to make pre-emptive demand surges.[49] In this regard, critical minerals and metals represent the archetypical example of industrial raw materials exposed to sudden and significant demand surges.

In such cases, the expansion of demand may easily translate into an export surge due to the heavy geographical concentration of mine production. Moreover, when top producers have also established and/or are in the process of establishing downstream processing industries using the critical materials in question, increased exports may jeopardise availability of supply domestically, causing critical shortages and/or severe price increases. In such a context, it has been argued, the use of export restrictions may in principle alleviate the effect of export surges.

China has contended in *China–Rare Earths* that its export quotas on rare earth elements were safeguard tools. Accordingly, it claimed that such instruments were meant to 'help manag[e] the limited supply [of rare earth elements] and provide reasonable certainty on the quantities that will be supplied to the domestic and foreign users', thus 'moderating speculative surges that could upset market balance and certainty sought by China'.[50] While China's intention was to protect its domestic industries from the disruption of supply caused by speculative foreign demand surges, it defended them under Article XX(g) under the proposition that preventing demand surges would also contribute to conservation in addition to serving the sustainable economic development needs of the country.[51] Based on its interpretation of Article XX(g) GATT, however, the Panel reiterated that export safeguards aimed at mitigating the effects of speculative demand surges in the world market were measures

[47] M. Radetzki, *A Handbook of Primary Commodities in the Global Economy* (Cambridge University Press, 2010), pp. 139–42.

[48] Ibid., pp. 100–10. [49] Ibid., pp. 127–30.

[50] Panel Reports, *China–Rare Earths*, para. 7.449. [51] Ibid.

of industrial policy that could not be considered as 'relating to' conservation.[52]

China's attempt to stretch Article XX(g) to defend export safeguards by insisting on the principle of sustainable development and on the developmental orientation of the principle of sovereignty over natural resources may be explained by the absence of an economic emergency exception *à la* Article XIX in the GATT.[53] The GATT does provide for a series of 'industrial' policy related exceptions such as Article XX(i) and (j) but, as explained above, these exceptions are construed to apply in narrowly circumscribed cases – the existence of a governmental stabilization plan in one case and a situation of local or short supply in the other case.[54] Article XI:2(a) GATT, moreover, allows WTO Members to adopt export restrictions to prevent or relieve a 'critical' shortage of an essential product.[55] These exceptions can thus be invoked in emergency situations putting in danger the physical availability of a product to domestic consumers. They are not aimed, however, at protecting the domestic industry from the injuries that may arise out of a surge of exports, which are not necessarily limited to supply disruptions, but may also encompass changes in the level of downstream product sales, losses of profits due to exaggerated price increases of domestic inputs caused by foreign demand surges, and related effects on domestic employment.

Without entering into the merits of China's specific claims in *China–Rare Earths*, it should be noted that the unavailability of an Article XIX GATT-like exception clause allowing export safeguards when serious harm to a domestic industry is caused by unforeseen developments represents a further element of asymmetry in the WTO legal treatment of trade measures applied on the import and the export sides and may also encourage two types of dynamics: on the one hand, WTO Members that, like China, are unable to use export duties on a certain product by virtue of an accession protocol provision, would be left with the choice to apply Article XI:1 GATT-inconsistent export restrictions and 'force' their defence under Article XX exceptions or Article XI:2 (a) GATT while they could have given preference to tariff instruments under an Article XIX-like exception. On the other hand, other WTO Members may be incentivized to abstain from scheduling and binding export duty rates on products that are or could be important for their own industrial fabric inasmuch as they would neither be able to

[52] Ibid., paras. 7.451–2. [53] See above, Chapter 6, Section 6.3.2. [54] Ibid.
[55] See above, Chapter 5, Section 5.3.1.

increase them nor be allowed to impose export restrictions in the event of an export surge under the GATT.

7.3.3 *Export restrictions as instruments to relieve commodity dependence and vulnerability to price volatility*

According to the United Nations Conference on Trade and Development (UNCTAD), two-thirds of all developing countries rely on primary resources (mineral, agricultural and energy commodities) for more than 50 per cent of their export earnings. This percentage grows to 80 per cent for half of the developing countries located in the African continent.[56] Commodity-dependent developing countries are particularly vulnerable to strong price fluctuations over the business cycle. Instability of export revenues due to commodity price volatility exposes them to large trade imbalances, which, in turn, may lead to balance of payments disequilibria and exchange rate fluctuations ultimately harmful for economic diversification.[57]

Against this backdrop, export restrictions have frequently been used to prevent and/or remedy the adverse consequences of commodity price volatility on the economic growth of resources-based, mainly low-income, developing countries.[58] Commodity-dependent developing countries exposed to large trade imbalances caused by boom-bust price cycles and/or by speculative attacks[59] often resort to export restrictions and, in particular, export taxes, to stabilize export revenues. During periods of high commodity prices, these countries tend to raise the export tax rate in order to reap the benefits of the bonanza while at the same time offsetting the negative effects on external competitiveness that may be induced by

56 UNCTAD, 'State of Commodity Dependence 2012', www10.iadb.org/intal/intalcdi/PE/2012/12065.pdf (accessed January 2015).

57 H. Bargawi, E. Bova, B. Ferrarini, and S. Newman, 'Low-income Countries and Commodity Price Volatility', in T. Cottier and P. Delimatsis (eds.), *The Prospects of International Trade Regulation* (Cambridge University Press, 2010), pp. 464–71; Radetzki, *A Handbook of Primary Commodities*, pp. 188–215.

58 R. Piermartini, 'The Role of Export Taxes in the Field of Primary Commodities', WTO Discussion Paper, Economic Research and Statistics Division, 2004, www.wto.org/english/res_e/booksp_e/discussion_papers4_e.pdf (accessed May 2014), pp. 9–13; K. C. Fung and J. Korinek, 'Economics of Export Restrictions as Applied to Industrial Raw Materials', OECD Trade Policy Papers, No. 155 (OECD Publishing, 2014), pp. 31–5.

59 Speculation is widely recognized as having a critical role in driving up the price of commodities. C. R. Knittel and R. S. Pindyck, 'The Simple Economics of Commodity Price Speculation', MIT Center for Energy and Environmental Policy Research Working Paper No. 6/2013, http://web.mit.edu/ceepr/www/publications/workingpapers/2013-006.pdf (accessed January 2015), p. 31.

large surpluses in their balance of payments. In this scenario, the ultimate goal is avoiding Dutch disease-like implications undermining efforts toward economic diversification and development.[60]

The use of temporary measures on the export side is considered less problematic than other types of interventions such as exchange rate managements. Exchange rate policy would in fact determine an increase in the price of imported products and potentially destabilize other sectors of the economy. This is particularly true for mineral-based countries, which generally import large quantities of food commodities.[61] Moreover, in the case of small economies (i.e. in the case of many low-income CDDCs), devaluation may not have an appreciable impact on export competitiveness or, in any case, such impact may be short lived as the country would become exposed to inflationary pressures or currency speculation.[62] The use of export restrictions, in contrast, would also inherently control internal inflation and, in the case of export taxes, increase government revenues.[63]

While the use of export taxes may reflect such additional redistributional effects, the preference for export taxes may also derive from the lack of relevant exceptions applicable to export restrictions in the GATT. As discussed above, Article XVIII(b) does include a balance of payments exception, drafted in more flexible terms than Article XII GATT, for WTO Members with low standards of living. Yet, this exception refers solely to imports.[64] The same holds true for Article XVIII(c) GATT, which allows WTO Members at the early stages of development to take 'specific measures affecting imports' to assist the establishment of a particular industry.[65]

The specific vulnerabilities and development needs of low-income CDDCs notwithstanding, the justification of Article XI:1 GATT-inconsistent quantitative restrictions responding to balance of payments and economic development purposes is in other words solely contemplated on the import side. Again, this circumstance may have a bearing on the lack of engagement on the part of these WTO Members in the negotiation of export duty concessions to be included in their GATT schedules.

[60] Radetzki, *A Handbook of Primary Commodities*, pp. 207–10.

[61] Bargawi *et al.*, 'Low-income Countries', pp. 469–70.

[62] Ibid. See also Radetzki, *A Handbook of Primary Commodities*, p. 212.

[63] Many developing countries have employed so-called variable export taxes. Accordingly, they raise export taxes during commodity booms and relax or dismantle them when prices fall. Piermartini, 'The Role of Export Taxes', pp. 10–11.

[64] See above, Chapter 5, Section 5.3.3. [65] Ibid.

7.4 Arguments against export restrictions as beggar-thy-neighbour instruments

At the other end of the spectrum, export restrictions are looked at suspiciously as beggar-thy-neighbour policies that deteriorate conditions of access to raw materials and *de facto* subsidize downstream processing industries in the imposing countries at the expense of world prices and foreign consumers, thus disrupting global value chains. Due to the current configuration of the WTO disciplines on export restrictions, discussions on how best to address trade distortions caused by export restrictions have mainly focused on the use of export taxes. These instruments being lawful under Article XI:1 GATT in the absence of any corresponding export duty concessions inscribed in GATT schedules, the general feeling has been that WTO Members may easily evade the WTO rules on export quantitative restrictions (QRs) by shifting to high levels of export taxes. However, exports QRs such as export quotas have also come under close scrutiny due to their greater distortive effects and their rent-seeking implications.

7.4.1 Disruptions in global value chains caused by export restrictions

In a world of atomized production processes, export restrictions on raw materials may entail repercussions along the multiple processing stages composing the so-called global value chains (GVCs). Cattaneo *et al.* define GVCs as 'the full range of activities that are required to bring a good or service from conception through the different phases of production – provision of raw materials; the input of various components, subassemblies, and producer services; the assembly of finished goods – to delivery to final consumers, as well as disposal after use'.[66] They explain that GVCs are by now a structural component of the world economy, which sensibly transformed the way international trade occurs. The globalization of production, in particular, has led to an increase of trade in intermediate goods and to the emergence of various kinds of coordinated networks which, on a global level, carry out all the activities that compose the value chain.[67] In this context, resource-endowed developing countries have

[66] O. Cattaneo, G. Gereffy and C. Staritz, 'Global Value Chains in a Postcrisis World: Resilience, Consolidation, and Shifting End Markets', in O. Cattaneo, G. Gereffy and C. Staritz (eds.), *Global Value Chains in a Postcrisis World: A Development Perspective* (The World Bank, 2010), p. 4.

[67] Ibid., p. 7.

entered GVCs mainly through the attraction of resource-seeking investments.[68] Large emerging economies such as China have also participated in manufacturing GVCs, although predominantly in low value-added segments of the value chain such as the assembly phase.[69]

Export restrictions on raw materials introduce an element of disturbance in the furthest upstream segment of the value chain, thus compromising the effective functioning of the cascade of activities that constitute manufacturing GVCs. The use of export restrictions intervening in the extraction phase has indeed been identified as a main factor of risk and constraint in the value chain leading to the production of cutting-edge and environmentally friendly manufactured products by the US Department of Energy.[70] The lack of transparency in the use of such measures is, in this respect, another major point of concern.[71]

To the extent that globalization of production makes national economies increasingly integrated and interdependent, trade liberalization as a necessary precondition for the consolidation and profitability of GVCs should in theory be considered a priority by all countries participating in GVCs. Yet, as the revival of the resource nationalism narrative shows, countries are increasingly prone to privilege domestic industrial development objectives over trade liberalization. One reason for this lies again in the 'unequal partition of the total value-added and rewards along the value chain', with developing country suppliers usually left with 'little leverage' in the chain and embroiled in a pattern of resource dependence that does not produce much value for the economy.[72] WTO Members would thus resort to export restrictions on raw materials to encourage downstream (higher value-added) activities domestically as a way of upgrading their position in manufacturing GVCs.[73] Another reason that could explain the renewed engagement in domestic industrial policy

[68] P. Draper and A. Freytag, *Who Captures the Value in the Global Value Chain? High Level Implications for the World Trade Organization*, E15 Initiative (International Centre for Trade and Sustainable Development (ICTSD) and World Economic Forum, 2014), p. 3.

[69] Ibid., p. 4.

[70] US Department of Energy, 'Critical Materials Strategy', 2010, http://energy.gov/sites/prod/files/edg/news/documents/criticalmaterialsstrategy.pdf (accessed December 2014), p. 9.

[71] See B. Fliess, C. Arriola and P. Liapis, 'Recent Developments in the Use of Export Restrictions in Raw Materials Trade', in OECD, *Export Restrictions in Raw Materials Trade: Facts, Fallacies and Better Practices* (OECD Publishing, 2014), p. 33.

[72] Cattaneo *et al.*, 'Global Value Chains in a Postcrisis World', p. 18.

[73] Draper and Freytag, *Who Captures the Value in the Global Value Chain*, p. 3.

is the awareness that GVCs have acted as conduits for the transmission of the economic crisis, with the weakest components being exposed to the effects of the shocks occurring in Western countries.[74] Developing countries with emerging economies have reacted by focusing on their home markets, and adopting industrial strategies to help local downstream producers compete with foreign producers.[75] The strong focus on industrial policy is thus aimed at reaping the benefits (and compensating for the negative spillovers) of resources exploitation domestically in view of achieving economic diversification.[76]

In this scenario, however, the divergence of economic policy orientations may eventually come to a close once governments realize they are not immune from disruptions in the value chain given the irreversible interdependence of the global economy and the exhaustibility of extractive resources.[77] In other words, as the European Commission highlighted in its last Report on Potentially Trade-Restrictive Measures, 'the unprecedented increase in the use of restrictions on exports of natural resources ... represents a great danger to the effective functioning of value chains, and can eventually harm its own perpetrators, as no country is self-sufficient in terms of resource needs'.[78]

7.4.2 *Subsidy-equivalent effects of export restrictions*

As explained above, export restrictions artificially reduce the domestic price of raw materials, thereby conferring an advantage to downstream processing industries that avail themselves of such materials. By reserving cheaper inputs for domestic consumers, export restrictions *de facto* subsidize national downstream industries.[79] In this respect, the fact that export restrictions, at a practical level, operate as subsidies does not have any implications under the GATT. Article XI:1 GATT bans all export

74 Cattaneo *et al.*, 'Global Value Chains in a Postcrisis World', pp. 6–11.

75 Ibid., pp. 12–13.

76 M. Wu and J. Salzman, 'The Next Generation of Trade and Environment Disputes: The Rise of Green Industrial Policy', *Northwestern University Law Review* 108 (2014), 401, pp. 426–30.

77 Cattaneo *et al.*, 'Global Value Chains in a Postcrisis World', pp. 10–11.

78 European Commission, 'Eleventh Report on Potentially Trade-Restrictive Measures Identified in the Context of the Financial and Economic Crisis', 2014, http://horizon.hozint.com/2014/11/dg-trade-11th-report-on-potentially-trade-restrictive-measures/ (accessed December 2014), p. 38.

79 See above, Chapter 3, Section 3.2.2.

restrictions, irrespective of whether they are functionally equivalent to a subsidy or not, as long as they have a limiting effect on the volume of exports. As per the same Article, moreover, export duties or taxes are lawful – to the extent that they are consistent with any existing export duty commitments assumed by the WTO Member in question under its GATT Schedule – irrespective of their rationale.[80] The use of export taxes has thus come under particular scrutiny as net importing WTO Members have started to fear that such instruments could be abused within the purview of Article XI:1 GATT to grant domestic industries an unfair advantage ultimately equivalent to a subsidy. Such concerns lie in the awareness of the difficulties of successfully challenging an export tax, lawful as per Article XI:1 GATT, under the WTO Agreement on Subsidies and Countervailing Measures (ASCM).[81] Only measures falling within the scope of the definition of 'subsidy' contained in the Agreement can in fact be subject to ASCM disciplines.

Under Article 1 ASCM, a subsidy within the meaning of the Agreement must fulfil two definitional elements: (i) it must constitute a financial contribution given by a government or any public body within the territory of a Member (Article1.1(a)(1) ASCM)[82] or any form of income or price support in the sense of Article XVI of GATT 1994 (Article1.1(a)(2) ASCM); and, (ii) this financial contribution must confer a benefit on the recipient (Article 1.1.(b) ASCM). A subsidy must then also be 'specific' as per Article 2 ASCM to fall within the remit of the ASCM.

The Appellate Body has not definitively clarified the status of export restrictions in general, and export taxes in particular, under the ASCM. In only one dispute, *US–Export Restraints*, has the Panel been confronted with the issue of whether certain export restraints constituted a 'subsidy' under the ASCM and, in particular, a government-entrusted or

[80] For a full account see Chapter 4, Section 4.3.3.

[81] Agreement on Subsidies and Countervailing Measures, Multilateral Agreements on Trade in Goods, Annex 1A to the Agreement Establishing the World Trade Organization, 1869 U.N.T.S. 14, entered into force 1 January 1995.

[82] According to Article 1.1(a)(1) ASCM, there is a financial contribution when: '(i) a government practice involves a direct transfer of funds (e.g. grants, loans, and equity infusion), potential direct transfers of funds or liabilities (e.g. loan guarantees); (ii) government revenue that is otherwise due is foregone or not collected (e.g. fiscal incentives such as tax credits); (iii) a government provides goods or services other than general infrastructure, or purchases goods; (iv) a government makes payments to a funding mechanism, or entrusts or directs a private body to carry out one or more of the type of functions illustrated in (i) to (iii) above which would normally be vested in the government, and the practice, in no real sense, differs from practices normally followed by governments'.

government-directed provision of goods in the sense of Article 1.1.(a)(1)(iv) ASCM. It observed that, although export restraints in the sense used in the dispute[83] were based on a government intervention having an effect on the marketplace, the 'entrusts or directs' standard was not satisfied to the extent that it required an 'explicit and affirmative action of delegation or command'.[84] Accordingly, the Panel concluded that the export restraints at issue did not constitute a 'financial contribution' within the meaning of Article 1.1 (a)(1)(iv) ASCM, and thus did not continue to assess whether the measures at issue conferred a benefit as required under Article 1.1(b) ASCM.

Although the Appellate Body has never explicitly ruled on the issue, the 'governmental financial contribution' requirement appears difficult to satisfy in the case of an export tax. The only category that may be relevant is the government 'provision of goods' under Article 1.1(a)(1)(iii) ASCM, associated with the last category, that is, 'the government entrusts or directs a private body to carry out one or more of the type of functions illustrated in (i) to (iii) above'. Although the government does not itself provide the cheaper raw material, the effect of an export tax is in fact that the government entices national economic operators to provide the material to domestic, and not foreign, consumers at a price that is below the prevailing international market prices.[85] The difficulty would be in demonstrating that the government's encouragement is a form of 'entrustment' or 'direction' within the meaning of Article 1.1(a)(1)(iv) ASCM. Existing relevant WTO case law seems to require that the government is in a position to control the private suppliers and to command them to sell the raw material to domestic users.[86] Depending on the factual circumstances of each dispute, this cannot be excluded in those cases where an export tax is associated with sales or purchases by domestic state-trading enterprises.[87]

[83] The definition of export restraint referred to by Canada for the purpose of the dispute was: 'a border measure that takes the form of a government law or regulation which expressly limits the quantity of exports or places explicit conditions on the circumstances under which exports are permitted, or takes the form of a government-imposed fee or tax on exports of the products calculated to limit the quantity of export'. Panel Report, *US–Export Restraints*, Section 4(a), para. VIII.3.

[84] Ibid., Section 4(d)(ii), para. VIII.5.

[85] D. Luff, H. Nguyen and N. Anh Thu, 'Export Control by WTO Members and Recommendations for Vietnam', EU-MUTRAP Project Activity Code: Supe-7, 2013 (on file with the author), pp. 75-80.

[86] See AB Report, *United States – Countervailing Duty Investigation on DRAMs*, paras. 141–98.

[87] See above, Chapter 2, Section 2.5.4.

Another way to establish the existence of a subsidy is if a measure provides 'any form of income or price support in the sense of Article XVI of GATT 1994' as per Article 1.1(a)(2) ASCM. Article XVI GATT does not specify the notion of 'income or price support', but Note 2 *Ad* Article XVI adds that the subsidies it addresses are those which 'operate directly or indirectly to increase exports of any product from [the Contracting Party which grants the subsidy], or to reduce imports of any product into [the Contracting Party which grants the subsidy]'.[88] This definitional element has not yet been interpreted by WTO dispute settlement bodies, but the language of Note 2 *Ad* Article XVI seems to contemplate measures inducing a price differential that is profitable for domestic producers rather than consumers.[89] Yet, it remains to be seen whether export taxes may be considered as falling within the ambit of this definitional element considering that, in the case of an export tax, the domestic consumers benefiting from the artificial advantage created by the tax are producers of downstream products.

Under the second definitional element of subsidy provided for in Article 1.1(b) ASCM, a financial contribution by a government or an income or price support measure must provide a 'benefit' to the recipient. According to Article 14 ASCM and based on a substantial body of consistent WTO case law, the existence of a benefit is defined in relation to normal commercial conditions applicable in a given market (i.e. a measure must improve the competitive position of the recipient compared with what its situation would be in the 'market' without the measure). Although export taxes confer a benefit on domestic downstream producers in the economic sense when they artificially decrease domestic input prices below world market levels, proving the existence of a 'benefit' under Article 1.1(b) ASCM requires the identification of an appropriate 'market benchmark'. This may be difficult when a domestic market is distorted by a government intervention, such as the case of an export tax.[90] Depending on the specific criteria used to determine the market benchmark, the benefit test may yield different results. If the WTO adjudicators determine that the benchmark is the price before the export tax was introduced, then there would be a benefit. However, if the

[88] See Note 2 *Ad* Article XVI, paragraph 1, included in Annex I to the GATT.

[89] V. Pogoretskyy, 'Energy Dual Pricing in International Trade: Subsidies and Anti-dumping Perspectives', in Y. Selivanova (ed.), *Regulation of Energy in International Trade Law: WTO, NAFTA and the Energy Charter* (Kluwer Law International, 2012), pp. 202–3.

[90] Luff *et al.*, 'Export Control by WTO Members', pp. 78–80.

benchmark is determined as the price existing in the market at the time the domestic inputs were purchased, all regulatory measures existing at the time, including the export tax, would be incorporated into the benchmark price and there could be no benefit.[91] WTO dispute settlement bodies may however decide to consider alternative cross-border benchmarks in situations where domestic market conditions cannot be of use due to the predominant role played by the government.[92] Under this scenario, alternative benchmarks may include proxies construed on the basis of the world or export price for similar goods, or the 'prices for the same inputs in economies at a similar stage of development with a similar resource base'.[93]

Finally, an export tax would need to be 'specific' within the meaning of Article 2 ASCM as it does not come under the category of prohibited subsidies under Article 3 ASCM.[94] In principle, an export tax would be unlikely to meet the *de jure* specificity criterion as it artificially reduces the price of a raw material for all domestic users. In this sense, it does not apply to specific enterprises or groups of enterprises, or industries or groups thereof, or specific sectors as required under Article 2 ASCM. Yet, it could arguably be considered a *de facto* specific subsidy under Article 2.1 (c) ASCM inasmuch as only certain enterprises or industries reap the benefits generated by its implementation.[95] In this regard, WTO adjudicators seem to retain a quite significant margin of manoeuvre.

In conclusion, while it cannot be excluded at the outset that export taxes, albeit lawful under Article XI:1 GATT, could nevertheless be challenged as actionable subsidies due to their effects on domestic downstream producers, chances to prove that they constitute a 'subsidy' within the meaning of the ASCM seem generally quite low and in any case much would depend on the specific circumstances of each case. The uncertainty

[91] Ibid.

[92] AB Report, *US–Softwood Lumber IV*, para. 90. For the Appellate Body, the role of the government is predominant when 'it effectively determines the price at which the private suppliers sell the same or like products'. Ibid., paras. 87–96 and 101.

[93] Y. Selivanova, *Energy Dual Pricing in WTO Law: Analysis and Prospects in the Context of Russia's Accession to the World Trade Organization* (Cameron May, 2008), p. 123.

[94] Provided they fall within the definition of subsidy under Article 1 ASCM, subsidies that are *de jure* or *de facto* contingent on export performance (Article 3.1(a) ASCM) or on the use of domestic goods over imported goods (Article 3.1(b) ASCM) fall under the category of prohibited subsidies and are deemed to be specific within the meaning of Article 2 ASCM. P. C. Mavroidis, *Trade in Goods: The GATT and the Other Agreements Regulating Trade in Goods*, 2nd edn (Cambridge University Press, 2012), p. 549.

[95] See Panel Report, *US–Softwood Lumber IV*, para. 7.116.

surrounding the possibility to challenge beggar-thy-neighbour export taxes adopted for industrial policy purposes under the ASCM have likely had a bearing on the definition of a clear position against export taxes on the part of net importing WTO Members. Such Members have repeatedly advocated for the elimination of such instruments altogether, provided specific flexibilities are conceded to least developed and developing countries.[96]

7.4.3 *Rent-seeking behaviours induced by export restrictions and absence of competition rules in the WTO*

Although efforts to encourage more ambitious rules on export restrictions in the WTO have focused on the need to strengthen WTO disciplines on export taxes, other types of quantitative restrictions on exports have also continued to be closely scrutinized given their strong impacts on trade. Export quotas, in particular, have been under the radar as measures that may critically hamper access to raw materials depending on the stringency of the ceiling set by governments, while also incentivizing rent-seeking behaviours on the part of exporting firms whose dynamics are similar to those behind the creation of cartels.[97] In other words, exporting firms may be encouraged to act as 'single sellers' in order to raise the price of the restricted products when destined for the international market and thus increase their revenues at the expense of weak foreign consumers.[98] These types of collusive behaviour are facilitated in the minerals and metals sector, where price elasticity is generally low and global supply geographically concentrated in a few countries, with a small number of large exporting firms enjoying 'substantial market power'.[99]

The tangible risk of opportunistic behaviour linked to the implementation of export quotas may thus amplify the harmful effects produced by such measures, irrespective of whether the government has introduced them for legitimate purposes. In this regard, the question becomes whether the country applying the restrictions has in place a sufficiently stringent legislation regulating anti-competitive conduct. As is known,

[96] See above, Section 7.2 and below, Chapter 8, Section 8.3.

[97] For a thorough overview of the preconditions and implications of cartel action see Radetzki, *A Handbook of Primary Commodities*, pp. 144–65.

[98] Fung and Korinek, 'Economics of Export Restrictions', p. 25.

[99] Ibid., p. 6. See also US National Research Council of the National Academies, *Minerals, Critical Minerals and the US Economy* (The National Academies Press, 2008), www.nma.org/pdf/101606_nrc_study.pdf (accessed December 2014), pp. 67, 136.

the WTO legal framework does not regulate anti-competitive business practices affecting international trade.[100] Such practices may thus evade regulation if they occur in a country with ineffective competition law. Inasmuch as domestic competition laws remain an expression of the sovereignty of states, cross-border competition regulation continues to lag behind, despite the efforts of influential players such as the United States and, to a limited extent, the European Union to assert jurisdiction over foreign firms whose conduct generates effects within their territory or impinges on the import/export commerce of their domestic firms.[101] As recognized by the English House of Lords, in fact, 'claims to extra-territorial jurisdiction are particularly objectionable in the field of [competition] legislation because, among other reasons, such legislation reflects national economic policy which may not coincide, and directly conflict, with that of other states'.[102]

Until the WTO specifically addresses competition issues,[103] and in light of the inherent difficulties in ensuring the extraterritorial enforcement of

[100] The draft Havana Charter did contain a chapter specifically devoted to 'Restrictive Business Practices' (Chapter V), according to which Members would have had to commit to take all possible and appropriate measures to ensure that private and public commercial enterprises did not engage in anti-competitive behaviours. After the failure of the International Trade Organization (ITO) project, however, the GATT abandoned any ambition to incorporate rules on competition.

[101] In the United States, the so-called effects doctrine in support of the extraterritorial application of US competition legislation has relied on an extensive interpretation of Section 1 of the Sherman Act, which reads: 'Every contract, combination . . . or conspiracy, and restraint of trade in commerce . . . with foreign nations is declared to be illegal'. For a full account of the genesis and evolution of the effects doctrine in the jurisprudence of the US Supreme Court see M. Taylor, *International Competition Law: A New Dimension for the WTO?* (Cambridge University Press, 2006), pp. 56–65. Similarly, albeit more prudently, the European Court of Justice has elaborated the so-called implementation doctrine. Ibid., pp. 64–5.

[102] Per Lord Frazer in *Rio Tinto Zinc Corporation* v. *Westinghouse Electric Corporation* [1978] 2 WLR 81, p. 125, cited ibid., p. 55.

[103] Following the Singapore Ministerial Declaration in 1996, a Working Group on the Interaction between Trade and Competition Policy (WGTCP) was established in the WTO with a mandate to study various aspects of the interaction between domestic competition policy instruments and international trade. See WTO Doc. WT/MIN(96)/DEC, 18 December 1996, para. 20. The Doha Ministerial Declaration acknowledged the work of the Working Group and envisaged a number of issues to be addressed on priority in its future activities. See WTO Doc. WT/MIN(01)/DEC/1, 20 November 2001, para. 25. However, in 2004 the General Council of the WTO decided to drop the competition item from the Doha Agenda and since then the Working Group has been inactive. WTO, 'Interaction between Trade and Competition Policy', www.wto.org/english/tratop_e/comp_e/comp_e.htm (accessed January 2015). Some authors have

national competition law,[104] any collateral effects caused by the rent-seeking incentives produced by quantitative restrictions on exports are likely to escape WTO disciplines. The extent to which such circumstances may have played a role in the drafting of asymmetrical exceptions for quantitative restrictions applied on the export side in the GATT is uncertain. Yet, it may likely affect the chances of net importing countries accepting a reform of existing WTO disciplines on the export side aimed at, *inter alia*, restoring symmetry in the legal treatment of import and export quantitative restrictions.

7.5 Conclusions

This chapter has attempted to provide a simple conceptual framework to understand the foundational paradigms at the basis of the different conceptions of export restrictions, which have opposed import-dependent and resource-endowed WTO Members from the 1970s onwards. It also aimed to explain how these paradigms have evolved since then, alongside the structural changes intervening in the world economy and the consolidation of international law principles relevant for global natural resources management. The objective was to identify the scope and targets that a proposal for reform of WTO rules on export restrictions should embrace in order to respond to the legitimate, but divergent, concerns of both sides.

The analysis showed that the specific issues to be addressed by any attempts at reform in the area stem from two fundamental stances: first, that export restrictions – export taxes included – should be strictly disciplined to avoid their abuse as beggar-thy-neighbour policies and/or industrial policy instruments creating an unfair advantage for domestic industries; and second, that the developmental function of export restrictions should be acknowledged in the WTO legal framework and reflected in the availability of specific flexibilities. These poles of tension have been shown to be greatest when concerning the use of export restrictions on minerals and metals and, in particular, on critical minerals and metals. Yet, as shown above, most of the arguments underlying such divergent views could apply more generally to industrial raw materials.

Due to the peculiar traits of the WTO legal framework on export restrictions, the main area of tension has been that of export taxes.

advocated the conclusion of a new plurilateral WTO agreement on competition. See Taylor, *International Competition Law*, pp. 423–82.

104 Ibid., pp. 65–9.

Import-dependent countries have supported the negotiation of strengthened rules on export duties, and advocated in particular a blanket elimination of export duties mitigated by additional flexibilities for least developed and developing countries. In contrast, developing countries have firmly reclaimed the right to use export duties as the only legitimate export measure available under the GATT to achieve development goals, and have contested the asymmetries in the rights and obligations of different WTO Members created by the fragmentation of current WTO disciplines on export duties. In particular, they have denounced the systemic imbalances generated by such fragmentation, as disclosed through the approach adopted by the Appellate Body in *China–Raw Materials* and *China–Rare Earths*. Accordingly, they have contended that the current WTO accession regime on export duties is fundamentally iniquitous, and contradicts not only 'constitutional' WTO principles such as the principles of non-discrimination and reciprocity but also general international law principles such as the principle of sovereignty over natural resources.

Against this backdrop, the elimination of the inconsistencies emanating from the fragmentation of existing WTO disciplines on export duties stands as a preliminary issue to address in any discussion on how to reform the existing WTO rules on export restrictions. The growing awareness that the WTO accession regime on export duties has *de facto* generated substantial inequalities among Members has deepened the divide among different categories of WTO Members and, in particular, eroded the trust of developing countries in any opportunities for reforming relevant WTO disciplines. Only by showing that reform could go in the direction of greater regulatory coherence and, under appropriate conditions, re-establish a 'level playing field' among WTO Members, could a climate of trust be restored as the necessary precondition for raising the level of commitments in the area. In this respect, as shown above, ensuring the greatest level of comparability in the WTO legal treatment of import and export restrictions in a way that preserves the policy space of any WTO Member to resort to export restrictions, provided certain conditions are met, could be the key to overcoming the reluctance of developing countries to assume export duty commitments. With this framework in mind, the next chapter will examine the existing Doha proposals addressing the topic of export restrictions.

8

Export restrictions in the Doha Round negotiations

8.1 Introduction

This chapter examines the proposals on export restrictions that were circulated in the Non-agricultural Market Access (NAMA) Negotiating Group in the course of the Doha Round. The aim of this chapter is to investigate their inspiration and specific features in light of the conceptual framework provided in Chapter 7. Accordingly, the analysis will attempt to identify any elements that might have determined the lack of support for such proposals on the part of the World Trade Organization (WTO) Membership.

This chapter starts by giving an account of the history behind the Doha proposals on export restrictions in light of the original Doha mandate (Section 8.2). It then provides an analysis of the two proposals submitted by the European Union on export taxes in the course of the Doha negotiations, the proposal on a WTO Agreement on Export Taxes and the Revised Submission on Export Taxes (Section 8.3). Section 8.4 discusses the submission on a Protocol on Transparency in Export Licensing to the General Agreement on Tariffs and Trade (GATT) instigated by Japan and the United States. Finally, Section 8.5 draws some conclusions on the probable reasons why the Doha proposals on export restrictions have failed to garner the support of developing country Members.

8.2 The Doha mandate on export restrictions

The Doha Ministerial Declaration does not contain a mandate on export restrictions as such. Paragraph 16 of the Declaration, labelled 'Market Access for Non-agricultural Products', however, reads:

> We agree to negotiations which shall aim, by modalities to be agreed, to reduce or as appropriate eliminate tariffs, including the reduction or elimination of tariff peaks, high tariffs, and tariff escalation, *as well as non-tariff*

> *barriers*, in particular on products of export interest to developing countries. Product coverage shall be *comprehensive and without a priori exclusions*. The negotiations shall take fully into account the special needs and interests of developing and least-developed country participants, including through less than full reciprocity in reduction commitments, in accordance with the relevant provisions of Article XXVIII bis of GATT 1994 and the provisions cited in paragraph 50 below (emphasis added).[1]

In line with the comprehensive scope of non-tariff barriers as interpreted in the multilateral trading system,[2] the issue of export restrictions was brought into the NAMA Negotiating Group (NGMA) after the very first submission was circulated in the Group in 2002. The submission was prepared by the European Union and affirmed that the mandate conferred as per paragraph 16 of the Doha Declaration was 'more ambitious than the negotiating mandates for previous rounds, and require[d] a correspondingly ambitious approach to the negotiations'.[3] Accordingly, the European Union argued that '[n]egotiations should also address export restrictions, including export duties' as 'non-tariff barriers which considerably hamper trade and have a serious impact on the activity of trade operators'.[4] This view was then endorsed by a number of other WTO Members,[5] but also substantially rejected by many developing countries including India, Egypt, India, Indonesia, Kenya, Malaysia, Mauritius, Nigeria, Tanzania, Uganda and Zimbabwe.[6]

8.3 The EU proposals on export taxes

8.3.1 The proposal for a WTO Agreement on Export Taxes

The European Union was the first WTO member to submit a comprehensive textual proposal for reforming WTO disciplines on export taxes

[1] See WTO Doc. WT/MIN(01)/DEC/1, 20 November 2001. Paragraph 50 of the Declaration states: 'The negotiations and the other aspects of the Work Programme shall take fully into account the principle of special and differential treatment for developing and least developed countries embodied in: Part IV of the GATT 1994; the Decision of 28 November 1979 on Differential and More Favourable Treatment, Reciprocity and Fuller Participation of Developing Countries; the Uruguay Round Decision on Measures in Favour of Least-Developed Countries; and all other relevant WTO provisions'.

[2] See above, Chapter 4, Section 4.2.

[3] WTO Doc. TN/MA/W/1, 24 June 2002, para. 3.

[4] Ibid., para 8.

[5] See, e.g. WTO Doc. TN/MA/W/5, 5 August 2002, para. 8; WTO Doc. TN/MA/W/15/Add.4/Rev.1, 11 April 2008.

[6] See, e.g. WTO Doc. TN/MA/W/10, 22 October 2002, para. 8; WTO Doc. TN/MA/W/31, 25 March 2003, para. 16.

to the NAMA Negotiating Group.[7] The solution proposed by the EU was the conclusion of a new WTO Agreement on Export Taxes on non-agricultural goods.[8] The underlying objective of the EU submission, as clarified in the 'General Remarks' section of the proposals was:

> ... to address distortions to international trade caused by those export taxes used for the purpose of (or otherwise having the effect of):
> - artificially transferring gains from trade between WTO Members (beggar thy neighbour);
> - creating unfair advantages to domestic industries involved in international trade at the expense of other WTO Members' producers, including infant industries in developing countries; or
> - evading existing WTO disciplines on export restrictions by shifting to more or less prohibitive taxes on the exportation of goods.[9]

The core principle, established in Article 1 of the proposed Agreement, is the complete elimination of all export duties on non-agricultural products by any WTO Member.[10] In parallel, Articles 2 and 3 envisage the possibility for developing country Members and especially least developed country Members to maintain, and correspondingly list and bind in their GATT schedules, export taxes on a most favoured nation (MFN) basis[11] 'for a limited number of products, at low levels', provided that:

> (a) they are necessary, in conjunction with domestic measures, to maintain fiscal stability, to satisfy fiscal needs, or to facilitate economic diversification and avoid excessive dependence on the export of primary products; and
> (b) they do not adversely affect international trade by limiting the availability of goods to WTO Members in general or by raising world market prices of any goods beyond the prices that would prevail in the absence of such measures, or otherwise cause serious prejudice to the interest of developing country Members.[12]

[7] WTO Doc. TN/MA/W/11/Add.6, 27 April 2006. Following the entry into force of the Lisbon Treaty on 1 December 2009, the European Union replaced the European Communities in the WTO. WTO Doc. WT/TPR/G/248, 1 June 2011, p. 6.

[8] The proposal aims at covering all NAMA products as defined in the Doha mandate. See Article 1 of the proposed WTO Agreement on Export Taxes annexed to WTO Doc. TN/MA/W/11/Add.6.

[9] Ibid., p. 1.

[10] Article 1 of the proposed WTO Agreement on Export Taxes states: 'No duties, taxes or other charges imposed on or in connection with the exportation of non-agricultural goods destined for the territory of any other Member, as well as internal taxes and other charges on products exported to any other Member that are in excess of those imposed on like products destined for internal sale, shall be instituted or maintained'.

[11] Ibid., Article 3.2. [12] Ibid., Article 3.1.

These Articles therefore prescribe a number of preconditions incumbent on developing and least developed WTO Members for scheduling and binding export duties: first, the export taxes must be 'necessary' to address a number of situations considered legitimate and applied 'in conjunction with domestic measures'; second, they must not alter the price and/or the availability of a particular product. These conditions do not have a parallel in existing GATT rules on tariff concessions[13] and clearly aim at preventing any large supplier, whether or not a developing country Member, from maintaining export duties.[14] Article 4 of the proposed WTO Agreement, moreover, requires that the prohibition and, when applicable, the reduction of export taxes takes effect upon the entry into force of the Agreement for developed WTO Members, while developing countries and least developed countries (LDCs) enjoy a transition period of three and five years, respectively.[15] During such period, these Members must gradually implement their commitments through 'equal annual reductions'[16] and cannot introduce any new export taxes.[17]

A particularly interesting article of the proposed WTO Agreement is Article 9, which clarifies the applicability of Article XXVIII GATT to export duty concessions coming under the scope of Articles 2 and 3:

> GATT Article XXVIII and the Understanding on the Implementation of Article XXVIII of GATT 1994 shall apply *mutatis mutandis* to the modification of a schedule within the meaning of Article 2.

Although the limited policy space left to developing countries and LDCs to schedule and bind export duties under Articles 2 and 3 somewhat deprives Article 9 from achieving its full potential, this provision remains important as it provides an indication of how WTO Members interpret Article XXVIII GATT with respect to export duty concessions included in GATT schedules.

In the same way, it is noteworthy that Article 5 of the proposed WTO Agreement formally confirms the applicability of a number of GATT exceptions to export taxes adopted in breach of the Agreement by *any*

[13] See above, Chapter 4, Section 4.3.

[14] Although Article 3.1 does not give any indication as to what constitutes 'low level' of export duty rate, the wording of Article 3.1(a) seems to suggest that any export tax that has the effect of limiting the availability of a product to WTO Members, increasing its world price, or affecting the interests of developing country Members may be deemed not to be 'low' within the meaning of Article 3.1. See proposed WTO Agreement on Export Taxes annexed to WTO Doc. TN/MA/W/11/Add.6.

[15] Ibid., Article 4.1. [16] Ibid. [17] Ibid., Article 4.2.

WTO Member, namely the general exceptions and the security exceptions under Articles XX and XXI, the economic exceptions under Article XII and Article XVIII, and Article XV:9 on Exchange Arrangements.[18] Export taxes adopted 'in accordance with the conditions, rules and procedures' laid out in Articles XII, XV:9 and XVIII GATT should, moreover, be applied 'temporarily and under specific circumstances'. Article 5(c) further specifies that this avenue is available to 'any Member, and especially developing Members and least-developed Members'.[19] In case of any dispute arising with respect to the use of export taxes under the Agreement, Article 8 of the proposed Agreement confirms that any violation of its provisions can be challenged before the WTO dispute settlement bodies.

Article 7 of the proposed Agreement, moreover, attempts to reinforce the transparency and notification requirements applicable to export taxes:

> Members reaffirm, with respect to export taxes, their commitment to obligations on transparency and notification in Article X of GATT 1994 and in the Ministerial Decision on Notification Procedures adopted on 15 April 1994. Any new export taxes and any increase in existing export taxes must be notified to the WTO Secretariat 60 days before their entry into force. The notification shall contain a detailed description of the export taxes in question, their product and trade coverage, and their applied levels. Upon request, the Member seeking to institute new or raise existing export taxes shall afford adequate and prompt opportunity for consultations and provide information on the reasons for the export taxes, on their potential effects and on other matters of interest or concern to any other Member. The Member shall also allow a reasonable interval between the adoption of the measure instituting new or raising existing export taxes and its entry into force.

Overall, the architecture of the proposed WTO Agreement on Export Taxes is thus centred on the principle of general elimination of export duties. Limited flexibilities are provided for developing countries and LDCs to residually maintain export duties in the form of export tariff bindings included in their GATT schedules. While the applicability of Article XXVIII GATT to such commitments is explicitly recognized under Article 9, they are nevertheless to be gradually phased out in accordance with Article 10 of the Agreement:

[18] Ibid., Article 5(b) and (c).

[19] Article 5(a) also clarifies that fees and charges imposed on or in connection with the exportation of goods in accordance with Article VIII GATT do not run afoul of the proposed WTO Agreement.

> No later than 5 years after the entry into force of this Agreement, Members shall begin negotiations with the aim of gradually phasing out all remaining export taxes listed in Members' schedules. Priority should be given to those export taxes of concern to developing countries and particularly least-developed countries.

Article 5 mitigates the general elimination obligation by allowing WTO Members to invoke a range of GATT-specific exceptions, including not only those already applicable to both imports and exports (e.g. Article XX) but also a number of import-specific GATT exceptions (e.g. Article XVIII).

In the intention of the European Union, this balancing exercise was meant to 'establish a workable compromise in the area of export taxes between those many countries affected by "beggar thy neighbour" measures adopted by a few major supplier and other large economies, and the use of export taxes by small economies, which includes the majority of developing countries'.[20] Yet, the founding concept of the proposed Agreement, which strongly condemned the use of export duties as beggar-thy-neighbour instruments, fundamentally challenged the view of export restrictions as developmental tools consistently defended by developing country Members, including large and resource-abundant emerging economies.[21] In this respect, the incorporation of additional flexibilities did not prove sufficient to overcome the resistance of developing countries, which felt that the proposal reflected a strong bias towards the needs of import-dependent WTO Members to improve the conditions of access to supplies, while leaving very limited policy space to lawfully introduce and/or maintain export duties for legitimate policy reasons.[22] The European Union accordingly revised its submission, eliminating its most controversial elements.

8.3.2 The EU Revised Submission on Export Taxes

On 16 January 2008, the European Union submitted a revised, less ambitious proposal on export taxes aimed at meeting the requests of developing country Members while still providing 'appropriate remedies to the specific problem related to the use of export taxes as "beggar thy

20 See WTO Doc. TN/MA/W/103/Rev.3/Add.1, 21 April 2011, p. 68.

21 See above, Chapter 7, Section 7.3.

22 WTO, 'World Trade Report 2010: Trade in Natural Resources', www.wto.org/english/res_e/booksp_e/anrep_e/world_trade_report10_e.pdf (accessed December 2014), p. 184.

neighbour" instruments'.[23] The underlying rationale inspiring the second EU proposal, in this sense, does not differ from that of the first submission. The European Union explicitly reiterated the urgency of addressing the 'distortive effects on global commodity trade' generated by export taxes, especially when applied by large suppliers and/or 'when used for industrial or trade policy purposes'.[24] In particular, the European Union stressed that 'the recent proliferation in the use of these instruments', on the one hand, and, on the other hand, 'the short global supply of some specific commodities, despite their abundance in a few countries – a situation that is aggravated by export taxes in key supplying countries', were crucial factors to take into consideration in designing the core features of any revised proposals on export taxes.[25]

Starting from this premise, the European Union maintained that 'any negotiated solution for export taxes would have to build upon existing GATT concepts and rules', and recommended that any revised approach ensures, '*as a minimum,* increased transparency and predictability'.[26]

As to the transparency component, the European Commission admitted that the 1993 Ministerial Decision on Notification Procedures[27] 'has had little, if any, practical effects on Members' level of transparency'.[28] Accordingly, it advocated for stricter transparency provisions on export duties in view of 'ensur[ing] that existing obligations are made operational and enforced in a satisfactory manner'.[29] More specifically, it called on WTO Members to agree on transparency commitments along the lines of the Uruguay Round Understanding on the Interpretation of Article XVII GATT.[30]

The cited Understanding sets out detailed notification requirements for WTO Members on state trading enterprises. It requires each WTO Member to 'conduct a review of its policy with regard to the submission of notifications on state trading enterprises to the Council for Trade in

23 WTO Doc. TN/MA/W/101, 17 January 2008, para. 6. 24 Ibid., para. 1.

25 Ibid., para. 2. The European Union also underlined that export taxes may harm developing countries themselves by frustrating their aspiration 'to build new (infant) processing industries in specific sectors where export taxes by other [developing] countries are prevalent on the raw materials or other inputs'. Ibid., para. 1. See also C. Michalopoulos, 'Trade Policy and Market Access Issues for Developing Countries: Implications for the Millenium Round', The World Bank Policy Research Working Paper No. 2214, 1999, http://elibrary.worldbank.org/content/workingpaper/10.1596/1813-9450-2214 (accessed January 2015).

26 WTO Doc. TN/MA/W/101, para. 6 (emphasis added).

27 See above, Chapter 4, Section 4.5. 28 WTO Doc. TN/MA/W/101, para. 7.

29 Ibid. 30 Ibid.

Goods, taking account of the provisions of this Understanding' and specifies that '[i]n carrying out such a review, each Member should have regard to the need to ensure the maximum transparency possible in its notifications so as to permit a clear appreciation of the manner of operation of the enterprises notified and the effect of their operations on international trade'.[31] Moreover, the Understanding contemplates the possibility for each Member that has reason to believe that another Member has not adequately met its notification obligations to raise the matter and, under certain conditions, make a counter-notification to the Council for Trade in Goods.[32] Both notifications and counter-notifications are to be regularly reviewed, on behalf of the Council for Trade in Goods, by a working party that may make recommendations with regard to the adequacy of notifications and the need for further information.[33]

The choice of the Uruguay Round Understanding on the Interpretation of Article XVII of GATT 1994 as a model for any new transparency provisions on export taxes indicates the intention to raise the level of ambition in comparison with Article 7 of the former proposed Agreement on Export Taxes. Accordingly, the European Union has clarified that developing countries and LDCs should be granted specific flexibilities as per their transparency commitments.[34]

As to the predictability component, the EC replaced the general prohibition approach to export taxes with a 'negotiated solution'[35] consisting in scheduling and binding export taxes in WTO Members' GATT schedules:

> WTO Members should undertake to schedule export taxes on non-agricultural products in their Schedules of Concessions and bind the export taxes at a level to be negotiated, except that:
>
> (a) Least-developed countries would undertake to schedule export taxes on non-agricultural products but may maintain these export taxes unbound; and
> (b) [Small and vulnerable] countries would schedule export taxes but may maintain these export taxes unbound for a certain number of tariff lines (the number is to be negotiated), in reflection of their specific developmental interests and concerns.[36]

[31] Uruguay Round Understanding on the Interpretation of Article XVII of GATT 1994, para. 2. The notifications have to be compiled in accordance with an apposite questionnaire on state trading adopted on 24 May 1960. See GATT Doc. BISD 9S/184–185.

[32] Uruguay Round Understanding on the Interpretation of Article XVII of GATT 1994, para. 4.

[33] Ibid., para 5.

[34] WTO Doc. TN/MA/W/101, para 7.

[35] Ibid., para. 8.

[36] Ibid., para. 9.

Apart from the specific flexibilities envisaged for LDCs and small and vulnerable economies, export taxes would 'have to be bound at a level that "reduce or eliminate tariff peaks, high tariffs, and tariff escalation"' in pursuance of the Doha mandate. In the view of the European Union, such a negotiated option would still ensure that Members 'can reasonably expect what measures any other Member may impose that influence trade'.[37]

The revised submission of the European Union on export taxes has the merit of mitigating the general prohibition approach in the spirit of seeking that common ground with other WTO Members which was missed at the time of the first submission. In this regard, the proposed negotiated solution goes in the direction of a greater comparability between GATT rules on import and export duties, *de facto* seeking the implementation of Article XXVIII(*bis*) GATT on the export side as admitted by virtue of Article II:2(a) GATT.

However, unlike the first proposal, the revised submission does not specifically address the question of the applicability of GATT-specific flexibilities to export taxes but simply states that the suggested approach would be without prejudice to 'the right of WTO Members to apply export taxes when exceptional circumstances under GATT rules are invoked'.[38] While it is reasonable to assume that Article XXVIII GATT would be deemed applicable to export duty concessions included in Members' GATT schedules,[39] it remains open to question whether the expression 'exceptional circumstances under GATT rules' is intended to refer to the whole range of flexibilities expressly mentioned in the proposed Agreement on Export Taxes or only to those GATT exceptions conceived of as applying to both imports and exports (e.g. Article XX) and not including import-specific exceptions (e.g. Article XVIII).

The revised proposal on export taxes of the European Union was inserted into the fourth revision of draft modalities for non-agricultural market access.[40] It also appears among the Textual Proposals concerning non-tariff barriers (NTBs) annexed to the most recent Textual Report of the NGMA Chairman on the State of Play of the NAMA Negotiations.[41] Yet, the inclusion of such a proposal 'does not presume a consensus around it'.[42] Within the general stalemate of the Doha negotiations and

[37] Ibid., para. 8. [38] Ibid., para. 9. [39] See above, Chapter 4, Section 4.3.1.
[40] WTO Doc. TN/MA/W/103/Rev.3, 6 December 2008, pp. 45–7.
[41] WTO Doc. TN/MA/W/103/Rev. 3/Add.1, 21 April 2011, pp. 75–7. [42] Ibid., p. 43.

the limited progress registered in the NAMA negotiations, there have been no advances in the area of export taxes in the last few years.[43]

8.4 The proposed Protocol on Transparency in Export Licensing to the GATT

On 11 June 2008, Japan and the United States submitted to the NAMA Negotiating Group a communication on enhanced transparency in export licensing. The communication aimed to introduce a textual proposal for a new Protocol on Transparency in Export Licensing to the GATT 1994.[44] The proposal then gained the support of other WTO Members such as the Separate Customs Territory of Taiwan, Penghu, Kinmen and Matsu,[45] the Republic of Korea,[46] Ukraine,[47] Chile,[48] and Costa Rica.[49]

In the view of the sponsor Members, the Protocol would serve 'to fill up a gap by providing a set of procedural rules for Members to follow in order to achieve greater transparency in running their respective export licensing regimes in conformity with WTO obligations'.[50] In this regard, the Protocol does not diminish the right of WTO Members to adopt export licensing schemes in accordance with relevant GATT provisions, but merely intends to instruct WTO Members on how to meet the transparency requirements set out in the GATT provisions (e.g.

[43] WTO, 'Non-Agricultural Market Access Negotiations News Archive', www.wto.org/english/news_e/archive_e/nama_arc_e.htm (accessed January 2015).

[44] WTO Doc. TN/MA/W/15/Add.4/Rev.1. Japan had first launched the idea of a new Protocol to the WTO Agreement on the subject in 2006. In the initial proposal of Japan, the text of the Agreement was almost integrally based on the WTO Agreement on Import Licensing Procedures, as a way of bridging the gap between transparency disciplines on import licensing and analogous ones on the export side. See WTO Doc. TN/MA/W/15/Add.4, 18 April 2006. Interestingly, Japan noted: 'The issue was initially identified with respect to export restrictions on mineral resources, but it actually had a horizontal expanse since similar experiences of difficulty in importation in a stable manner due to export restrictions have been pointed out by other Members, in the field such as wood and wood products. It later turned out that it was a systemic issue since the WTO agreement appears to lack sufficient rules for transparency on export restrictions'. Ibid., para. 1.

[45] WTO Doc. TN/MA/W/15/Add.4/Rev.2, 7 July 2008.

[46] WTO Doc. TN/MA/W/15/Add.4/Rev.3, 18 March 2009.

[47] WTO Doc. TN/MA/W/15/Add.4/Rev.5, 7 September 2009.

[48] WTO Doc. TN/MA/W/15/Add.4/Rev.6, 23 September 2010.

[49] WTO Doc. TN/MA/W/15/Add.4/Rev.7, 27 September 2010.

[50] WTO Doc. TN/MA/W/130, 11 December 2009, p. 2.

Articles X:2, XIII:1 and XIII:5 GATT)[51] in order to 'secure effectiveness' of such provisions and 'give substance to the ideal of transparency' established therein.[52] This goal is reflected in the language of the Preamble to the proposed Agreement, which – similar to the Preamble to the WTO Agreement on Import Licensing Procedures – affirms, among others, the desire to 'ensure that export licensing procedures are not utilized in a manner contrary to the principles and obligations of the GATT 1994' and to 'bring transparency to the procedures and practices related to export licensing so as to inform traders and Members and facilitate trade in these products'.[53]

The text of the proposed Protocol is composed of five articles. Article 1 proposes a definition of export licensing, inspired by the WTO Agreement on Import Licensing Procedures. Under this definition, export licensing refers to:

> any administrative procedures involving the submission of an application or other documentation (i.e. other than that required for customs purposes) to the relevant administrative bodies as a prior condition for exportation from the customs territory of the exporting Member.

In contrast to the Agreement on Import Licensing, the proposed Protocol on Transparency in Export Licensing does not distinguish between automatic and non-automatic licensing. In this regard, it is neutral regarding the permitted and not permitted forms of export licensing.[54]

Article 2 poses a general obligation of notification whereby WTO Members are required, within 60 days after the entry into force of the Protocol, to notify in writing to the Committee on Market Access all existing measures and, subsequently, any new measures or modification of existing measures on export licensing within 60 days after the entry into force of such measures.[55] In both cases, the Member is required to contextually provide a copy of the relevant measure.[56] The notification shall indicate, similar to what is required by the relevant provision in the WTO Agreement on Import Licensing, all relevant information for importing WTO Members, including the list of products subject to the measure, the description of the procedures and eligibility criteria, the

[51] See above, Chapter 5, Sections 5.2.2 and 5.4. [52] WTO Doc. TN/MA/W/130, p. 2.

[53] Two more recitals were eventually removed from the text of the Preamble. One affirmed that WTO Members were 'convinced that access to information on export licensing measures benefits traders in both developed and developing country Members'; the other recognized that 'the obligations set out in this Protocol are without prejudice to a Member's rights and obligations under Article XX of the GATT 1994'. Ibid., p. 1.

[54] Ibid., p. 2. [55] Ibid., Article 2.1. [56] Ibid., Article 2.3.

expected duration of the measure, the indication of whether it is implemented in connection with other export restrictions (e.g. a quota), its rationale, and the contact details of the pertinent administrative bodies.[57] This 'standard' requirement is reinforced by the inclusion of a counter-notification requirement along the lines of that contemplated in the proposed WTO Agreement on Export Taxes.[58]

Under Article 3 of the proposed Protocol, moreover, each WTO Member is required to provide all relevant information concerning the administration of its export licensing scheme, the export licences granted and any restrictions implemented in conjunction with the export licensing measure to any other WTO Members who may request it; where available, moreover, WTO Members are also to provide information on the distribution of the export licences among importing countries and statistics concerning the expected and/or actual exports of the product subject to the export licensing scheme. Article 4, however, clarifies that Members should not be required as per the Protocol to disclose 'confidential information' when such disclosure would 'impede law enforcement or otherwise be contrary to the public interest or would prejudice the legitimate commercial interests of particular enterprises, public or private'. Finally, Article 5 provides that the Committee on Market Access should establish a registry of notifications submitted in pursuance of the Protocol, to be made available to the public, and periodically review (at least once every two years) the implementation and operation of the Protocol having due regard to its objectives as well as the rights and obligations established therein.

The proposed Protocol on Transparency in Export Licensing was inserted into the fourth draft modalities for non-agricultural market access.[59] Similarly to the EU Revised Submission on Export Taxes, it also appears among the NTB Textual Proposals annexed to the latest Textual Report of the NGMA Chairman on the State of Play of the NAMA Negotiations.[60] Although the matter has not advanced within

[57] Ibid., Article 2.2.

[58] Article 2.4 of the proposed Protocol reads: 'Any interested Member which considers that another Member has not notified a new or existing measure on export licensing or modification thereto in accordance with the provisions of paragraph 1 and 2 may bring the matter to the attention of such other Member. If notification is not made promptly thereafter, the interested Member may itself notify the measure on export licensing or changes therein, including all relevant information'.

[59] WTO Doc. TN/MA/W/103/Rev. 3, pp. 75–7.

[60] WTO Doc. TN/MA/W/103/Rev. 3/Add.1, pp. 67–9.

the NAMA in the past few years, significant progress in the direction of enhanced transparency in export licensing, as well as other types of export restrictions (including export taxes) will nevertheless be achieved after the entry into force of the WTO Agreement on Trade Facilitation.[61]

8.5 Conclusions

The analysis of the proposals on export restrictions tabled to date in the course of the Doha negotiations has shown that the discussions within the NGMA have focused on three main issues: (i) the question of transparency of export restrictions; (ii) the question of establishing a standard overarching principle which shall govern the use of export taxes on a system-wide level as a way to ensure predictability; (iii) the question of establishing the purposes for and the conditions under which export restrictions might be used as a way to preserve flexibility. The Doha proposals have thus identified and addressed the main shortcomings of existing WTO provisions on export restrictions, namely the lack of sufficiently stringent and enforceable notification requirements and the lack of a system-wide discipline on export taxes.[62]

Regarding transparency, the revised EU submission on export taxes advocates for stronger notification requirements, along the lines of the Understanding on the Interpretation of Article XVII GATT, in order to ensure that existing transparency obligations are 'made operational and enforced in a satisfactory manner'.[63] In the same spirit, the proposed Protocol on Transparency in Export Licensing sets out procedural rules aimed at ensuring the effectiveness of and giving substance to GATT rules on transparency in running export licensing schemes.[64] The two initiatives address the problem of transparency in the use of the most frequently reported types of export restrictions on industrial raw materials and are thus laudable.[65]

[61] See above, Chapter 4, Section 4.5 and Chapter 5, Section 5.4.

[62] See above, Chapter 6, Section 6.2.3.

[63] WTO Doc. TN/MA/W/101, para 7.

[64] WTO Doc. TN/MA/W/130, p. 2.

[65] See above, Chapter 2, Section 2.4.3 and Section 2.5.3.

As to the reform of WTO disciplines on export taxes, the EU proposals attempted to find an approach that could ensure an optimal trade-off between increased predictability in the use of such instruments and political feasibility. In the first proposal, the European Union advocated a 'radical' solution consisting of the adoption of a new WTO Agreement on Export taxes that prohibited export taxes as a general principle, leaving limited flexibilities for developing and least developed countries. The proposed Agreement confirmed that all WTO Members could still resort to export taxes under specific circumstances in accordance with a number of GATT exceptions, including some relevant exceptions originally intended to be import specific in the GATT. However, the elimination of export taxes altogether as a general rule was certain to generate strong resistance among LDCs and developing countries, and especially among those large suppliers responsible for the vertical proliferation of export taxes which were manifestly the main target of the proposal.

The need to garner the support of developing countries required a shift to less ambitious solutions. As a result, the revised submission on export taxes of the European Union proposes the system-wide scheduling and binding of export taxes at a level to be negotiated, as already contemplated under Article XXVIII(*bis*) GATT. Similar to the first submission, specific flexibilities are envisaged for LDCs and small and vulnerable economies. This proposal goes in the direction of greater comparability between WTO rules on export taxes and import tariffs and, by envisaging the negotiation of export duty concessions in WTO Members' GATT schedules, it *de facto* 'corrects' the inconsistencies linked to the fragmentation of current WTO rules on export duties. Nevertheless, it seems to go a step back with respect to the question of flexibility, as it avoids confirming, in contrast to the first proposal, the availability of relevant import-specific exceptions provided for in the GATT.

While the described proposals on export restrictions are likely to remain only on paper, given the lack of consensus on whether the Doha mandate on NAMA even includes export restrictions, the scarce progress in the work of the NAMA Negotiating Group and the more general downgrading of the Doha Round work programme, they remain an important point of reference for future attempts to reform WTO rules on export restrictions. They identify the elements of weakness in the existing WTO disciplines to be addressed most urgently, and shed light on the orientation of both import-dependent and net exporting developing country Members as to 'how far' new (or strengthened) rules could go to improve transparency and predictability. Thus, the analysis of the Doha proposals, particularly those on export taxes, offers useful insights into designing new proposals to achieve balanced and equitable rules on export restrictions.

9

The way forward

9.1 Introduction

This study has shown that the existence of clashing paradigms on the role and function that export restrictions should play within the World Trade Organization (WTO) legal system has so far compromised efforts towards a reassessment of the relevant WTO disciplines. Resistance to change has been particularly strong among developing countries, determined to defend the right to use export restrictions as developmental tools. In light of the current configuration of WTO rules, those developing country Members bound exclusively by the standard provisions of the General Agreement on Tariffs and Trade (GATT) have consistently rejected any solutions capable of restricting the policy space available under the Agreement to resort to export duties – as the only type of export restrictive measure still lawful, as per Article XI:1 GATT, regardless of its rationale. At the same time, newly acceded Members bound by WTO-plus obligations on export duties have advocated a more flexible approach in the interpretation of the additional commitments assumed under individual provisions of their accession protocols before the WTO dispute settlement bodies. The direction pointed to by the Appellate Body in *China–Raw Materials* and *China–Rare Earths* has, however, reinforced the belief in those countries, China *in primis*, that they have already given up too much. The result has been a somewhat ideological opposition towards any proposals aimed at addressing, and limiting, the use of export taxes for industrial or trade purposes.

Yet, the analysis of existing WTO provisions on export restrictions, as arising out of general GATT disciplines and country-specific export duty commitments contained in accession protocol provisions, has shown that WTO Members would gain from a system-wide reform in the area, which could enhance transparency and predictability but still correct the most evident inconsistencies emanating from the current fragmentation, that is, re-establish a 'level playing field' among the current four 'tiers' of WTO

Members. This is true for import-dependent WTO Members contesting export restrictions as beggar-thy-neighbour instruments, but even more so for newly acceded WTO Members already constrained in their ability to resort to export taxes by virtue of more or less *in*flexible WTO-plus obligations. Such reform would, moreover, serve the interests of many developing country Members whose aspirations to economic diversification may be hampered by abuses of such instruments by other countries.[1]

This chapter discusses possible avenues for designing more coherent and balanced disciplines on export restrictions. These proposals address the specific shortcomings identified in Part Two of this study, taking into account the lessons of the Tokyo and Doha Rounds. Because of the gaps and inconsistencies arising out of the current WTO disciplines on export duties, the options discussed focus in particular on the area of export taxes. Section 9.2 attempts to identify the 'building blocks' of a reform which could represent a workable compromise between the legitimate concerns of import-dependent WTO Members and resource-endowed developing countries' claims over policy space. It builds on the conclusions drawn from the analysis of existing WTO provisions and starts from the premise that, in light of the current configuration of relevant WTO rules, greater levels of commitment can only be achieved by ensuring greater levels of flexibility. Accordingly, it identifies a number of specific solutions that could be incorporated, either selectively or comprehensively, in a possible reform of relevant WTO rules. Section 9.3 discusses a number of potential options for implementing the proposed solutions in the WTO legal system. Section 9.4 provides some final considerations on the opportunity for renewing engagement in reforming WTO rules on export restrictions.

9.2 Building blocks

9.2.1 Enhanced transparency in the use of export restrictions

Lack of transparency is one of the main shortcomings of the WTO legal regime on export restrictions and is responsible for the uncertainty

[1] WTO Doc. TN/MA/W/101, 17 January 2008, para. 1. As noted by the European Commission, although many developing countries may depend on exports of one or more commodities, there is no country endowed with all the resources necessary for its economic development. European Commission, 'Eleventh Report on Potentially Trade-Restrictive Measures Identified in the Context of the Financial and Economic Crisis', 2014, http://horizon.hozint.com/2014/11/dg-trade-11th-report-on-potentially-trade-restrictive-measures/ (accessed December 2014), p. 38.

produced by such measures in international commodity markets.[2] As explained above, this uncertainty is particularly detrimental in the minerals and metals sector due to the high capital costs and long-term payback perspectives characterizing investment in the extractive industry.[3] Not surprisingly, transparency has featured prominently in the discourse on export restrictions on industrial raw materials in the WTO, figuring as a core objective in all of the relevant Doha proposals.[4] Such proposals have focused on strengthening notification requirements as a way of enhancing transparency in the implementation of export taxes and export licensing schemes – the two most frequently used types of export restrictions.

Action on transparency would therefore appear to be essential for any WTO reform in the area of export restrictions, whether or not confined to the regulation of export taxes. Detailed and enforceable notification requirements remain in fact the minimum common denominator to give substance to WTO Members' commitments on the export side. In the specific area of export taxes, moreover, strengthened rules remains necessary both in scenarios where Members reach agreement on a system-wide regime to govern the use of such measures and where the *status quo* is maintained. In the latter scenario, greater transparency would be needed to ensure that Members are fully informed of measures taken by any other Member that may influence trade.[5] In the former case, it would constitute a necessary precondition for ensuring that the new regime is made operational and enforced in a satisfactory manner.[6]

To date, 'export duties' are subject to the general notification regime established under the Uruguay Round Ministerial Decision on Notification Procedures, whereas 'export restrictions' fall within the scope of the Decision on Notification Procedures for Quantitative Restrictions adopted by the Council for Trade in Goods in 1995 and recently revised in 2012.[7] Implementation of the former Decision has so far been quite modest.[8] Significant improvements have however been introduced through the 2012 revision of the Decision of the Council for Trade in Goods applicable to export quantitative restrictions (QRs),

[2] See above, Chapter 2, Section 2.3. [3] See above, Chapter 3, Section 3.2.3.

[4] See above, Chapter 8, Sections 8.3 and 8.4. [5] WTO Doc. TN/MA/W/101, para. 8.

[6] Currently, notification requirements on export taxes are generally weaker than those imposed on export quantitative restrictions. See above, Chapter 5, Section 5.4.

[7] See above, Chapter 4, Section 4.5 and Chapter 5, Section 5.4.

[8] WTO Doc. TN/MA/W/101, para. 7.

although the Decision does not specify the exact types of export restrictions subject to the new regime.

Against this backdrop, action on transparency could aim at putting in place a centralized regime for the notification of export restrictions, ideally encompassing both export taxes and any type of QRs having a limiting effect on exports (i.e. including export licensing schemes, minimum export prices, etc.). This regime could build on the model established by the 2012 revision of the Decision on Notification Procedures for Quantitative Restrictions, and introduce stricter time limits (currently six months after the entry into force of a measure) and content requirements for WTO Members, drawing from the existing Doha proposals. For instance, Article 7 of the proposed WTO Agreement on Export Taxes requires that Members provide notification of any new export taxes or any variations thereof within sixty days before the effective date of entry into force of such measures, indicating all relevant information for importing WTO Members.[9] In the revised submission on export taxes, moreover, the European Union points to the Understanding on the Interpretation of Article XVII GATT, according to which each WTO Member needs to ensure 'the maximum transparency possible in its notifications so as to permit a clear appreciation of . . . the effect . . . on international trade' of the notified measures.[10] Any WTO Member can also make a counter-notification to the Council for Trade in Goods, and both notifications and counter-notifications have to be regularly reviewed on behalf of the Council for Trade in Goods by a working party that can make recommendations on the adequacy of the notifications and the need for further information.[11]

WTO Members could thus be required to make complete notifications of any export restrictions in force every year, and to notify any new measures, or any modifications thereof, within a reasonable period of time (e.g. 60 days) before the effective date of entry into force. Counter-notification could also be allowed. The Council for Trade in Goods or a working party as envisaged in the Understanding on Article XVII GATT could then regularly review the notifications and counter-notifications on behalf of the Council for Trade in Goods and make recommendations

[9] In the proposed Protocol on Enhanced Transparency on Export Licensing, Article 2 envisages a time limit of 60 days after the entry into force of export licensing schemes. See above, Chapter 8, Section 8.4.

[10] See Uruguay Round Understanding on the Interpretation of Article XVII of GATT 1994, para. 2.

[11] See above, Chapter 8, Section 8.3.2.

on the adequacy of the notifications and the need for further information, including on the effect of the notified measures on trade. Alternatively, a specific Committee on Export Restrictions could also be established to this end.[12]

The notification regime could also be strengthened through the introduction of consultation requirements, also to be administered by the Committee. An interesting model in this respect could be the Committee on Balance of Payments (BoP) Restrictions.[13] Finally, all information provided by Members could be then made available to the public. A report could also be circulated every year based on such information.[14]

9.2.2 *Negotiation of export duty concessions in WTO Members' GATT schedules*

Another major shortcoming of the WTO legal regime applicable to export restrictions is the lack of a system-wide discipline on export taxes.[15] Any attempts at reform should thus address the inconsistencies arising from the current fragmentation of relevant WTO rules, particularly the WTO accession regime on export duties, and establish a general overarching principle to govern the use of such measures as a means of re-establishing a 'level playing field' among WTO Members.

Any such principle would need to ensure a reasonable balance between the need to prevent any abuses in the use of export taxes and that of ensuring a sufficient level of flexibility capable of keeping all the tiers of Members interested in endorsing a proposal that raises the level of commitments in this area, albeit for different reasons. In this regard, the history of the Tokyo and Doha Rounds shows that the vast majority of WTO Members are not ready to accept a blanket prohibition on the use of export taxes, even if mitigated by extended flexibilities.[16] However, they

[12] Pursuant to Article IV:7 of the Marrakesh Agreement, additional committees may be established by the Ministerial Conference 'with such function as it may deem appropriate'.

[13] For a full account see P. C. Mavroidis, *Trade in Goods: The GATT and the Other Agreements Regulating Trade in Goods*, 2nd edn (Cambridge University Press, 2012), pp. 76–80.

[14] WTO Doc. G/L/59/Rev.1, 22 June 2012, paras. 4 and 7.

[15] See above, Chapter 4.

[16] See above, Chapter 7, Section 7.2 and Chapter 8, Section 8.3.1. The approach based on the complete elimination of export duties has been successfully promoted by the European Union, along with a number of other net importing economies, in bilateral and regional trade agreements. See J. Korinek and J. Bartos, 'Multilateralising Regionalism: Disciplines on Export Restrictions in Regional Trade Agreements', in OECD, *Export Restrictions in*

might agree to negotiate export duty commitments on a product-by-product basis in the form of concessions to be listed in their GATT schedules, provided adequate flexibilities can be agreed upon.[17]

Such a 'negotiated solution'[18] is not an original one. As explained above, the GATT already contains a set of provisions that allow WTO Members to negotiate export duty concessions in the same way as they do for import duty concessions. Such provisions are Article XXVIII(*bis*) and Article II:1(a) GATT. The former encourages WTO Members to engage in such negotiations, while Article II:1(a) guarantees the legal enforceability of export duty concessions *once included in the GATT schedules* of WTO Members.[19] The Russian Federation's choice to create a new Part in its Schedule of Concessions specifically devoted to 'Export Duties' (Part V) *de facto* represents the first example of systematic incorporation of export duty concessions into a Member's GATT schedule in accordance with what is explicitly admitted by virtue of GATT provisions.[20]

WTO Members have so far lacked the incentive to schedule and bind export duties.[21] The proliferation of export taxes, on the one hand, and the consolidation of the Appellate Body approach to the relationship between export duty commitments assumed under accession protocol provisions and the GATT 1994, on the other hand, may however induce WTO Members to reconsider the opportunity for embracing a 'negotiated solution'. This approach would, in fact, bring along positive outcomes for different categories of WTO Members and the multilateral trading system in general. Firstly, a negotiated solution would be in keeping with the spirit of the GATT, which fundamentally opts for 'tariffs' as 'the preferred and acceptable form of protection'.[22] Secondly, it would

Raw Materials Trade: Facts, Fallacies and Better Practices (OECD Publishing, 2014), pp. 149–182. Yet, the countries more actively making use of export restrictions (e.g. China, the Russian Federation, etc.), at least in the minerals and metals sector, have not yet concluded free trade agreements with these players. Pursuing a multilateral solution alongside bilateral strategies remains therefore essential.

17 Ibid.

18 The expression 'negotiated solution' borrows from WTO Doc. TN/MA/W/101, para. 8.

19 See Chapter 4, Section 4.3.1.

20 The Russian Federation scheduled and bound more than 700 tariff lines. As noted above, Australia has also committed not to apply export duties on eleven products of mineral origin. However, Australia did not create a new Part in its GATT Schedule to incorporate such commitments but inserted an ad hoc note referring to the specific tariff lines concerned in Section 2 of Part I of its Schedule on 'MFN import tariff commitments on non-agricultural products'. See above, Chapter 4, Section 4.3.1.

21 See Chapter 4, Section 4.3 and Chapter 5, Section 5.5.

22 Panel Report, *Turkey–Textiles*, para. 9.63.

represent a step forward in the WTO regulation of export duties inasmuch as, as also admitted by the European Union in its revised submission on export taxes, the 'scheduling and binding of Members' export taxes could offer an appropriate route of ensuring adequate predictability'.[23] Thirdly, by bringing all export duty commitments assumed by WTO Members into the GATT framework as per Article II:7 GATT,[24] a negotiated solution would harmonize the procedure used to assume obligations in the area of export duties by all WTO Members and thus make available relevant GATT adjustment procedures and exceptions across the board. The flexibility of export duty concessions would thus be assured by virtue of Article XXVIII GATT.[25] The applicability of Article XX exceptions would also follow automatically. Finally, such a solution would allow for specific flexibilities being envisaged for least developed countries (LDCs) and small and vulnerable developing countries, pursuant to the 1979 Decision on Differential and More Favourable Treatment Reciprocity and Fuller Participation of Developing Countries (the Enabling Clause).[26] Various approaches could be envisaged: one possibility would be to let these Members maintain scheduled export taxes unbound for a number of tariff lines to be negotiated based on their specific development needs, as suggested by the European Union in its revised submission on export taxes.[27] Another option could consist of according special and differential treatment depending on whether the Member in question is a significant exporter (say, for instance, supplying more than 5 per cent of world exports) of a specific product.

A negotiated solution would therefore appear to be the most adequate solution to foster commitment in the area of export duties on the part of

[23] WTO Doc. TN/MA/W/101, para. 8. The European Union is one of the WTO Members that has most strenuously contested export restrictions, both at a judiciary level – by challenging China's export regime before the WTO adjudicators – and at a diplomatic level – by proactively looking for negotiable solutions within the Non-agricultural Market Access (NAMA) Negotiating Group.

[24] See above, Chapter 4, Section 4.3.1. [25] Ibid., Section 4.3.2.

[26] See GATT Doc. L/4901, BISD 26S/203, 28 November 1979. Paragraph 5 of the Decision states: 'The developed countries do not expect reciprocity for commitments made by them in trade negotiations to reduce or remove tariffs and other barriers to the trade of developing countries, i.e., the developed countries do not expect the developing countries, in the course of trade negotiations, to make contributions which are inconsistent with their individual development, financial and trade needs. Developed contracting parties shall therefore not seek, neither shall less-developed contracting parties be required to make, concessions that are inconsistent with the latter's development, financial and trade needs'.

[27] See above, Chapter 8, Section 8.3.2.

WTO Members. Such an approach would raise the level of ambition of any attempts at reform, rather than frustrating it, as countries would be more prone to accept lower levels of export tariff bindings as long as they were left with the right to renegotiate or derogate from such commitments under legitimate circumstances. The modification or withdrawal of export duty concessions, at the same time, would still have to occur in accordance with the conditions set out under Article XXVIII, which guarantee predictability and ensure that the overall level of concessions remains balanced.[28] Likewise, detailed conditions are also set forth in GATT exceptions to prevent any abuses on the part of WTO Members.

As such, scheduling and binding export duties on a product-by-product basis may represent a win-win solution for all tiers of WTO Members. The first tier of Members would see its margin of manoeuvre restricted, but would profit from increased predictability in the use of export duties on the part of trade partners. This outcome is clearly a priority for those import-dependent industrialized countries that over the years have advocated a stronger approach towards beggar-thy-neighbour and subsidy-equivalent export duties. Large emerging economies belonging to the first tier (e.g. India) would also gain from taking part in such negotiations inasmuch as their participation would stimulate other Members to join, as in a virtuous cycle. Apart from the immediate interests in defending the legitimacy of export duties, such countries need to take into account their long-term interests as countries whose intensive industrialization requires, and will increasingly be dependent upon, imports of a wide range of raw materials. Finally, LDCs and small developing countries would have no practical reasons to object to such an approach, as long as they were provided with adequate flexibilities while still enjoying greater levels of predictability resulting from other WTO Members' commitments.

Depending on the coverage of the negotiations, the second tier of WTO Members could also in principle experience erosion of their margin for manoeuvre if asked to assume commitments in excess of those already inscribed in their GATT schedules. Yet, these Members have already scheduled and bound the export duty rates applied on those products of interest to trade partners.[29]

[28] The underlying principle of Article XXVIII GATT is that the renegotiation of concessions shall be accompanied by compensatory adjustments. Failing those, affected Members maintain the right to withdraw substantially equivalent concessions in order to 'maintain a general level of reciprocal and mutually advantageous concessions not less favourable to trade than before'.

[29] See above, Chapter 4, Section 4.3.1 and Section 4.1.

The third and fourth tiers of WTO Members would also gain from supporting a negotiated solution. In addition to the advantages brought about by increased overall predictability, this solution would in fact allow them to incorporate their WTO-plus obligations on export duties into their GATT schedules, thus 'correcting' the systemic imbalances produced by the current WTO accession regime on export duties. The integration of existing WTO-plus obligations on export duties in an appropriate part of newly acceded Members' schedules would not diminish their scope and coverage. Rather, depending on the level of ambition of the negotiations, these Members may be asked to schedule and bind export duties on a greater number of tariff lines and/or to reduce the maximum export duty rate applicable to those tariff lines that were already subject to export duty bindings in their accession protocols.[30] Yet, these Members would regain the right to modify and/or withdraw all their concessions in accordance with Article XXVIII GATT. Moreover, the fourth tier of WTO Members would also regain access to Article XX GATT general exceptions.

According to Qin, the incorporation of WTO-plus obligations on export duties into newly acceded Members' GATT schedules would not require amending the respective accession protocol provisions. In her view, 'adding Part V to a GATT schedule is a matter of amending the GATT schedule'.[31] Accordingly, inasmuch as the export duty commitments assumed under individual accession protocol provisions would constitute trade-liberalizing concessions, if compared with a situation of previously unbound duty rates, their inclusion in GATT schedules could be accommodated through recourse to the internal GATT procedures used to record unilateral, bilateral and multilateral trade-liberalizing concessions.[32] Yet, given that such incorporation would practically alter the terms of accession of the third and fourth tiers of WTO

[30] This would likely be the case for those newly acceded WTO Members that have agreed on narrowly tailored commitments on export duties (e.g. Saudi Arabia, Vietnam and Ukraine). No significant expansion of coverage could arise in the case of countries such as China, Montenegro or Tajikistan, which have already committed to a general elimination obligation with very limited exceptions. See above, Chapter 4, Section 4.4.1. In the latter case, Qin suggests that Part V of such Members' schedules would not need to reproduce the whole list of products subject to a 0 per cent binding, but could contain a textual reference to relevant accession protocol provisions. Qin, 'Reforming WTO Discipline', p. 1182.

[31] Ibid., p. 1183.

[32] Ibid., p. 1183, citing the Decision on Procedures for Modification and Rectification of Schedules of Tariff Concessions, 26 March 1980, L/4962, BISD 27S/25. According to this Decision, the procedure set out in Article XXX:1 GATT, which requires the acceptance of all WTO Members, does not apply to reductions in the export duty rate on a product below the maximum bound rate indicated for such product in a schedule.

Members, Qin has suggested that the General Council could adopt 'a decision to approve the incorporation . . . following the same procedures for approving the terms of accession ... under the Decision-Making Procedures Under Article IX and XII of the WTO Agreement'.[33]

In conclusion, the negotiation of export duty concessions among WTO Members on a system-wide basis represents an opportunity for all WTO Members to enhance predictability in the use of export taxes while not necessarily depriving developing countries of the policy space to resort to such measures under legitimate circumstances. At the same time, it may be advantageous for newly acceded WTO Members inasmuch as it would allow the most evident inconsistencies arising out of the fragmentation of the WTO accession regime on export duties to be corrected.[34] Accordingly, to the extent that a negotiated solution would not espouse any purposive differentiation in the treatment of export duties but rather provide for system-wide rules governing the scheduling and binding of such measures in GATT schedules, such an option could have a chance of garnering broad support among WTO Members.

9.2.3 *Expansion of GATT flexibilities*

Finding agreement among WTO Members as to what principle should govern, on a system-wide basis, the use of export taxes constitutes a priority in order to raise the level of commitments in the area and resolve the inconsistencies resulting from the 'unlevel playing field' created by the WTO accession regime on export duties. In this regard, a negotiated solution building on GATT provisions and re-establishing a broader comparability in the treatment of import and export tariffs appears to be the best way to gaining the consent of different categories of WTO Members.

As explained above, prospects for reform could increase if additional flexibilities were incorporated into a proposal which adequately acknowledged the legitimate developmental function of export duties under specific circumstances. In this respect, it is noteworthy that the first EU submission on export taxes circulated within the Non-agricultural Market Access

[33] Ibid., p. 1184, citing WTO Doc. WT/L/93, 24 November 1995. Qin has also posited that the General Council could contextually confirm the applicability of Article XXVIII GATT to Part V of GATT schedules. Ibid., p. 1184.

[34] According to Qin, even just the incorporation of WTO-plus obligations on export duties in the respective GATT schedules of newly acceded Members would reconcile the WTO discipline on export duties with the international law principle of sovereignty over natural resources. Ibid., p. 1186. See above, Chapter 7, Section 7.3.1.1.

(NAMA) Negotiating Group in the course of the Doha negotiations not only clarified the applicability of GATT adjustment procedures relevant to tariff concessions and GATT exceptions designed to apply to imports and exports alike, but also confirmed that 'WTO Members, and in particular developing countries, may resort to export taxes by reference to GATT Article XVIII on governmental assistance to economic development and GATT Article XII on safeguarding the balance of payment'.[35] In other words, the proposal recognized the legitimacy of export duties when addressing balance of payments disequilibria and infant industry protection purposes, that is, situations that the GATT already acknowledges on the import side.[36]

The extension of relevant import-specific exceptions available in the GATT was contemplated as part of a more comprehensive agreement on export taxes whose underlying principle was the general elimination of such measures.[37] In this regard, it might have been perceived by the European Union as a necessary compromise in order to garner support for such an ambitious approach among developing country Members and, indeed, it was not reiterated in the revised submission on export taxes advocating for a negotiated solution. Yet, as shown above, the unavailability of economic development exceptions may have contributed to the lack of engagement in the negotiation of export duty concessions on the part of WTO Members as admitted in the GATT.

In light of the conflicting interests characterizing the debate over export restrictions in the multilateral trading system,[38] confirming the availability of relevant GATT economic exceptions may further contribute to a climate of trust on the part of developing countries and raise the level of ambition in the negotiation of export duty concessions. In principle, depending on the scope of the reform, the applicability of relevant import-specific GATT exceptions may not necessarily be confined to violations of WTO rules on export taxes, but could be extended to the use of Article XI:1 GATT-inconsistent export restrictions as already envisaged in the GATT on the import side.[39] Yet, conditioning the availability of Article XII and Article XVIII GATT upon the negotiation of export duty concessions would incentivise WTO Members to privilege export taxes over more trade-distortive types of export restrictions in keeping with the spirit of the multilateral trading system.[40]

35 WTO Doc. TN/MA/W/11/Add. 6, pp. 1–2.

36 See above, Chapter 4, Section 4.3.2.

37 See above, Chapter 8, Section 8.3.1.

38 See above, Chapter 7, Sections 7.3 and 7.4.

39 See above, Chapter 5, Section 5.3.3.

40 Article XII and Article XVIII GATT already set out specific conditions and procedures aimed at ensuring that WTO Members preferably use measures consistent with GATT

Finally, it is noteworthy that both Article XII and Article XVIII GATT laid down specific conditions, rules and procedures, which aim to ensure that recourse to import tariffs in excess of those set forth in a Member's GATT schedule (or import restrictions in derogation of Article XI:1 GATT), occurs only after the WTO Member concerned has entered into negotiations or notified and consulted with affected WTO Members. In the case of BoP provisions, in particular, the Committee on Balance of Payment Restrictions specifically presides over the administration of Article XII and Article XVIII(b) GATT in order to ensure transparency in the use of such exceptions on the part of WTO Members.[41]

9.2.4 *Envisaging an export safeguard mechanism*

In the spirit of encouraging the highest level of ambition in the negotiation of export duty concessions, one may wonder whether the introduction of a safeguard mechanism on the export side could also be envisaged as a confidence-building means. As explained above, export restrictions may in principle be used by WTO Members as safety valves to relieve situations of excessive surge of exports, either caused by world price increases or speculative behaviour in the international markets that could disrupt domestic supply or contribute to price volatility in the domestic market to the detriment of downstream processing industries and, therefore, economic diversification.[42]

The question here is whether and to what extent exporting countries should be entitled to protect domestic downstream sectors from market forces which would prevent them from competing with foreign buyers of raw materials. In this regard, the GATT does contain a safeguard clause under Article XIX GATT but it applies to the import side only.[43] Moreover, this exception only allows WTO Members to use safeguards when a surge of imports causes or threatens serious injury to domestic producers *of like or directly competitive products*. This condition could not apply, *mutatis*

provisions to achieve legitimate economic objectives. Under both exceptions, WTO Members are thus encouraged to use tariff-like measures as the preferred tools to respond to balance of payments and economic development needs. In this respect, it appears unlikely that import-dependent WTO Members support the extension of GATT economic exceptions to export restrictions other than export duties.

41 For a full account see Mavroidis, *Trade in Goods*, pp. 75–80.

42 See above, Chapter 7, Sections 7.3.2. and 7.3.3. For a full account see F. Piérola, *The Challenge of Safeguards in the WTO* (Cambridge University Press, 2014).

43 See above, Chapter 5, Section 5.3.3.

mutandis, to export safeguards, as in such case the domestic producers to be protected would be downstream processors using a particular raw material.

Of better use on the export side could in principle be an export safeguard mechanism modelled on the so-called special agricultural safeguards (SSGs) envisaged under Article 5 of the WTO Agreement on Agriculture (AoA). SSGs consist of import tariff increases based on price and quantity triggers; such increases follow automatically after an increase in imports above a certain level or a fall in domestic prices below a certain level. Reproduced on the export side, this logic could offer WTO Members some leeway for shielding domestic producers from surges of exports or price spikes of critical raw materials caused by external upward shifts in world demand.

Getting agreement on appropriate and equitable thresholds for export safeguards may prove quite controversial. Discussions within the Agriculture Negotiations of the WTO are enlightening in this respect.[44] In theory, developed countries may reasonably be expected to be excluded from export safeguards. However, the question of whether developing countries, including large suppliers of raw materials, should be entitled to resort to such mechanisms under the same price and/or quantity triggers as LDCs and small developing economies may arise. Thresholds could be varied to reflect differences in the development stage of the imposing country or in the size of exports subject to taxes (and thus the expected impact of such measures). For example, they could vary between developing countries excessively exposed to commodity export dependence and unable to influence world prices of the taxed products and emerging economies detaining large shares of world exports of the taxed materials.

Similar to the case of SSGs, export safeguards could be made available only to those products for which export duty rates were scheduled and bound in Members' GATT schedules and could take the form of export tariff increases.[45] Members resorting to export safeguards could also be required to give notice to either the Council for Trade in Goods or a specific Committee, depending on the features of the transparency component of the WTO reform.[46] Members scheduling and binding

[44] WTO, 'Market Access: Special Agricultural Safeguards (SSGs)', www.wto.org/english/tratop_e/agric_e/negs_bkgrnd11_ssg_e.htm (accessed January 2015). See above, Introduction, Section I.2.1.

[45] SSGs can only be employed with respect to those products that were tariffed under Article 4 AoA. For a full account see Mavroidis, *Trade in Goods*, pp. 755–8.

[46] Under Article 5.7 AoA, WTO Members implementing SSGs shall give notice 'in writing, including relevant data' to the Committee on Agriculture 'as far in advance as may be practicable and in any event within 10 days of the implementation of such action'.

export duties into Part V of their GATT schedules could then reserve therein the right to use export safeguards.[47] Accordingly, Members would in principle be further encouraged to engage in the negotiation of export duty concessions on a wider range of products.

9.3 Implementing options

The section above shows that, to the extent that transparency, predictability and flexibility inform efforts towards reforming WTO disciplines on export restrictions, all WTO Members, including traditional objectors, would have to gain from supporting such efforts. Evidently, the current configuration of relevant WTO rules makes the regulation of export taxes a key area to address in view of achieving these core objectives. Nothing, however, precludes enhanced rules, particularly in the area of transparency, also being developed with respect to other types of export restrictions.

Although the proposals circulated in the NAMA Negotiating Group are highly informative in this respect, the lack of consensus on the existence of a mandate on export restrictions under Paragraph 16 of the Doha Declaration has fundamentally compromised progress on the topic since the inception of the negotiations. These discussions would thus have to be resumed, independently of the outcome of the Doha Round, in the context of a new post-Doha agenda. Alternatively, they could be de-linked from the next round of multilateral trade negotiations and form the basis of a 'critical mass' agreement between a subset of WTO Members or a 'plurilateral' agreement, depending on the scope of the reform and the level of its ambition.

9.3.1 A deal on transparency

WTO Members may decide to take action on transparency on a priority basis, assuming that agreement over other regulatory aspects require longer and more complex negotiations. A deal on transparency could either cover all types of export restrictions or be limited to, for instance, export taxes and/or other specific types of export restrictions, depending on the scope and particular features of the WTO reform.

Strengthened notification requirements could be set forth in a decision adopted by the Ministerial Conference. According to Article IV:1 of the Marrakesh Agreement, the Ministerial Conference has in fact 'the

[47] See Article 5.1 AoA.

authority to take decisions on all matters under any of the Multilateral Trade Agreements, if so requested by a Member'.[48] A decision on transparency of export restrictions, including or limited to export duties, would fall within the scope of this provision as coming under the GATT. The decision could set out notification and documentation requirements of the kind discussed above and perhaps also specific consultation procedures to be administered by an ad hoc Committee.

Finally, these rules could also in principle form part of a more comprehensive agreement laying down the general principles governing the use of selected types of export restrictions (i.e. export taxes).

9.3.2 *Launching negotiations on export duty concessions: options and modalities*

Pursuant to Article XXVIII (*bis*), WTO Members have traditionally exchanged tariff concessions in the context of the so-called rounds of multilateral negotiations. Given that Article XXVIII (*bis*) explicitly encourages tariff negotiations on the import and export sides alike, negotiations on export duty concessions could in principle be launched in the context of the next multilateral round without the need for formally amending the GATT or concluding a new WTO agreement, provided that Members are willing to do so. For the sake of clarity, however, an interpretative understanding could be useful to definitively confirm the applicability of the scheduling and binding procedure *à la* Article II:1(b) GATT to export duty concessions in light of Article II:1(a) GATT and Article XXVIII(*bis*) GATT, including the deconsolidation procedure under Article XXVIII GATT.

The multilateral negotiation of export duty concessions, moreover, should follow and/or give effect to the incorporation of WTO-plus commitments on export duties assumed under accession protocol provisions by newly acceded WTO Members.[49] Specific flexibilities for LDCs and small developing countries, along the lines of those envisaged by the European Union in its revised submission on export taxes, would also need to be reflected in the conduct of negotiations.

WTO Members could agree on specific modalities governing the negotiation of export duty concessions to be included in GATT

[48] According to Article IV:2, the General Council may conduct the functions of the Ministerial Conference in the intervals between the biennial meetings of the Ministerial Conference.

[49] See above, Section 9.2.2 for a discussion on how to implement such incorporation.

schedules. Considering that a relatively limited number of WTO Members is currently responsible for a large share of the export taxes in place in the NAMA sector, with some of them already bound by specific obligations in the area, modalities based on linear or tiered cuts would not necessarily provide such Members with the necessary incentive to engage in the negotiations, whereas the 'request-offer' modalities appear in principle more appropriate. [50] Another viable solution could be to opt for a 'sector approach' modality, with the product coverage limited to a specific list of raw materials.[51] In theory, the larger the product coverage, the greater the benefit WTO Members would get from negotiating export tariff bindings multilaterally,[52] although it might be difficult to reach an agreement on which raw materials to include.

Should consensus on the reduction of export taxes through a negotiated solution fail multilaterally, a subset of WTO Members representing a 'critical mass'[53] could in principle still commit to schedule and bind export taxes on an agreed list of (mineral) raw materials similarly to what happened in the case of the Information Technology Agreement (ITA).[54] In such a case, it would suffice for a group of WTO Members to sign a Ministerial Declaration, whose provisions would come into effect upon acceptance by a number of Members collectively accounting for a sufficiently large percentage of world trade in raw materials as agreed upon in the Declaration itself.[55] Following the conclusion and signing of this Declaration, participants would have to modify their GATT schedules by means of creating a

[50] For a full account see Mavroidis, *Trade in Goods*, pp. 99–100.

[51] See WTO Doc. TN/MA/S/13, 24 January 2005, pp. 3-4. See also WTO Doc. TN/MA/W/37/Add.5, 25 July 2008.

[52] This is true for both large suppliers of raw materials and small economies dependent on the exportation of a few commodities. For instance, while China is a top exporter of non-fuel minerals and metals, it relies extensively on the importation of energy resources.

[53] According to Patrick Low, 'a critical mass may be said to exist when a sufficient number of parties that do not represent the entire membership agree upon a common course of cooperative action to be taken under the auspices of the WTO'. P. Low, 'WTO Decision-Making for the Future', WTO Economic Research and Statistics Division, Staff Working Paper ERSD-2011-05, www.wto.org/ENGLISH/res_e/reser_e/ersd201105_e.pdf (accessed January 2015), p. 6. For a comprehensive account of the decision-making challenges currently faced by the WTO see T. Cottier, 'A Two-Tier Approach to WTO Decision Making', NCCR Working Paper No. 2009/06, www.peacepalacelibrary.nl/ebooks/files/NCCR_Cottier_Two-Tier-Approach.pdf (accessed January 2015).

[54] WTO Doc. WT/MIN(96)/16, 13 December 1996. For a full account see Mavroidis, *Trade in Goods*, pp. 101–4.

[55] In the case of the ITA, the original signatories of the Ministerial Declaration had agreed that the participants would have to represent around 90 per cent of world trade in information technology products. Ibid., para. 4.

Part V devoted to 'Export Duties' and incorporating therein all agreed concessions in order to give legal effect to the agreement.[56] The export tariff concessions would then have to be applied on a non-discriminatory basis in accordance with the most favoured nation (MFN) clause.[57]

Finally, to the extent that WTO Members include export duty concessions in their GATT schedules, which are an integral part of the GATT, they would remain free to modify or withdraw such concessions in accordance with the procedures laid out in Article XXVIII GATT and could invoke Article XX general exceptions to justify export tariff increases. Such a negotiated solution, however, would not permit flexibilities to extend beyond those envisaged under the GATT for imports and exports alike.

9.3.3 A new WTO Agreement on Export Taxes

WTO Members could also conclude a new Agreement on Export Taxes. This solution would allow them to go beyond the mere implementation of the existing GATT disciplines. In other words, a new Agreement on Export Taxes could not only lay down a standard procedure *à la* Article II:1(b) GATT in order for WTO Members to assume obligations in the area of export duties, and clarify the applicability of any GATT provisions relevant to export duty concessions. It could also contain specific provisions confirming the availability of GATT import-specific exceptions, such as Article XII and Article XVIII GATT, to such concessions, in addition to setting out the procedures and conditions for resorting to export safeguards.

Such an Agreement could only enter into force upon acceptance by a majority of two-thirds of WTO Members in accordance with Article X:3 of the Marrakesh Agreement. Alternatively, a subset of WTO Members could in principle conclude a plurilateral agreement on export taxes to be added to Annex 4 to the WTO Agreement by the Ministerial Conference

56 The ITA was given legal effect through the certification of modification of schedules on the part of participating Members in accordance with the 1980 Decision on Procedures for Modification and Rectification of Schedules of Tariff Concessions. See Panel Reports, *EC–IT Products*, paras. 7.375–84. Notably, this is the same decision that Qin indicates as the basis for incorporating WTO-plus obligations on export duties into the respective schedules of newly acceded WTO Members. See above, Section 9.2.2. It could also be relied upon to incorporate export duty concessions agreed by means of a Ministerial Declaration.

57 The principle at the basis of critical mass agreements, as recalled by Low, is that '[w]hen the critical mass has defined itself . . . those left outside it are presumably considered too small in the market to undermine the agreement'. P. Low, 'WTO Decision-Making for the Future', p. 9.

'exclusively by consensus'.[58] Such an Agreement would bind only adhering Members, and third parties would remain excluded from any benefits accruing from it.[59]

A new WTO Agreement on Export Taxes remains however the preferable option.[60] For the reasons explained above, an agreement of this kind could in fact represent a workable compromise for different categories of WTO Members, inasmuch as it would correct the inconsistencies inherent to the current WTO regime and ensure that greater predictability and transparency are matched with confidence-building flexibilities currently not envisaged under GATT provisions. Finally, by limiting the availability of such additional flexibilities to products scheduled and bound in Members' GATT schedules, such an agreement could provide Members with the right incentives to privilege export taxes over more trade-restrictive types of export restrictions in keeping with the spirit of the multilateral trading system.

9.4 Conclusions

Access to supplies is one of the fundamental challenges of our era. The scarcity of natural resources, on the one hand, and the epochal geopolitical and geoeconomic transformations affecting traditional international trade patterns, on the other hand, call for a reconsideration of the role that the WTO should play in the regulation of trade barriers on the export side. However, any efforts in this area could not succeed without addressing the growing concerns among developing country Members that further action would irrevocably limit their policy space, ultimately depriving them of the right to use export restrictions to pursue their own economic and sustainable development needs in accordance with the principle of sovereignty over natural resources.

It is argued that a workable compromise in this respect could be accommodated under the auspices of the WTO to the extent that greater levels of transparency and predictability are combined with greater levels of flexibility for Members. Such an approach would build on GATT rules and, ideally, would expand the flexibilities provided therein to encourage participation in a negotiated solution that contemplates the scheduling and binding of export duties as the standard procedure to assume obligations in the area

[58] See Article X:9 of the Marrakesh Agreement. [59] Ibid., p. 7.

[60] As noted by Kennedy, '[t]here is a degree of mistrust among some developing countries regarding the GPA-type of optional agreement'. M. Kennedy, *Legal Options for a Sustainable Energy Trade Agreement* (Geneva: International Centre for Trade and Sustainable Development, 2012).

by all WTO Members. Such a system-wide approach would be in keeping with the spirit (and the provisions) of the GATT while still representing a step forward in the WTO regulation of export taxes. Moreover, it would re-establish coherence in existing disciplines by dissipating the inconsistencies produced by the WTO accession regime on export duties as revealed in *China–Raw Materials* and *China–Rare Earths*, thus re-establishing a 'level playing field' among WTO Members.

In light of the Appellate Body's approach to the relationship between export duty commitments assumed under accession protocol provisions and the GATT 1994, it is argued that developing country Members that have traditionally resisted the calls for reform (China *in primis*) would have a tangible interest in rebuilding momentum in the WTO and proactively engaging in the definition of more coherent and equitable disciplines on export restrictions, either horizontal or sectoral.[61] In the latter case, priority would have to be given to raw materials and, particularly, to critical minerals and metals in light of the peculiar mixture of practical and legal considerations underlying their proliferation.

In conclusion, the issue of export restrictions should be resumed in the post-Doha agenda of the WTO and action should be led by developing country Members. This could mark the start of a new phase in the negotiations, where the bias towards access to supplies would be substituted with a more balanced approach that takes into account the development concerns of resource-endowed and resource-dependent WTO Members. In the meantime, countries negotiating the terms of their accession to the WTO, and assuming export duty commitments in this context, should endeavour to list such commitments in their GATT schedules of concessions by adding therein a Part V specifically devoted to 'Export Duties'. The consolidation of a model based on the Russian precedent, which *de facto* builds on already existing GATT provisions, would reassure resource-rich Members that meaningful WTO obligations on export duties might not necessarily deprive them of the right to use such measures for sustainable economic development purposes. Under these premises, confidence and engagement in the topic of export restrictions in the context of a post-2015 agenda could be restored and the level of ambition of a system-wide reform raised.

[61] In this sense, see also Qin, 'Reforming WTO Discipline', p. 1184.

APPENDIX 1

Export taxes applied to critical minerals and metals (2008–2014)

Minerals and metals	Forms	Country	WTO Status	Applied export duty	Date of introduction and/or modification of a measure	Any relevant developments since 2008
Antimony	Ores and concentrates	China	*Newly acceded WTO Member*	*20%*	1 Jan. 2008[1]	
		Vietnam	*Newly acceded WTO Member*	20%	21 Jun. 2008	Increased in 2008
	Ash and residues (excl. from the manufacture of iron/steel) containing antimony/beryllium/cadmium/chromium/their mixtures	Argentina	Original WTO Member	5%	2002	
		China	*Newly acceded WTO Member*	10%	1 Jan. 2008[1]	
		Zambia	Original WTO Member	10%	1 Jan. 2012–30 Sept. 2014	
	Antimony unwrought	China	*Newly acceded WTO Member*	5%	1 Jan. 2008[1]	
	Antimony powders	China	*Newly acceded WTO Member*	*20%*	1 Jan. 2008[1]	

	Antimony and articles thereof	Vietnam	*Newly acceded WTO Member*	5%	1 Jan. 2009	
	Waste and Scrap	Afghanistan	Candidate WTO Member	5%	2010	
		Argentina	Original WTO Member	5%	2002	40% from 17 Jan. 2008 to 6 Jan. 2010
		China	*Newly acceded WTO Member*	*20%*	1 Jan. 2008[1]	
		Pakistan	Original WTO Member	25%	9 Jun. 2007	
		Russian Federation	*Newly acceded WTO Member*	6.5%	1 Jan. 2007	
		Sri Lanka	Original WTO Member	25%	17 Mar. 2006	
		Vietnam	*Newly acceded WTO Member*	*22%*	1 Jan. 2009	Decreased yearly from 37% in 2009 to 22% in 2012
Beryllium	Ash and residues (excl. from the manufacture of iron/steel) containing antimony/beryllium/cadmium/chromium/their mixtures	Argentina	Original WTO Member	5%	2002	
		China	*Newly acceded WTO Member*	***10%***	1 Jan. 2008[1]	
		Zambia		10%		

(*cont.*)

Minerals and metals	Forms	Country	WTO Status	Applied export duty	Date of introduction and/or modification of a measure	Any relevant developments since 2008
			Original WTO Member		1 Jan. 2012–30 Sept. 2014	
	Waste and scrap	Afghanistan	Candidate WTO Member	5%	2010	
		Argentina	Original WTO Member	5%	2002	40% from 17 Jan. 2008 to 6 Jan. 2010
		Pakistan	Original WTO Member	25%	9 Jun. 2007	
		Russian Federation	*Newly acceded Member*	6.5%	1 Jan. 2007	
		Sri Lanka	Original WTO Member	25%	17 Mar. 2006	
		Vietnam	*Newly acceded WTO Member*	*22%*	1 Jan. 2009	Decreased yearly from 37% in 2009 to 22% in 2012
Borates	Natural sodium borates and concentrates thereof (whether/not calcined)	Argentina	Original WTO Member	10%	2002	
	Natural borates and concentrates thereof (excl. sodium borates), whether/not calcined but not	Argentina	Original WTO Member	5%	2002	10% until 11 Jan. 2012

	incl. borates separated from natural brine; natural boric acid containing not >85% of H3BO3 calc. on the dry weight					
	Boron; tellurium	Argentina	Original WTO Member	5%	2002	
	Oxides of boron; boric acids	Argentina	Original WTO Member	5%	2002	
	Disodium tetraborate (refined borax), other than anhydrous	Argentina	Original WTO Member	5%	2002	
	Borates other than disodium tetraborate (refined borax)	Argentina	Original WTO Member	5%	2002	
	Peroxoborates (perborates)	Argentina	Original WTO Member	5%	2002	
Cobalt	Cobalt ores and concentrates	Argentina	Original WTO Member	10%	2002	
		China	*Newly acceded WTO Member*	15%	1 Jan. 2008[1]	
		Democratic Republic of the Congo	Original WTO Member	10%		
		Vietnam	*Newly acceded WTO Member*	20%	21 Jun. 2008	Increased in 2008
	Cobalt oxides and hydroxides; commercial cobalt oxides	Argentina	Original WTO Member	5%	2002	
		China		***15%***	1 Jan. 2011[1]	

(*cont.*)

Minerals and metals	Forms	Country	WTO Status	Applied export duty	Date of introduction and/or modification of a measure	Any relevant developments since 2008
			Newly acceded WTO Member			
	Cobalt mattes and other intermediate products of cobalt metallurgy; unwrought cobalt; powders	Argentina	Original WTO Member	5%	2002	
		Vietnam	*Newly acceded WTO Member*	5%	1 Jan. 2009	
	Waste and scrap	Argentina	Original WTO Member	5%	2002	40% from 17 Jan. 2008 to 6 Jan. 2010
		Pakistan	Original WTO Member	25%	9 Jun. 2007	
		Russian Federation	*Newly acceded WTO Member*	30% but not less than 1200 EUR/ tonne	23 Dec. 2006	
		Sri Lanka	Original WTO Member	25%	17 Mar. 2006	
		Ukraine	*Newly acceded WTO Member*	***21%***	16 May 2008	Decreased from 27% in 2008 to 21% in 2012

		Vietnam	*Newly acceded WTO Member*	22%	1 Jan. 2009	Decreased yearly from 37% in 2009 to 22% in 2012
	Cobalt and articles thereof	Argentina	Original WTO Member	5%	2002	
		Vietnam	*Newly acceded WTO Member*	5%	1 Jan. 2009	
Chromium	Chromium ores and concentrates	India	Original WTO Member	3,000 INR/ tonne	1 Mar. 2008–16 Mar. 2012	Increasing
				30%	17 Mar. 2012	
		Zimbabwe	Original WTO Member	20%	1 Jan. 2010–1 Apr. 2011	Increased from 15% to 20% on 1 Aug. 2010
	Waste and scrap	Afghanistan	Candidate WTO Member	5%	2010	
		Argentina	Original WTO Member	5%	2002	40% from 17 Jan. 2008 to 6 Jan. 2010
		Pakistan	Original WTO Member	25%	9 Jun. 2007	
		Russian Federation	*Newly acceded Member*	6.5%	1 Jan. 2007	
		Sri Lanka	Original WTO Member	25%	17 Mar. 2006	
		Vietnam	*Newly acceded WTO Member*	22%	1 Jan. 2009	Decreased yearly from 37% in 2009 to 22% in 2012

(cont.)

Minerals and metals	Forms	Country	WTO Status	Applied export duty	Date of introduction and/or modification of a measure	Any relevant developments since 2008
	Ash and residues (excl. from the manufacture of iron/steel) containing antimony/beryllium/cadmium/chromium/their mixtures	Argentina	Original WTO Member	5%	2002	
		China	*Newly acceded WTO Member*	***10%***	1 Jan. 2008[1]	
		Zambia	Original WTO Member	10%	1 Jan. 2012–30 Sept. 2014	
	Ferro-chromium, containing by weight >4% of carbon, in granular/powder form	China	*Newly acceded WTO Member*	*20%*	1 Jan. 2008[1]	
	Ferro-chromium, other than that containing by weight >4% of carbon, in granular/powder form	China	*Newly acceded WTO Member*	*20%*	1 Jan. 2008[1]	
Coke	Coke and semi-coke made from coal whether or not agglomerated	China	*Newly acceded WTO Member*	***40%***	1 Dec. 2008–31 Dec. 2012[1]	Eliminated since 1 Jan. 2013[2]
		Russian Federation	*Newly acceded WTO Member*	*6.5%*	23 Dec. 2006	Eliminated since 12 Jun. 2006

	Coke and semi-coke of coal/lignite/peat, whether/not agglomerated; retort carbon					
Copper	Copper ores and concentrates	Argentina	Original WTO Member	10%	2002	
		China	*Newly acceded WTO Member*	***10%***	1 Jan. 2008[1]	
		Democratic Republic of the Congo	Original WTO Member	10%		
		Indonesia	Original WTO Member	25%	11 Jan. 2014	Further increase (up to 60% by 2016) announced in May 2014
		Vietnam	*Newly acceded WTO Member*	30%	1 Jan. 2012	20% from 21 Jun. 2008 to 31 Dec. 2011
		Zambia	Original WTO Member	15%	28 Mar. 2008–27 Dec. 2008	
				10%	Early 2013	
	Ash and residues (excl. from the manufacture of iron/steel) containing mainly copper	Argentina	Original WTO Member	5%	2002	
		China	*Newly acceded WTO Member*	***10%***	1 Jan. 2008[1]	

(cont.)

Minerals and metals	Forms	Country	WTO Status	Applied export duty	Date of introduction and/or modification of a measure	Any relevant developments since 2008
		Zambia	Original WTO Member	10%	1 Jan. 2012–30 Sept. 2014	Intermittently eliminated and reintroduced
	Copper oxides and hydroxides	Argentina	Original WTO Member	5%	2002	
	Chloride oxides and chloride hydroxides, of copper	Argentina	Original WTO Member	5%	2002	
	Sulphates of copper	Argentina	Original WTO Member	5%	2002	
	Copper mattes; cement copper (precipitated copper)	Argentina	Original WTO Member	5%	2002	
		China	*Newly acceded WTO Member*	***15%***	1 Jan. 2008[1]	
		Malaysia	Original WTO Member	5%	1996	
		Russian Federation	*Newly acceded WTO Member*	*10% (as a percentage of customs value or in euro)*	1 Jan. 2007	
		Vietnam		*20%*	1 Jan. 2011	Increased in 2008

		Newly acceded WTO Member			
	Zambia	Original WTO Member	15%	28 Mar. 2008	
Unrefined copper; copper anodes	Argentina	Original WTO Member	5%	2002	
	China	*Newly acceded WTO Member*	*15%*	1 Jan. 2008[1]	
	Malaysia	Original WTO Member	5%	1996	
	Russian Federation	*Newly acceded WTO Member*	*10% (as a percentage of customs value or in euro)*	1 Jan. 2007	
	Zambia	Original WTO Member	15%	28 Mar. 2008	
Cathodes and sections of cathodes, of refined copper, unwrought	Argentina	Original WTO Member	5%	2002	
	China	*Newly acceded WTO Member*	*10%*	1 Jan. 2008[1]	
	Malaysia	Original WTO Member	5%	1996	
	Russian Federation	*Newly acceded WTO Member*	*10% (as a percentage of customs*	1 Jul. 2007– 4 Jun. 2012; 23 Aug.	Eliminated from 21 Jan. 2010 to 19 Dec. 2010

(*cont.*)

Minerals and metals	Forms	Country	WTO Status	Applied export duty	Date of introduction and/or modification of a measure	Any relevant developments since 2008
				value or in euro)	2012–22 Aug. 2014	
				893 USD/ tonne	***5 Jun. 2012–22 Aug. 2012***	
		Vietnam	*Newly acceded WTO Member*	20%	1 Jan. 2011	Increased from 15% to 20% on 3 Sept. 2011
	Wire-bars of refined copper, unwrought	Argentina	Original WTO Member	5%	2002	
		China	*Newly acceded WTO Member*	*10%*	1 Jan. 2008[1]	
		Malaysia	Original WTO Member	5%	1996	
		Russian Federation	*Newly acceded WTO Member*	*10% (as a percentage of customs value or in euro)*	1 Jan. 2007–4 Jun. 2012; 23 Aug. 2012	
				893 USD/ tonne	***5 Jun. 2012–22 Aug. 2012***	

	Vietnam	*Newly acceded WTO Member*	20%	1 Jan. 2011	Increased from 15% to 20% on 3 Sept. 2011
Unwrought products of refined copper (excl. of 7403.11–7403.13)	Argentina	Original WTO Member	5%	2002	
	China	*Newly acceded WTO Member*	*10%*	1 Jan. 2008[1]	
	Malaysia	Original WTO Member	5%	1996	
	Russian Federation	*Newly acceded WTO Member*	10% *(as a percentage of customs value or in euro)*	1 Jan. 2007–4 Jun. 2012; 23 Aug. 2012	
			893 USD/ tonne	***5 Jun. 2012–22 Aug. 2012***	
	Vietnam	*Newly acceded WTO Member*	20%	1 Jan. 2011	Increased from 15% to 20% on 3 Sept. 2011
Billets of refined copper, unwrought	Argentina	Original WTO Member	5%	2002	
	China	*Newly acceded WTO Member*	*10%*	1 Jan. 2008[1]	
	Malaysia	Original WTO Member	5%	1996	
	Russian Federation	*Newly acceded WTO Member*	*10%(as a percentage of customs*	1 Jan. 2007–4 Jun. 2012;	

(*cont.*)

Minerals and metals	Forms	Country	WTO Status	Applied export duty	Date of introduction and/or modification of a measure	Any relevant developments since 2008
				value or in euro)	23 Aug. 2012	
				893 USD/ tonne	***5 Jun. 2012– 22 Aug. 2012***	
		Vietnam	*Newly acceded WTO Member*	20%	1 Jan. 2011	Increased from 15% to 20% on 3 Sept. 2011
	Copper alloys	Argentina	Original WTO Member	5%	2002	
		China	*Newly acceded WTO Member*	*5%*	1 Jan. 2008[1]	
		Malaysia	Original WTO Member	5%	1996	
		Russian Federation	*Newly acceded WTO Member*	*10%(as a percentage of customs value or in euro)*	1 Jan. 2007	
		Vietnam	*Newly acceded WTO Member*	20%	1 Jan. 2011	Increased from 15% to 20% on 3 Sept. 2011
	Waste and scrap	Argentina		5%	2002	

	Original WTO Member			40% duty applied from 12 Jan. 2011 to 6 Jan. 2012
China	*Newly acceded WTO Member*	*15%*	1 Jan. 2008[1]	
Kazakhstan	Candidate WTO Member	30%, but not less than 330 EUR/ 1000 kg	28 Dec. 2007	
Malaysia	Original WTO Member	10%	1996	
Pakistan	Original WTO Member	25%	9 Jun. 2007	Initially declared temporary (to expire on 30 Jun. 2010)
Russian Federation	*Newly acceded WTO Member*	*50%, but not less than 420 EUR/ 1000 kg*	1 Jan. 2007	
Sri Lanka	Original WTO Member	50% FOB value	17 Mar. 2006	Renovated multiple times after expiration
Ukraine	*Newly acceded WTO Member*	***21%***	16 May 2008	Decreased from 27% in 2008 to 21% in 2012

(*cont.*)

Minerals and metals	Forms	Country	WTO Status	Applied export duty	Date of introduction and/or modification of a measure	Any relevant developments since 2008
		Vietnam	*Newly acceded WTO Member*	*22%*	1 Jan. 2009	Decreased yearly from 37% in 2009 to 22% in 2012
		Zambia	Original WTO Member	25% or 80,000 kwacha/tonne, whichever is greater	28 Mar. 2008	
	Bars, rods and profiles, of refined copper and copper alloys	Argentina	Original WTO Member	5%	2002	
		Vietnam	*Newly acceded WTO Member*	10%	1 Jan. 2011	5% until 31 Dec. 2010
	Master alloys of copper	Argentina	Original WTO Member	10%	2002	
		China	*Newly acceded WTO Member*	***10%***	1 Jan. 2008[1]	
		Malaysia	Original WTO Member	10%	1996	
		Russian Federation	*Newly acceded WTO Member*	10%	1 Jan. 2007	

		Vietnam	*Newly acceded WTO Member*	15%	1 Jan. 2008	
	Powders of non-lamellar structure, of copper	Vietnam	*Newly acceded WTO Member*	15%	1 Jan. 2010	
	Powders and flakes	Argentina	Original WTO Member	5%	2002	
		Vietnam	*Newly acceded WTO Member*	15%	1 Jan. 2010	
	Copper wires and wires of copper alloys	Argentina	Original WTO Member	5%	2002	
	Plates, sheets and strip and copper foil	Argentina	Original WTO Member	5%	2002	
Fluorspar	Fluorspar, containing by weight 97% or less of calcium fluoride	China	*Newly acceded WTO Member*	***15%***	1 Dec. 2008–31 Dec. 2012[1]	Eliminated since 1 Jan. 2013[2]
	Fluorspar, containing by weight >97% of calcium fluoride	China	*Newly acceded WTO Member*	***15%***	1 Dec. 2008–31 Dec. 2012[1]	Eliminated since 1 Jan. 2013[2]
Gallium, germanium indium, niobium, rhenium and vanadium	Germanium, vanadium, gallium, hafnium, indium, niobium (columbium), rhenium and articles of these metals, excl. waste and scrap, powder, unwrought	Vietnam	*Newly acceded WTO Member*	5%	1 Jan. 2009	
	Germanium, vanadium, gallium, hafnium, indium, niobium	Argentina	Original WTO Member	5%	2002	40% from 17 Jan. 2008 to 6 Jan. 2010

(*cont.*)

Minerals and metals	Forms	Country	WTO Status	Applied export duty	Date of introduction and/or modification of a measure	Any relevant developments since 2008
	(columbium), rhenium, and articles of these metals, incl. waste and scrap, powder, unwrought					
		China	*Newly acceded Member*	*2%*	1 Jan. 2008[1]	15% from 1 Jan. 2008 to 1 Jul. 2009; 5% until 31 Dec. 2013
		Pakistan	Original WTO Member	25%	9 Jun. 2007	
		Russian Federation	*Newly acceded Member*	*6.5%*	1 Jan. 2007	
		Sri Lanka	Original WTO Member	25%	17 Mar. 2006	
Germanium	Germanium oxides and hydroxides	China	*Newly acceded WTO Member*	*5%*	1 Jan. 2008[1]	
Graphite	Natural graphite, in powder/flakes	China	*Newly acceded WTO Member*	***20%***	1 Jan. 2009[1]	
Lithium	Lithium oxide and hydroxide	Argentina	Original WTO Member	5%	2002	
	Lithium carbonates	Argentina	Original WTO Member	5%	2002	
Magnesite		China		***5%***	1 Jan. 2008[1]	

	Natural magnesium carbonate (magnesite)		*Newly acceded WTO Member*			
Magnesium	Waste and scrap	Argentina	Original WTO Member	5%	2002	40% from 17 Jan. 2008 to 6 Jan. 2010
		China	*Newly acceded WTO Member*	***10%***	1 Jan. 2008–31 Dec. 2012[1]	Eliminated since 1 Jan. 2013[2]
		Pakistan	Original WTO Member	25%	9 Jun. 2007	
		Russian Federation	*Newly acceded WTO Member*	20% but not less than 138 EUR/1000 kg	8 Nov. 2009	Increased from 5% to 20% in Nov. 2009
		Sri Lanka	Original WTO Member	25%	17 Mar. 2006	
		Vietnam	*Newly acceded WTO Member*	*22%*	1 Jan. 2009	Decreased yearly from 37% in 2009 to 22% in 2012
	Unwrought magnesium, containing at least 99.8% by weight of magnesium	China	*Newly acceded WTO Member*	***10%***	1 Jan. 2008–31 Dec. 2012[1]	Eliminated since 1 Jan. 2013[2]
	Unwrought magnesium (excl. of 8104.11)	China	*Newly acceded WTO Member*	***10%***	1 Jan. 2008–31 Dec. 2012[1]	Eliminated since 1 Jan. 2013[2]
		Vietnam		5%	1 Jan. 2009	

(*cont.*)

Minerals and metals	Forms	Country	WTO Status	Applied export duty	Date of introduction and/or modification of a measure	Any relevant developments since 2008
	Magnesium raspings, turnings and granules, graded according to size; magnesium powders		*Newly acceded WTO Member*			
Manganese	Waste and scrap	Argentina	Original WTO Member	5%	2002	40% from 17 Jan. 2008 to 6 Jan. 2010
	Manganese ores and concentrates, incl. ferruginous manganese ores and concentrates with a manganese content of 20% or more, calc. on the dry weight	China	*Newly acceded WTO Member*	***15%***	1 Jan. 2008–31 Dec. 2012[1]	Eliminated since 1 Jan. 2013[2]
		Gabon	Original WTO Member	3.5% of their reference value	2009	
		India	Original WTO Member	4 INR/tonne	11 Sept. 2001	
		Indonesia	Original WTO Member	20%	11 Jan. 2014	Further increase (up to 60% by 2016) announced in May 2014
		Vietnam	*Newly acceded WTO Member*	15%	1 Jan. 2010	20% until 31 Dec. 2009
		China		***20%***		

	Manganese and articles thereof, incl. waste and scrap		*Newly acceded WTO Member*		1 Jan. 2008–31 Dec. 2012[1]	Eliminated since 1 Jan. 2013[2]
		Pakistan	Original WTO Member	25%	9 Jun. 2007	
		Russian Federation	*Newly acceded Member*	6.5%	1 Jan. 2007	
		Sri Lanka	Original WTO Member	25%	17 Mar. 2006	
		Vietnam	*Newly acceded WTO Member*	*22%*	1 Jan. 2009	Decreased yearly from 37% in 2009 to 22% in 2012
	Ferro-manganese, containing by weight >2% of carbon, in granular/powder form	China	*Newly acceded WTO Member*	*20%*	1 Jan. 2008[1]	
	Ferro-manganese, other than that containing by weight >2% of carbon, in granular/powder form	China	*Newly acceded WTO Member*	*20%*	1 Jan. 2008[1]	
Molybdenum	Molybdenum ores and concentrates, roasted	China	*Newly acceded Member*	***15%***	1 Jan. 2008[1,3]	To be eliminated by 2 May 2015[4]
		Russian Federation	*Newly acceded Member*	6.5%	1 Jan. 2007	
		Vietnam	*Newly acceded WTO Member*	20%	21 Jun. 2008	
	Molybdenum ores and concentrates, other than roasted	China	*Newly acceded Member*	***15%***	1 Jan. 2008[1,3]	To be eliminated by 2 May 2015[4]

(cont.)

Minerals and metals	Forms	Country	WTO Status	Applied export duty	Date of introduction and/or modification of a measure	Any relevant developments since 2008
		Vietnam	*Newly acceded WTO Member*	20%	21 Jun. 2008	Increased in 2008
	Molybdenum oxides and hydroxides	China	*Newly acceded Member*	**5%**	1 Jan. 2008[1,3]	To be eliminated by 2 May 2015[4]
	Molybdates	China	*Newly acceded WTO Member*	**5%**	1 Jan. 2008[1,3]	To be eliminated by 2 May 2015[4]
	Ferro-molybdenum in granular/ powder form	China	*Newly acceded WTO Member*	**20%**	1 Jan. 2008[1,3]	To be eliminated by 2 May 2015[4]
	Molybdenum powders	China	*Newly acceded Member*	**5%**	1 Jan. 2008[1,3]	To be eliminated by 2 May 2015[4]
		Vietnam	*Newly acceded WTO Member*	5%	1 Jan. 2009	
	Unwrought molybdenum, incl. bars and rods obt. simply by sintering	China	Newly acceded Member	**5%**	1 Jan. 2008[1,3]	To be eliminated by 2 May 2015[4]
		Vietnam	*Newly acceded WTO Member*	5%	1 Jan. 2009	
	Molybdenum bars and rods (excl. those obt. simply by sintering), profiles, plates, sheets, strip and foil	Vietnam	*Newly acceded WTO Member*	5%	1 Jan. 2009	

	Molybdenum wire	Vietnam	*Newly acceded WTO Member*	5%	1 Jan. 2009	
	Waste and scrap	Argentina	Original WTO Member	5%	2002	40% from 17 Jan. 2008 to 6 Jan. 2010
		China	*Newly acceded Member*	***15%***	1 Jan. 2008[1,3]	To be eliminated by 2 May 2015[4]
		Pakistan	Original WTO Member	25%	9 Jun. 2007	
		Russian Federation	*Newly acceded WTO Member*	6.5%	1 Jan. 2007	
		Sri Lanka	Original WTO Member	25%	17 Mar. 2006	
		Vietnam	*Newly acceded WTO Member*	*22%*	1 Jan. 2009	Decreased yearly from 37% in 2009 to 22% in 2012
	Molybdenum and articles thereof	Vietnam	*Newly acceded WTO Member*	5%	1 Jan. 2009	
Nickel	Nickel ores and concentrates	Vietnam	*Newly acceded WTO Member*	20%	21 Jun. 2008	Increased in 2008
	Waste and scrap	Argentina	Original WTO Member	5%		40% from 17 Jan. 2008 to 6 Jan. 2010
		Pakistan	Original WTO Member	25%	9 Jun. 2007	
		Russian Federation	*Newly acceded WTO Member*	*30%, but not less than*	1 Jan. 2007	

(cont.)

Minerals and metals	Forms	Country	WTO Status	Applied export duty	Date of introduction and/or modification of a measure	Any relevant developments since 2008
				720 EUR/ 1000 kg		
		Sri Lanka	Original WTO Member	25% FOB value	17 Mar. 2006	Renovated several times after expiration
		Ukraine	*Newly acceded WTO Member*	***21%***	16 May 2008	Decreased from 27% in 2008 to 21% in 2012
		Vietnam	*Newly acceded WTO Member*	*22%*	1 Jan. 2009	Decreased yearly from 37% in 2009 to 22% in 2012
	Nickel powders and flakes	Vietnam	*Newly acceded WTO Member*	5%	1 Jan. 2010	
	Nickel mattes, nickel oxide sinters and other intermediate products of nickel metallurgy	Malaysia	Original WTO Member	10%	1996	
		Russian Federation	*Newly acceded WTO Member*	*5%*	1 Jan. 2007	
	Nickel, not alloyed, unwrought	Russian Federation	*Newly acceded WTO Member*	*5%*	21 Dec. 2009–22 Aug. 2014	Eliminated from 30 Jan. 2009 to 21 Dec. 2009; 10% from 12 Dec. 2010 to 22 Aug. 2012

		Vietnam	*Newly acceded WTO Member*	5%	1 Jan. 2008	
	Nickel alloys, unwrought	Malaysia	Original WTO Member	10%	1996	
		Russian Federation	*Newly acceded WTO Member*	5%	21 Dec. 2009	
		Vietnam	*Newly acceded WTO Member*	5%	1 Jan. 2008	
	Bars, rods, profiles and wire, of nickel, not alloyed	Vietnam	*Newly acceded WTO Member*	5%	1 Jan. 2008	
	Bars, rods, profiles and wire, of nickel alloys	Vietnam	*Newly acceded WTO Member*	5%	1 Jan. 2008	
	Ferro-nickel, in granular/powder form	China	*Newly acceded WTO Member*	**20%**	1 Jan. 2008[1]	
Niobium	Ores and concentrates	China	*Newly acceded WTO Member*	30%	1 Jan. 2008[1]	
	Ferro-niobium, in granular/powder form	China	*Newly acceded WTO Member*	20%	1 Jan. 2008[1]	
Phosphates	Natural calcium phosphates, natural aluminium calcium phosphates and phosphatic chalk, unground	China	*Newly acceded WTO Member*	**35%**	1 Jan. 2008[1]	
	Yellow phosphorus	China	*Newly acceded WTO Member*	20%	1 Dec. 2010–31 Dec. 2014[1]	50% special duty removed since 1 Jun. 2009

(*cont.*)

Minerals and metals	Forms	Country	WTO Status	Applied export duty	Date of introduction and/or modification of a measure	Any relevant developments since 2008
	Phosphorus, n.e.s.	China	*Newly acceded Member*	*10%*	1 Jan. 2008[1]	
	Phosphoric acid and polyphosphoric acids, whether/not chemically defined	China	*Newly acceded WTO Member*	***300RMB/ tonne***	1 Jan. 2013[1]	10% in 2009; 7% from 2010 to 2012
PGMs	Ores and concentrates	China	*Newly acceded WTO Member*	***10%***	1 Jan. 2008[1]	10% in 2009
		Vietnam	*Newly acceded WTO Member*	20% to 30%	21 Jun. 2008	Increased in 2008 and 2011
	Ash containing precious metal/ precious metal compounds	Argentina	Original WTO Member	5%	2002	
		Gambia	Original WTO Member	5%	2010	
		Russian Federation	*Newly acceded WTO Member*	6.5%	1 Jan. 2007	
	Waste and scrap (platinum)	Argentina	Original WTO Member	5%	2002	
		Gambia	Original WTO Member	5%	2010	
		Russian Federation	*Newly acceded WTO Member*	6.5%	1 Jan. 2007	
		Malaysia		5%	12 Sept. 2003	

	Platinum, palladium, rhodium, iridium, osmium, ruthenium unwrought/in powder form		Original WTO Member			
		Russian Federation	*Newly acceded WTO Member*	*6.5%*	1 Jan. 2007	
	Platinum, palladium, rhodium, iridium, osmium, ruthenium in semi-manufactured forms	Malaysia	Original WTO Member	5%	12 Sept. 2003	
		Russian Federation	*Newly acceded WTO Member*	6.5%	1 Jan. 2007	
	Base metals/silver/gold, clad with platinum, not further worked than semi-manufactured	Malaysia	Original WTO Member	5%	12 Sept. 2003	
		Russian Federation	*Newly acceded WTO Member*	*6.5%*	1 Jan. 2007	
Rare earth elements	Ores of rare earth metals	China	*Newly acceded Member*	***15%***	1 Jan. 2008[1,3]	Various forms raised from 20% to 25% on 1 Jan. 2011; To be eliminated by 2 May 2015[4]
	Thorium ores and concentrates	China	*Newly acceded Member*	***10%***	1 Jan. 2008[1,3]	Various forms raised from 20% to 25% on 1 Jan. 2011; To be eliminated by 2 May 2015[4]

(cont.)

Minerals and metals	Forms	Country	WTO Status	Applied export duty	Date of introduction and/or modification of a measure	Any relevant developments since 2008
	Neodymium, dysprosium, terbium, lanthanum, cerium, praseodymium, yttrium	China	*Newly acceded Member*	***25%***	1 Jan. 2008[1,3]	Various forms raised from 20% to 25% on 1 Jan. 2011; To be eliminated by 2 May 2015[4]
	Other rare earth metals, scandium and yttrium, not intermixed or interalloyed	China	*Newly acceded Member*	***25%***	1 Jan. 2008[1,3]	Various forms raised from 20% to 25% on 1 Jan. 2011; To be eliminated by 2 May 2015[4]
	Other rare earth metals, scandium and yttrium, intermixed or interalloyed	China	*Newly acceded Member*	***25%***	1 Jan. 2008[1,3]	Various forms raised from 20% to 25% on 1 Jan. 2011; To be eliminated by 2 May 2015[4]
	Europium, terbium, dysprosium, yttrium as oxides, carbonates, chlorides or fluoride (incl. praseodymium oxide)	China	*Newly acceded Member*	***25%***	1 Jan. 2008[1,3]	Various forms raised from 20% to 25% on 1 Jan. 2011; To be eliminated by 2 May 2015[4]

	All other rare earth oxides, carbonates and chlorides	China	*Newly acceded Member*	***15%***	1 Jan. 2008[1,3]	Various forms raised from 20% to 25% on 1 Jan. 2011; To be eliminated by 2 May 2015[4]
	Cerium compounds	China	*Newly acceded Member*	***15%***	1 Jan. 2008[1,3]	Various forms raised from 20% to 25% on 1 Jan. 2011; To be eliminated by 2 May 2015[4]
	Other compounds of rare earth metals/yttrium/scandium these metals, other than cerium compounds	China	*Newly acceded Member*	***25%***	1 Jan. 2008[1,3]	Various forms raised from 20% to 25% on 1 Jan. 2011; To be eliminated by 2 May 2015[4]
	Ferro-alloy containing rare earths with weight of more than 10%	China	*Newly acceded Member*	***25%***	1 Jan. 2008[1,3]	Various forms raised from 20% to 25% on 1 Jan. 2011; To be eliminated by 2 May 2015[4]
Silicon	Silicon, containing by weight <99.99% of silicon	China	*Newly acceded Member*	***15%***	1 Jan. 2008–31 Dec. 2012[1]	Eliminated since 1 Jan. 2013[2]
		China	*Newly acceded WTO Member*	*25%*	1 Jan. 2008[1]	

(*cont.*)

Minerals and metals	Forms	Country	WTO Status	Applied export duty	Date of introduction and/or modification of a measure	Any relevant developments since 2008
	Ferro-silicon, containing by weight >55% of silicon, in granular/ powder form					
	Ferro-silicon, other than that containing >55% of silicon, in granular/powder form	China	*Newly acceded WTO Member*	*25%*	1 Jan. 2008[1]	
	Ferro-silico-manganese, in granular/powder form	China	*Newly acceded WTO Member*	*20%*	1 Jan. 2008[1]	
	Ferro-silico-chromium, in granular/powder form	China	*Newly acceded WTO Member*	***20%***	1 Jan. 2008[1]	
Silver	Ores and concentrates	China	*Newly acceded WTO Member*	***10%***	1 Jan. 2008[1]	
		Vietnam	*Newly acceded WTO Member*	30%	21 Jun. 2008	Increased in 2008
	Silver (incl. silver plated with gold/ platinum), in powder form	Fiji	Original WTO Member	3% FOB value	19 Feb. 2009	
		Malaysia	Original WTO Member	5%	12 Sept. 2003	
		Vietnam	*Newly acceded WTO Member*	5%	1 Jan. 2009	
	Silver (incl. silver plated with gold/ platinum), unwrought	Fiji	Original WTO Member	3% FOB value	19 Feb. 2009	

		Malaysia	Original WTO Member	5%	12 Sept. 2003	
	Silver (incl. silver plated with gold/ platinum), in semi-manufactured forms	Malaysia	Original WTO Member	5%	12 Sept. 2003	
	Base metals clad with silver, not further worked than semi-manufactured	Fiji	Original WTO Member	3% FOB value	19 Feb. 2009	
		Malaysia	Original WTO Member	5%	12 Sept. 2003	
	Waste and scrap	Argentina	Original WTO Member	5%	2002	
		Gambia	Original WTO Member	5%	2010	
		Russian Federation	*Newly acceded WTO Member*	6.5%	1 Jan. 2007	
Tantalum	Ores and concentrates	China	*Newly acceded WTO Member*	30%	1 Jan. 2008[1]	
		Democratic Republic of the Congo	Original WTO Member	10%		
		Vietnam	*Newly acceded WTO Member*	30%	21 Jun. 2008	Increased in 2008
	Waste and scrap	Argentina	Original WTO Member	5%	2002	40% from 17 Jan. 2008 to 6 Jan. 2010

(*cont.*)

Minerals and metals	Forms	Country	WTO Status	Applied export duty	Date of introduction and/or modification of a measure	Any relevant developments since 2008
		China	*Newly acceded WTO Member*	***10%***	1 Jan. 2008[1]	
		Pakistan	Original WTO Member	25%	9 Jun. 2007	
		Russian Federation	*Newly acceded WTO Member*	6.5%	1 Jan. 2007	
		Sri Lanka	Original WTO Member	25%	17 Mar. 2006	
		Vietnam	*Newly acceded WTO Member*	*22%*	1 Jan. 2009	Decreased yearly from 37% in 2009 to 22% in 2012
Titanium	Ores and concentrates	China	*Newly acceded WTO Member*	***10%***	1 Jan. 2008[1]	
		Vietnam	*Newly acceded WTO Member*	30%	1 Jan. 2011	20% until 12 Dec. 2010
	Unwrought titanium; powders	Vietnam	*Newly acceded WTO Member*	5%	1 Jan. 2009	
	Titanium and articles thereof	Vietnam	*Newly acceded WTO Member*	5%	1 Jan. 2009	
		China		*20%*	1 Jan. 2008[1]	

	Ferro-titanium and ferro-silico-titanium, in granular/powder form		*Newly acceded WTO Member*			
	Waste and scrap	Argentina	Original WTO Member	5%	2002	40% from 17 Jan. 2008 to 6 Jan. 2010
		Pakistan	Original WTO Member	25%	9 Jun. 2007	
		Russian Federation	*Newly acceded WTO Member*	30%, but not less than 225 EUR/1000 kg	1 Jan. 2007	
		Sri Lanka	Original WTO Member	25%	17 Mar. 2006	
		Ukraine	*Newly acceded WTO Member*	***21%***	16 May 2008	Decreased yearly from 27% in 2008 to 21% in 2012
		Vietnam	*Newly acceded WTO Member*	*22%*	1 Jan. 2009	Decreased from 37% in 2009 to 22% in 2012
Tungsten	Tungsten ores and concentrates	China	*Newly acceded WTO Member*	*20%*	1 Jan. 2008[1]	
		Russian Federation	*Newly acceded WTO Member*	10%	19 Dec. 2012	
		Vietnam	*Newly acceded WTO Member*	20%	21 Jun. 2008	Increased in 2008

(*cont.*)

Minerals and metals	Forms	Country	WTO Status	Applied export duty	Date of introduction and/or modification of a measure	Any relevant developments since 2008
	Tungsten acid, trioxide, oxides and hydroxides	China	*Newly acceded WTO Member*	**5%**	1 Jan. 2008[1,3]	To be eliminated by 2 May 2015[4]
	Tungstates (wolframates)	China	*Newly acceded WTO Member*	**5%**	1 Jan. 2008[1,3]	To be eliminated by 2 May 2015[4]
	Tungsten (wolfram) powders	China	*Newly acceded WTO Member*	**5%**	1 Jan. 2008[1,3]	To be eliminated by 2 May 2015[4]
		Vietnam	*Newly acceded WTO Member*	5%	1 Jan. 2009	
	Unwrought tungsten (wolfram), incl. bars and rods obt. simply by sintering	China	*Newly acceded WTO Member*	**5%**	1 Jan. 2008[1,3]	To be eliminated by 2 May 2015[4]
		Vietnam	*Newly acceded WTO Member*	5%	1 Jan. 2009	
	Tungsten (wolfram) wire	Vietnam	*Newly acceded WTO Member*	5%	1 Jan. 2009	
	Ferro-tungsten and ferro-silico-tungsten, in granular/powder form	China	*Newly acceded WTO Member*	**20%**	1 Jan. 2008[1,3]	To be eliminated by 2 May 2015[4]
	Waste and scrap	Argentina	Original WTO Member	5%	2002	40% from 17 Jan. 2008 to 6 Jan. 2010

		China	*Newly acceded WTO Member*	***15%***	1 Jan. 2008[1,3]	To be eliminated by 2 May 2015[4]
		Pakistan	Original WTO Member	25%	9 Jun. 2007	
		Russian Federation	*Newly acceded WTO Member*	6.5%	1 Jan. 2007	
		Sri Lanka	Original WTO Member	25%	17 Mar. 2006	
		Ukraine	*Newly acceded WTO Member*	***21%***	16 May 2008	Decreased from 27% in 2008 to 21% in 2012
		Vietnam	*Newly acceded WTO Member*	*22%*	1 Jan. 2009	Decreased yearly from 37% in 2009 to 22% in 2012
Vanadium	Vanadium oxides and hydroxides	China	*Newly acceded WTO Member*	5%	1 Jan. 2008[1]	
	Ferro-vanadium, in granular/ powder form	China	*Newly acceded WTO Member*	*20%*	1 Jan. 2008[1]	

Sources: OECD Inventory (2014); WTO Trade Monitoring Reports (various years); WTO Trade Policy Reviews (various countries, various years); Eleventh Report on Potentially Trade-Restrictive Measures of the European Commission (2014); China's annual Tariff Implementation Plans.

Note: Newly acceded WTO Members are those WTO Members that have joined the Organization after 1994. When italics are used, the Members have assumed export duty commitments in their accession protocols and/or related Working Party reports. When the rate of the export duty is in italics, the WTO Member has undertaken specific obligations on the use of export duties applied on that particular material. If the rate is in italics and bold, the export duty in place is in breach of the WTO obligations assumed by that WTO Member.

1 Announced yearly in the annual Tariff Implementation Plan released in December.

2 Condemned by the Appellate Body in *China–Raw Materials*. Eliminated since 1 January 2013 (see WT/DS394/19/Add.1, WT/DS395/18/Add.1 and WT/DS398/17/Add.1, 18 January 2013)

3 Condemned by the Appellate Body in *China–Rare-Earths*. Reasonable period of time expires on 2 May 2015 (see WT/DS431/16, WT/DS432/14 and WT/DS433/14, 10 December 2014)

4 Elimination announced in the summer of 2014 but not confirmed in the Tariff Implementation Plan for 2015 released on 17 December 2014.

APPENDIX 2

Export licensing schemes applied to critical minerals and metals (2008–2014)

Minerals and metals	Forms	Country
Antimony	Ores and concentrates	China, Grenada, Malaysia, Philippines, Russian Federation
	Semi-processed forms	China, Tajikistan[1]
	Waste and scrap	Afghanistan, Algeria, Argentina, Belarus, China, Guyana, Jamaica, Japan, Mauritius, Morocco, Russian Federation,[2] South Africa, Sri Lanka, Tajikistan,[1] Trinidad and Tobago, Turkmenistan, Zambia
Beryllium	Semi-processed forms	China
	Waste and scrap	Afghanistan, Algeria, Argentina, China, Guyana, Jamaica, Japan, Mauritius, Morocco, Russian Federation,[2] South Africa, Sri Lanka, Tajikistan,[1] Trinidad and Tobago, Turkmenistan, Zambia
Chromium	Ores and concentrates	Grenada, India, Malaysia, Philippines
	Semi-processed forms	South Africa
	Waste and scrap	Afghanistan, Algeria, Argentina, China, Guyana, Jamaica, Japan, Mauritius, Morocco, Russian

(*cont.*)

Minerals and metals	Forms	Country
		Federation,[2] South Africa, Sri Lanka, Tajikistan,[1] Trinidad and Tobago, Turkmenistan, Zambia
Coke	Unprocessed	China, Russian Federation[2]
Copper	Ores and concentrates	Argentina, Grenada, Malaysia, Philippines, Russian Federation
	Semi-processed forms	Argentina, Belarus, Japan, Indonesia, Malaysia, Paraguay, Russian Federation,[2] Sri Lanka, South Africa, Ukraine
	Waste and scrap	Algeria, Belarus, Guyana, Indonesia, Jamaica, Malaysia, Mauritius, Morocco, Paraguay, Russian Federation,[2] South Africa, Sri Lanka, Tajikistan,[1] Trinidad and Tobago, Turkey, Turkmenistan, Zambia
Cobalt	Ores and concentrates	Argentina, Grenada, Malaysia, Philippines, Russian Federation[2]
	Semi-processed forms	China, Russian Federation[2]
	Waste and scrap	Algeria, China, Guyana, Jamaica, Mauritius, Morocco, Russian Federation,[2] Sri Lanka, Tajikistan,[1] Trinidad and Tobago, Turkmenistan, Zambia
Germanium, vanadium, gallium, indium, niobium, rhenium	Ores and concentrates (niobium, tantalum and vanadium)	Grenada, Philippines, Russian Federation,[2] Rwanda, South Africa (tantalum)
	Semi-processed forms	

(*cont.*)

Minerals and metals	Forms	Country
		China (germanium and vanadium), Russian Federation[2] (tantalum and vanadium)
	Waste and scrap	Algeria, China, Guyana, Jamaica, Mauritius, Morocco, Russian Federation,[2] South Africa, Sri Lanka, Tajikistan,[1] Turkmenistan, Trinidad and Tobago, Zambia
Fluorspar	Unprocessed	Malaysia
Lithium	Ores and concentrates	South Africa
Magnesite	Unprocessed	China, Malaysia
	Semi-processed forms	Malaysia
Magnesium	Unprocessed	Russian Federation[2]
	Semi-processed forms	Brazil, Russian Federation[2]
	Waste and scrap	Algeria, Brazil, Guyana, Jamaica, Mauritius, Morocco, Russian Federation,[2] South Africa, Sri Lanka, Tajikistan,[1] Trinidad and Tobago, Turkmenistan and Zambia
Manganese	Ores and concentrates	Grenada, India, Philippines
	Semi-processed forms	China, Guyana, Jamaica, Malaysia, Mauritius, Morocco, South Africa, Sri Lanka, Tajikistan,[1] Trinidad and Tobago, Zambia
	Waste and scrap	Algeria, Guyana, Jamaica, Mauritius, Morocco, Russian Federation,[2] South Africa, Sri Lanka, Tajikistan,[1] Trinidad and Tobago, Turkmenistan, Zambia

(*cont.*)

Minerals and metals	Forms	Country
Molybdenum	Ores and concentrates	China, Grenada, Malaysia, Philippines, Russian Federation,[2] South Africa
	Semi-processed forms	China, Russian Federation[2]
	Waste and scrap	Algeria, China, Guyana, Jamaica, Mauritius, Morocco, Russian Federation,[2] South Africa, Sri Lanka, Tajikistan,[1] Turkmenistan, Trinidad and Tobago, Zambia
Nickel	Ores and concentrates	Grenada, Malaysia, Philippines, Russian Federation[2]
	Semi-processed forms	Russian Federation[2]
	Waste and scrap	Belarus, Guyana, Indonesia, Malaysia, Mauritius, Morocco, Russian Federation,[2] South Africa, Sri Lanka, Tajikistan,[1] Zambia
Phosphates	Unprocessed	China, Malaysia
PGMs	Ores and concentrates	Ghana, Grenada, Malaysia, Russian Federation, Philippines
	Semi-processed forms	Argentina, Belarus, China, Colombia, Japan, Indonesia, Russian Federation, South Africa
	Waste and scrap	Argentina, Belarus, China, Colombia, Russian Federation, South Africa, Tajikistan,[1] Turkmenistan
Rare earth elements	Ores and concentrates	China
	Semi-processed forms	China
Silicon	Semi-processed forms	China
Silver	Ores and concentrates	Grenada, Malaysia, Philippines

(*cont.*)

Minerals and metals	Forms	Country
	Semi-processed forms	Belarus, China, Fiji, Indonesia
	Waste and scrap	Argentina, Belarus, China, Colombia, Russian Federation, South Africa, Tajikistan,[1] Turkmenistan
Titanium	Ores and concentrates	Grenada, Philippines
	Semi-processed forms	China, Philippines
	Waste and scrap	Algeria, China, Guyana, Jamaica, Mauritius, Morocco, Russian Federation,[2] Sri Lanka, Tajikistan,[1] Turkmenistan, Trinidad and Tobago, Zambia
Tungsten	Tungsten ores and concentrates	China, Grenada, Malaysia, Philippines, South Africa
	Semi-processed forms	China, Russian Federation[2]
	Waste and scrap	Algeria, China, Guyana, Jamaica, Mauritius, Morocco, Russian Federation,[2] South Africa, Sri Lanka, Tajikistan,[1] Turkmenistan, Trinidad and Tobago, Zambia

Source: OECD Inventory (2014).

[1] Phased out since 3 July 2012.
[2] Phased out since 28 January 2012.

BIBLIOGRAPHY

WTO documents

G/L/59, Decision on Notification Procedures for Quantitative Restrictions adopted by the Council for Trade in Goods on 1 December 1995, 10 January 1996.

G/L/59/Rev.1, Decision on Notification Procedures for Quantitative Restrictions adopted by the Council for Trade in Goods on 22 June 2012, 3 July 2012.

G/STR/N/9/CHN/Add.1, Working Party on State Trading Enterprises – Updating Notification Pursuant to Article XVII:4(a) of the GATT 1994 and Paragraph 1 of the Understanding on the Interpretation of Article XVII, 14 July 2003.

TN/AG/W/4/Rev.4, Committee on Agriculture – Revised Draft Modalities for Agriculture, 6 December 2008.

TN/MA/S/13, Negotiating Group on Market Access – Sector Specific Discussions and Negotiations on Goods in the GATT and WTO, Note by the Secretariat, 24 January 2005.

TN/MA/W/1, Negotiating Group on Market Access – Market Access for Non-agricultural Products, Communication from the European Communities, 24 June 2002.

TN/MA/W/5, Negotiating Group on Market Access – Market Access for Non-agricultural Products, Contribution Paper from Japan, 5 August 2002.

TN/MA/W/10, Negotiating Group on Market Access – Market Access for Non-Agricultural Products, Submission by India, 22 October 2002.

TN/MA/W/11/Add.6, Negotiating Group on Market Access – Negotiating Proposal on Export Taxes, Communication from the European Communities, 27 April 2006.

TN/MA/W/15/Add.4, Negotiating Group on Market Access – Text-Based Proposal for Negotiation on Enhanced Transparency on Export Restrictions, Communication from Japan, 18 April 2006.

TN/MA/W/15/Add.4/Rev.1, Negotiating Group on Market Access – Enhanced Transparency on Export Licensing, Communication from Japan and the United States, 11 April 2008.

TN/MA/W/15/Add.4/Rev.2, Negotiating Group on Market Access – Enhanced Transparency on Export Licensing, Communication from the Separate

Customs Territory of Taiwan, Penghu, Kinmen and Matsu; Japan and the United States, 7 July 2008.

TN/MA/W/15/Add.4/Rev.3, Negotiating Group on Market Access – Enhanced Transparency on Export Licensing, Communication from the Separate Customs Territory of Taiwan, Penghu, Kinmen and Matsu; Japan; Korea and the United States, 18 March 2009.

TN/MA/W/15/Add.4/Rev.5, Negotiating Group on Market Access – Enhanced Transparency on Export Licensing, Communication from the Separate Customs Territory of Taiwan, Penghu, Kinmen and Matsu; Japan; Republic of Korea; Ukraine and the United States, 7 September 2009.

TN/MA/W/15/Add.4/Rev.6, Negotiating Group on Market Access – Enhanced Transparency on Export Licensing, Communication from Chile; Japan; Republic of Korea, the Separate Customs Territory of Taiwan, Penghu, Kinmen and Matsu; Ukraine and the United States, 23 September 2010.

TN/MA/W/15/Add.4/Rev.7, Negotiating Group on Market Access – Enhanced Transparency on Export Licensing, Communication from Chile; Costa Rica; Japan; Republic of Korea; the Separate Customs Territory of Taiwan, Penghu, Kinmen and Matsu; Ukraine and the United States, 27 September 2010.

TN/MA/W/31, Negotiating Group on Market Access – Market Access for Non-agricultural Products, Communication from Egypt, India, Indonesia, Kenya, Malaysia, Mauritius, Nigeria, Tanzania, Uganda and Zimbabwe, 25 March 2003.

TN/MA/W/37/Add.5, Negotiating Group on Market Access – Draft Modalities for the Sectoral Tariff Elimination in the Raw Material Sector, Communication from the United Arab Emirates, 25 July 2008.

TN/MA/W/101, Negotiating Group on Market Access – Revised Submission on Export Taxes, Communication from the European Communities, 17 January 2008.

TN/MA/W/103/Rev.3, Negotiating Group on Market Access – Revision of Draft Modalities for Non-Agricultural Market Access, 6 December 2008.

TN/MA/W/103/Rev.3/Add.1, Negotiating Group on Market Access – Textual Proposal by the Chairman, Ambassador Luzius Wasescha, on the State of Play of the NAMA Negotiations, 21 April 2011.

TN/MA/W/130, Negotiating Group on Market Access – Enhanced Transparency on Export Licensing, Communication from the Separate Customs Territory of Taiwan, Penghu, Kinmen and Matsu; Japan; Republic of Korea and the United States, 11 December 2009.

WT/ACC/CGR/38, Report of the Working Party on the Accession of Montenegro to the World Trade Organization, 5 December 2011.

WT/ACC/HRV/59, Report of the Working Party on the Accession of Croatia to the World Trade Organization, 29 June 2000.

WT/ACC/LAO/45, Report of the Working Party on the Accession of Lao People's Democratic Republic to the World Trade Organization, 1 October 2012.

WT/ACC/LVA/32, Report of the Working Party on the Accession of Latvia to the World Trade Organization, 19 February 1999.

WT/ACC/MNG/9, Report of the Working Party on the Accession of Mongolia to the World Trade Organization, 27 June 1996.

WT/ACC/RUS/70, Report of Working Party on the Accession of the Russian Federation to the World Trade Organization, 17 November 2011.

WT/ACC/SAU/61, Report of the Working Party on the Accession of Saudi Arabia to the World Trade Organization, 11 December 2005.

WT/ACC/TJK/3, Accession of the Republic of Tajikistan, Memorandum on the Foreign Trade Regime, 21 February 2003.

WT/ACC/TJK/7/Rev.7, Accession of the Republic of Tajikistan, Legislative Action Plan, 4 July 2012.

WT/ACC/TJK/24, Accession of the Republic of Tajikistan, Additional Questions and Replies, 20 June 2012.

WT/ACC/TJK/30, Report of the Working Party on the Accession of Tajikistan to the World Trade Organization, 6 November 2012.

WT/ACC/UKR/152, Report of the Working Party on the Accession of Ukraine to the World Trade Organization, 16 May 2008.

WT/ACC/VNM/48, Report of the Working Party on the Accession of Viet Nam to the World Trade Organization, 11 January 2007.

WT/DS431/16, China – Measures Related to the Exportation of Rare Earths, Tungsten, and Molybdenum – Agreement under article 21.3(b) of the DSU, 10 December 2014.

WT/DS432/14, China – Measures Related to the Exportation of Rare Earths, Tungsten, and Molybdenum – Agreement under article 21.3(b) of the DSU, 10 December 2014.

WT/DS433/14, China – Measures Related to the Exportation of Rare Earths, Tungsten, and Molybdenum – Agreement under article 21.3(b) of the DSU, 10 December 2014.

WT/L/93, Decision-Making Procedures under Articles IX and XII of the WTO Agreement, Statement by the Chairman as Agreed by the General Council on 15 November 1995, 24 November 1995.

WT/L/432, Protocol on the Accession of the People's Republic of China, 23 November 2001.

WT/L/579, Doha Work Programme, Decision adopted by the General Council on 1 August 2004, 2 August 2004.

WT/L/940, Protocol Amending the Marrakesh Agreement Establishing the World Trade Organization, General Council Decision of 27 November 2014, 28 November 2014.

WT/MIN(01)/DEC/1, Doha Ministerial Declaration, 14 November 2001.

WT/MIN(11)/2/Add.1, Schedule CLXV – The Russian Federation, 17 November 2011.

WT/MIN(11)/11, Eighth Ministerial Conference, Chairman's Concluding Statement, 17 December 2011.

WT/MIN(11)/ST/19, Statement by H.E. Mr Alfredo Chiaradia, Head of the Argentinean Delegation, 16 December 2011.

WT/MIN(11)/W/2, Elements for Political Guidance, 1 December 2011.

WT/MIN(13)/36, WT/L/911, Agreement on Trade Facilitation, Ministerial Decision of 7 December 2013, 11 December 2013.

WT/MIN(96)/16, Ministerial Declaration on Trade in Information Technology Products, 13 December 1996.

WT/MIN(96)/DEC, Singapore Ministerial Declaration, 18 December 1996.

WT/TPR/G/248, Trade Policy Review Report by European Union, 1 June 2011.

WT/TPR/M/249/Add.2, Trade Policy Review on India, Record of the Meeting, 28 October 2011.

WT/TPR/OV/14, Overview of Developments in the International Trading Environment, Annual Report by the Director-General, 21 November 2011.

WT/TPR/OV/15, Overview of Developments in the International Trading Environment, Annual Report by the Director-General, 29 November 2012.

WT/TPR/OV/16, Overview of Developments in the International Trading Environment, Annual Report by the Director-General, 31 January 2014.

WT/TPR/OV/17, Overview of Developments in the International Trading Environment, Annual Report by the Director-General, 24 November 2014.

WT/TPR/OV/W/1, Report to the TPRB from the Director-General on the Financial and Economic Crisis and Trade-Related Developments, 20 April 2009.

WT/TPR/OV/W/2, Report to the TPRB from the Director-General on the Financial and Economic Crisis and Trade-Related Developments, 15 July 2009.

WT/TPR/OV/W/3, Report to the TPRB from the Director-General on the Financial and Economic Crisis and Trade-Related Developments, 14 June 2010.

WT/TPR/OV/W/4, Symposium: The Financial and Economic Crisis and the Role of the WTO, Background Note by the Secretariat, 14 June 2011.

WT/TPR/OV/W/5, Report to the TPRB from the Director-General on the Financial and Economic Crisis and Trade-Related Developments, 9 June 2011.

WT/TPR/OV/W/6, Report to the TPRB from the Director-General on the Financial and Economic Crisis and Trade-Related Developments, 28 June 2012.

WT/TPR/OV/W/7, Report to the TPRB from the Director-General on the Financial and Economic Crisis and Trade-Related Developments, 5 July 2013.

WT/TPR/OV/W/8, Report to the TPRB from the Director-General on the Financial and Economic Crisis and Trade-Related Developments, 27 June 2014.

WT/TPR/S/176, Trade Policy Review on Argentina, Report by the Secretariat, 8 January 2007.

WT/TPR/S/199/Rev.1, Trade Policy Review on China, Report by the Secretariat, 12 August 2008.

WT/TPR/S/240, Trade Policy Review on the Democratic Republic of the Congo, Report by the Secretariat, 20 October 2010.

WT/TPR/S/247, Trade Policy Review on Nigeria, Report by the Secretariat, 24 May 2011.

WT/TPR/S/256, Trade Policy Review on Saudi Arabia, Report by the Secretariat, 21 December 2011.

WT/TPR/S/264/Rev.1, Trade Policy Review on China, Report by the Secretariat, 20 July 2012.

WT/TPR/S/277, Trade Policy Review on Argentina, Report by the Secretariat, 13 February 2013.

WT/TPR/S/279, Trade Policy Review on Mexico, Report by the Secretariat, 27 February 2013.

WT/TPR/S/292, Trade Policy Review on Malaysia, Report by the Secretariat, 27 January 2014.

WT/TPR/S/300, Trade Policy Review on China, Report by the Secretariat, 27 May 2014.

Other sources

Abi-Saab, G., 'Progressive Development of the Principles and Norms of International Law Relating to the New International Economic Order', UN Doc.A/39/504/Add.1, 23 October 1984.

Agatiello, O.R. and Fliess, B., 'Export Restrictions: Benefits of Transparency and Good Practices', OECD Trade Policy Papers, No. 146 (OECD Publishing, 2013).

Anderson, A. and Nelgen, N., 'Trade Barrier Volatility and Agricultural Price Stabilization', *World Development* 40 (2012) 36.

Appleyard, D., Field, A. and Cobb, S., *International Economics*, 7th edn (McGraw Hill, 2010).

Australian Government, Department of Foreign Affairs and Trade, 'Australia's WTO Tariffs Schedule: Consolidated Loose-leaf Schedule', www.dfat.gov.au/trade/negotiations/schedule/schedule.pdf (accessed January 2015).

Backer and McKenzie, 'Changes to Export Requirements for Rare Earths from China', 9 January 2014 (on file with the author).

Barbier, E., *Scarcity and Frontiers: How Economies Have Developed Through Natural Resource Exploitation* (Cambridge University Press, 2011).

Bargawi, H., Bova, E., Ferrarini, B. and Newman, S., 'Low-income Countries and Commodity Price Volatility', in T. Cottier and P. Delimatsis (eds.), *The Prospects of International Trade Regulation* (Cambridge University Press, 2010), pp. 452–482.

Baroncini, E., 'The Applicability of GATT Article XX to China's WTO Accession Protocol in the Appellate Body Report of the *China–Raw Materials* case: Suggestions for a Different Interpretative Approach', *China-EU Law Journal* 1 (2012) 1.

Baroncini, E., 'The *China-Rare Earths* WTO Dispute: A Precious Chance to Revise the *China–Raw Materials* Conclusions on the Applicability of GATT Article XX to China's WTO Accession Protocol', *Cuadernos de Derecho Internacionales* 4 (2012) 49.

Baroncini, E., 'Obblighi WTO-Plus, tutela dell'ambiente, della salute e preservazione delle risorse naturali: il Report del Panel nel caso *China–Raw Materials*', *Diritto Comunitario e degli Scambi Internazionali* 50 (2011) 627.

Bellmann, C. and Wilke, M., 'Trade Policies for Resource Security: Rethinking Export Restrictions' in R. Meléndez-Ortiz, C. Bellmann and M. Rodriguez Mendoza (eds.), *The Future and the WTO: Confronting the Challenges* (ICTSD, 2012), pp. 197–205.

Bhagwati, J., 'On the Equivalence of Tariffs and Quotas', in R. Caves, H.G. Johnson and P.B. Kenen (eds.), *Trade, Growth, and the Balance of Payments Payments – Essays in Honor of Gottfried Haberler* (Rand McNally, 1965), pp. 53–67.

Bhala, R., *Modern GATT Law: A Treatise on the General Agreements on Tariffs and Trade* (Sweet & Maxwell, 2005).

Bhala, R., 'Saudi Arabia, the WTO, and American Trade Law and Policy', *The International Lawyer* 34 (2004) 741.

Bonarriva, J., Koscielski, M. and Wilson, E., 'Export Controls: An Overview of their Use, Economic Effects, and Treatment in the Global Trading System', US International Trade Commission, Office of Industries Working Papers, August 2009.

Bouet, A. and Laborde Debucquet, D., 'The Economics of Export Taxes in the Context of Food Security', in OECD, *The Economic Impact of Export Restrictions on Raw Materials* (OECD Publishing, 2010).

Butler, N., 'After the Oil Price Fall, Is Natural Gas Next?' *Financial Times*, 4 January 2015.

Carter, C.A., Rasser, G.C. and Smith, A., 'Commodity Booms and Busts', *Annual Review of Resource Economics* 3 (2011) 87.

Cattaneo, O., Gereffy, G. and Staritz, C., 'Global Value Chains in a Postcrisis World: Resilience, Consolidation, and Shifting End Markets', in O. Cattaneo, G. Gereffy and C. Staritz (eds.), *Global Value Chains in a Postcrisis World: A Development Perspective* (The World Bank, 2010), pp. 3–20.

Charnovitz, S., 'Mapping the Law of the WTO Accession', in M.E. Janow, V. Donaldson and A. Yanovich (eds.), *The WTO: Governance, Dispute Settlement and Developing Countries* (Juris Publishing, 2008), pp. 855–920.

Chelliah, A., 'Tax and the Upstream O&G Sector', *Malaysian Business*, 16 August 2011.

Chi, M., '"Exhaustible Natural Resource" in WTO Law: GATT Article XX(g) Disputes and Their Implications', *Journal of World Trade* 48 (2014) 939.

Cline, W., *Trade Policy in the 1980s* (Peterson Institute for International Economics, 1983).

Cooper, R.N. and Lawrence, R.Z., 'The 1972–1975 Commodity Boom', *Brooking Papers on Economic Activity* 3 (1975) 671.

Cossy, M., 'Energy Trade and WTO Rules: Reflections on Sovereignty Over Natural Resources, Export Restrictions and Freedom of Transit', *European Yearbook of International Economic Law* 3 (2012) 281.

Cottier, T., 'A Two-Tier Approach to WTO Decision Making', NCCR Working Paper No. 2009/06, www.peacepalacelibrary.nl/ebooks/files/NCCR_Cottier_Two-Tier-Approach.pdf (accessed January 2015).

Cottier, T., Malumfashi, G., Matteotti-Berkutova, S., Nartova, O., de Sépibus, J. and Bigdeli, S.Z., 'Energy in WTO Law and Policy', in T. Cottier and P. Delimatsis (eds.), *The Prospects of International Trade Regulation: From Fragmentation to Coherence* (Cambridge University Press, 2011), pp. 211–244.

Coulibaly, A.L., 'The Food Price Increase of 2010–2011: Causes and Impacts', Publication No. 2013-02-E, Ottawa, Canada, Library of Parliament (2013), www.parl.gc.ca/Content/LOP/ResearchPublications/2013-02-e.pdf (accessed November 2014).

Critchlow, A., 'US Oil Production Surge to Break Saudi Arabia's Grip on World Energy', *The Telegraph*, 9 November 2014.

Crosby, D., 'WTO Legal Status and Evolving Practice of Export Taxes', *Bridges* 12, No. 5 (October–November 2008) 3.

Davies, G., 'Learning to Love the Dutch Disease: Evidence from the Mineral Economies', *World Development* 23 (1995) 1765.

Davies, G., 'Trade in Mineral Resources', WTO Economic Research and Statistics Division, Staff Working Paper ERSD-2010-01, www.wto.org/english/res_e/reser_e/ersd201001_e.pdf (accessed December 2014).

De Medeiros, J.F., 'Global Trade Law - China's Export Restraints Found to be Inconsistent with its Obligations as a Member of The World Trade Organization - *China - Measures Related to the Exportation of Various Raw Materials*', *Suffolk Transnational Law Review* 35 (2012) 203.

Desta, M.G., 'OPEC, the WTO, Regionalism and Unilateralism', *Journal of World Trade* 37 (2003) 523.

Devarajan, S., Go, D., Schiff, M. and Suthiwart-Narueput, S., 'The Whys and Why Nots of Export Taxes', The World Bank Policy Research Working Paper No. 1684, 1996, http://siteresources.worldbank.org/INTAFROFFCHIECO/Resources/The_Whys_and_Why_Nots_of_Export_Taxation.pdf (accessed November 2014).

Do, Q.-T., Levchenko, A. and Ravallion, M., 'Trade Insulation as Social Protection', The World Bank Policy Research Working Paper No. 6448,

2013, http://elibrary.worldbank.org/doi/book/10.1596/1813-9450-6448 (accessed November 2014).

Draper, P. and Freytag, A., *Who Captures the Value in the Global Value Chain? High Level Implications for the World Trade Organization*, E15 Initiative (ICTSD and World Economic Forum, 2014).

ECT, 'Putting a Price on Energy: International Pricing Mechanisms for Oil and Gas', 2007, www.encharter.org/fileadmin/user_upload/document/Oil_and_Gas_Pricing_2007_ENG.pdf (accessed November 2014).

Edminster, L.R., 'Control of Exports of Raw Materials: An International Problem', *The Annals of the American Academy of Political and Social Science* 150 (1930) 89.

Ehring, L. and Chianale, G.F., 'Export Restrictions in the Field of Energy', in Y. Selivanova (ed.), *Regulation of Energy in International Trade Law: WTO, NAFTA and Energy Charter* (Kluwer Law International, 2012), pp. 109–147.

Espa, I., 'The Appellate Body Approach to the Applicability of Article XX GATT in the Light of *China – Raw Materials*: A Missed Opportunity?', *Journal of World Trade* 46 (2012) 1399.

Espa, I. and Holzer, K., 'Negotiating an Energy Deal Under TTIP: Drivers And Impediments to US Shale Exports to Europe', *Denver Journal of International Law and Policy* 43 (2015) 358.

European Commission, 'Annex V to the Report of the Ad-hoc Working Group on Defining Critical Raw Materials', 30 July 2010, http://ec.europa.eu/enterprise/policies/raw-materials/files/docs/annex-v_en.pdf (accessed December 2014).

European Commission, 'Critical Raw Materials for the EU: Report of the Ad-hoc Working Group on Defining Critical Raw Materials', 30 July 2010, http://ec.europa.eu/enterprise/policies/raw-materials/files/docs/report-b_en.pdf (accessed November 2014).

European Commission, 'Eleventh Report on Potentially Trade-Restrictive Measures Identified in the Context of the Financial and Economic Crisis', 2014, http://horizon.hozint.com/2014/11/dg-trade-11th-report-on-potentially-trade-restrictive-measures/ (accessed December 2014).

European Commission, 'EU Trade Policy for Raw Materials, Second Activity Report', 2012, http://trade.ec.europa.eu/doclib/docs/2012/may/tradoc_149515.pdf (accessed November 2014).

European Commission, 'Fifth Report on Potentially Trade-Restrictive Measures Identified in the Context of the Financial and Economic Crisis', 2009, http://trade.ec.europa.eu/doclib/docs/2009/november/tradoc_145270.pdf (accessed November 2014).

European Commission, 'Report of the Ad-hoc Working Group on Defining Critical Raw Materials', May 2014, http://ec.europa.eu/enterprise/policies/raw-materials/files/docs/crm-report-on-critical-raw-materials_en.pdf (accessed December 2014).

European Commission, 'Report on Critical Raw Materials for the EU: Critical Raw Materials Profiles', May 2014, http://ec.europa.eu/enterprise/policies/raw-materials/files/docs/crm-critical-material-profiles_en.pdf (accessed December 2014).

European Commission, 'Report on Critical Raw Materials for the EU: Non-Critical Raw Materials Profiles', May 2014, http://ec.europa.eu/enterprise/policies/raw-materials/critical/index_en.htm (accessed December 2014).

European Commission, 'Tenth Report on Potentially Trade-Restrictive Measures Identified in the Context of the Financial and Economic Crisis', 2013, http://trade.ec.europa.eu/doclib/docs/2013/september/tradoc_151703.pdf (accessed November 2014).

European Commission, 'Ninth Report on Potentially Trade-Restrictive Measures Identified in the Context of the Financial and Economic Crisis', 2012, http://trade.ec.europa.eu/doclib/docs/2012/june/tradoc_149526.pdf (accessed December 2014).

European Commission, 'The Raw Materials Initiative – Meeting our Critical Needs for Growth and Jobs in Europe' (COM (2008) 699 final), 4 November 2008, http://eur-lex.europa.eu/LexUriServ/LexUriServ.do?uri=COM:2008:0699:FIN:en:PDF (accessed December 2014).

FAO, 'Draft Declaration of the World Summit on Food Security' (Rome, 16–18 November 2009), www.fao.org/fileadmin/templates/wsfs/Summit/Docs/Declaration/WSFS09_Draft_Declaration.pdf (accessed November 2014).

FAO, 'Report of the Thirty-fifth Section of the Committee on World Food Security' (Rome, 14-15 and 17 October 2009), www.fao.org/fileadmin/templates/cfs/Docs0910/ReformDoc/CFS_2009_2_Rev_2_E_K7197.pdf (accessed November 2014).

FAO, 'Soaring Food Prices: Facts, Perspectives, Impacts and Actions Required' (2008), HLC/08/INF/1, www.fao.org/fileadmin/user_upload/foodclimate/HLCdocs/HLC08-inf-1-E.pdf (accessed November 2014).

FAO, IFAD, IMF, OECD, UNCTAD, WFP, the World Bank, the WTO, IFPRI and the UN HLTF, 'Price Volatility in Food and Agricultural Markets: Policy Responses', 2 June 2011, www.oecd.org/tad/agricultural-trade/48152638.pdf (accessed November 2014).

FAO, WFP and IFAD, 'The State of Food Insecurity in the World', 2014, www.fao.org/3/a-i4030e.pdf (accessed November 2014).

Farooki, M. and Kaplinsky, R., *The Impact of China on Global Commodities Prices. The Global Reshaping of the Resource Sector* (Routledge Studies in Global Competition, 2012).

Findlay, R. and Lundhal, M., 'Resource-Led Growth – A Long-term Perspective: The Relevance of the 1870–1914 Experience for Today's Developing Economies', World Institute for Development Economics Research Working Paper No. 162/1999.

Finger, J.M. and Harrison, A., 'Import Protection for U.S. Textiles and Apparel: Viewed from the Domestic Perspective', in A.O. Krueger (ed.), *The Political Economy of Trade Protection* (University of Chicago Press, 1996), pp. 43–50.

Fliess, B., Arriola, C. and Liapis, P., 'Recent Developments in the Use of Export Restrictions in Raw Materials Trade', in OECD, *Export Restrictions in Raw Materials Trade: Facts, Fallacies and Better Practices* (OECD Publishing, 2014).

Fliess, B., Gou, H. and Mard, T., 'Taking Stock of Measures Restricting the Export of Raw Materials: Analysis of OECD Inventory Data', OECD Trade Policy Working Papers, No. 140 (OECD Publishing, 2012).

Fung, K.C. and Korinek, J., 'Economics of Export Restrictions as Applied to Industrial Raw Materials', OECD Trade Policy Papers, No. 155 (OECD Publishing, 2014).

G8, 'L'Aquila Joint Statement on Global Food Security: L'Aquila Food Security Initiative' (L'Aquila, 10 July 2009), www.ifad.org/events/g8/statement.pdf (accessed November 2014).

G20, 'Action Plan on Food Price Volatility and Agriculture adopted by the G20 Agriculture Ministers' (Paris, 22-23 July 2011), www.g8.utoronto.ca/g20/agriculture/index.html (accessed October 2014).

G20, 'Declaration of the Summit on Financial Markets and the World Economy' (Washington, DC, 15 November 2008), www.g20.utoronto.ca/2008/2008declaration1115.html (accessed December 2014).

G20, 'Food Security and Nutrition Framework' (Brisbane, Australia, 16 November 2014), www.g20.utoronto.ca/2014/2014-1116-communique.html (accessed November 2014).

G20, 'Leaders' Communiqué' (Brisbane, Australia, 16 November 2014), www.g20.utoronto.ca/2014/2014-1116-communique.html (accessed November 2014).

G20, 'Leaders' Declaration' (St. Petersburg, 6 September 2013), www.g20.utoronto.ca/2013/2013-0906-declaration.html (accessed November 2014).

G20, 'Report of the Agriculture Vice Ministers/Deputies Meeting' (Mexico City, 18 May 2012), www.g20.utoronto.ca/agriculture/index.html (accessed November 2014).

Gardner, R., *Sterling-Dollar Diplomacy in Current Perspective* (Columbia University Press, 1980).

Gilbert, C.L., 'How to Understand High Food Prices', *Journal of Agricultural Economics* 61 (2010) 398.

Giles, C., 'Winners and Losers of Oil Price Plunge', *Financial Times*, 15 December 2014.

Giordani, P., Rocha, N. and Ruta, M., 'Food Prices and the Multiplier Effect of Export Policy', WTO Economic Research and Statistics Division, Staff Working Paper ERSD-2012-08, https://www.wto.org/english/res_e/reser_e/ersd201208_e.pdf (accessed December 2014).

Gomes-Sabaini, J.C., 'Role of Export Taxes', in V. Tanzi (ed.), *Fiscal Policy in Open Developing Economies* (International Monetary Fund, 1998), pp. 42–53.

Goode, R. and Lent, G., 'Role of Export Taxes in Developing Countries', *Staff Papers (International Monetary Fund)* 13 (1966) 453.

Gorton, J., 'British Preferential Export Taxes', *American Economic Association* 14 (1924) 56.

Gu, B., 'Applicability of GATT Article XX in *China–Raw Materials*: A Clash within the WTO Agreement', *Journal of International Economic Law* 15 (2012) 1007.

Gu, B., 'Mineral Export Restraints and Sustainable Development – Are Rare Earths Testing the WTO's Loopholes?', *Journal of International Economic Law* 14 (2011) 765.

Häberli, C., 'God, the WTO – and Hunger', in K. Nadakavukaren Schefer (ed.), *Poverty and the International Economic Legal System: Duties to the World's Poor* (Cambridge University Press, 2013), pp. 79–106.

Hatch, G., 'The Second Round of Chinese Rare-Earth Export-Quota Allocations for 2014', Technology Metal Research, 19 July 2014, www.techmetalsresearch.com/2014/07/the-second-round-of-chinese-rare-earth-export-quota-allocations-for-2014/ (accessed December 2014).

Hausmann, R. and Rodrik, D., 'What You Export Matters', *Journal of Economic Growth* 12 (2007) 1.

Headey, D.D., 'Rethinking the Global Food Crisis: The Role of Trade Shocks', IFPRI Discussion Paper No. 958/2010, www.ifpri.org/sites/default/files/publications/ifpridp00958.pdf (accessed November 2014).

Hoda, A., *Tariff Negotiations and Renegotiations under the GATT and the WTO: Procedures and Practice* (Cambridge University Press, 2001).

Horn, H., Mavroidis, P.C. and Sapir, A., 'Beyond the WTO? An Anatomy of EU and US Preferential Trade Agreements', *The World Economy* 33 (2010) 1565.

Howse, R. and Josling, T., 'Agricultural Export Restrictions and International Trade Law: A Way Forward', IPC Position Paper, 2012, http://www.agritrade.org/Publications/documents/Howse_Josling_Export_Restriction_final.pdf (accessed November 2014).

Irwin, D., Mavroidis, P.C. and Sykes, A.O., *The Genesis of the GATT* (Cambridge University Press, 2008).

Ivanic, M., Martin, W. and Mattoo, A., 'Welfare and Price Impacts of Price-Insulating Policies', Presented at the 14th Annual Conference on Global Economic Analysis (Venice, Italy, 2001), GTAP Resource No. 3651, www.gtap.agecon.purdue.edu/resources/download/5580.pdf (accessed November 2014).

Jacks, D. S., 'From Boom to Bust: A Typology of Real Commodity Prices in the Long Run', NBER Working Paper No. 18874, 2013 (Cambridge, MA: National Bureau of Economic Research), http://www.nber.org/papers/w18874.pdf (accessed July 2014).

Jackson, J.H., *The Jurisprudence of GATT & the WTO: Insights on Treaty Law and Economic Relations* (Cambridge University Press, 2000).

Jackson, J.H., *World Trade and the Law of GATT* (Bobbs-Merrill Co., Inc., 1969).

Jensen, E.M., 'The Export Clause', *Florida Tax Review* 6 (2003) 6.

Jones, K., 'The Political Economy of WTO Accession: the Unfinished Business of Universal Membership', *World Trade Review* 8 (2009) 279.

Karapinar, B., 'China's Export Restriction Policies: Complying With "WTO-Plus" Or Undermining Multilateralism', *World Trade Review* 10 (2011) 389.

Karapinar, B., 'Defining the Legal Boundaries of Export Restrictions: A Case Law Analysis', *Journal of International Economic Law* 15 (2011) 443.

Karapinar, B., 'Export Restrictions and the WTO Law: How to Reform the "Regulatory Deficiency"', *Journal of World Trade* 45 (2011)1139.

Karapinar, B. and Haberli, C., *Food Crises and the WTO* (Cambridge University Press, 2010).

Kennedy, M., *Legal Options for a Sustainable Energy Trade Agreement* (Geneva: ICTSD, 2012).

Kim, J., 'Recent Trends in Export Restrictions on Raw Materials', in OECD, *The Economic Impact of Export Restrictions on Raw Materials* (OECD Publishing, 2010).

Knittel, C.R. and Pindyck, R.S., 'The Simple Economics of Commodity Price Speculation', MIT Center for Energy and Environmental Policy Research Working Paper No. 6/2013, http://web.mit.edu/ceepr/www/publications/workingpapers/2013-006.pdf (accessed January 2015).

Korinek, J. and Bartos, J., 'Multilateralising Regionalism: Disciplines on Export Restrictions in Regional Trade Agreements' in OECD, *Export Restrictions in Raw Materials Trade: Facts, Fallacies and Better Practices* (OECD Publishing, 2014).

Korinek, J. and Kim, J., 'Export Restriction on Strategic Raw Materials and their Impact on Trade and Global Supply', in OECD, *The Economic Impact of Export Restrictions on Raw Materials* (OECD Publishing, 2010).

Krugman, P. and Obstfeld, M., *International Economics: Theory and Policy*, 6th edn (Pearson Addison-Wesley, 2003).

Lamy, P., 'Global Commodities Trade: Perspective of the WTO', Speech given at the 9th Annual Global Conference on Commodities Finance held in Geneva on 14 June 2012, www.wto.org/english/news_e/sppl_e/sppl236_e.htm (accessed December 2014).

Lamy, P., 'Multilateralism Is Struggling', Inaugural speech given at the World Trade Forum 2012, www.wto.org/english/news_e/sppl_e/sppl244_e.htm (accessed November 2014).

Latina, J., Piermartini, R. and Ruta, M., 'Natural Resources and Non-Cooperative Trade Policy', *International Economics and Economic Policy* 8 (2012) 177.

Leamer, E.E., *Sources of International Comparative Advantage: Theory and Evidence* (MIT Press, 1984).

Lerner, A.P., 'The Symmetry between Import and Export Taxes', *Economica* 3 (1936) 306.

Levinson, A., 'Environmental Kuznets Curve', in *The New Palgrave Dictionary of Economics*, 2nd edn (Palgrave Macmillan, 2008).

Liapis, P., 'How Export Restrictive Measures Affect Trade in Agricultural Commodities', OECD Food, Agriculture and Fisheries Papers, No. 63 (OECD Publishing, 2013).

Liu, J., 'Accession Protocols: Legal Status in the WTO Legal System', *Journal of World Trade* 48 (2014) 751.

Liu, H.W. and Maughan, J., 'China's Rare Earths Export Quotas: Out of the China–Raw Materials Gate But Past the WTO's Finish Line?', *Journal of International Economic Law* 15 (2012) 971.

Low, P., 'WTO Decision-Making for the Future', WTO Economic Research and Statistics Division, Staff Working Paper ERSD-2011–05, www.wto.org/ENGLISH/res_e/reser_e/ersd201105_e.pdf (accessed January 2015).

Luff, D., Nguyen, H., and Anh Thu, N., 'Export Control by WTO Members and Recommendations for Vietnam', EU-MUTRAP Project Activity Code: Supe-7, 2013 (on file with the author).

Maddison, A., 'Explaining the Economic Performance of Nations', in W.J. Baumol, R. Nelson and E.N. Wolff (eds.), *Convergence of Productivity: Cross-National Studies and Historical Evidence* (Oxford University Press, 1994), pp. 20–61.

Mango, M.M. (ed.), 'Byzantine Trade, 4th–12th Centuries, The Archaeology of Local, Regional and International Exchange', Papers of the Thirty-eighth Spring Symposium of Byzantine Studies, St John's College, University of Oxford, March 2004 (Ashgate, 2004).

Martin, W. and Anderson, K., 'Export Restrictions and Price Insulation During Commodity Price Booms', The World Bank Policy Research Working Paper No. 5645, 2011, www.imf.org/external/np/seminars/eng/2011/trade/pdf/session1-martin-paper.pdf (accessed May 2014).

Martin, W. and Ivanic, M., 'The Food Price Crisis, Poverty and Agricultural Trade Policy', in B. Karapinar and C. Häberli (eds.), *Food Crises and the WTO* (Cambridge University Press, 2010), pp. 25–48.

Matsushita, M., 'Export Control of Natural Resources: WTO Panel Ruling on the Chinese Export Restrictions of Natural Resources', *Trade Law and Development* 3 (2011) 267.

Mavroidis, P.C., *Trade in Goods: The GATT and the Other Agreements Regulating Trade in Goods*, 2nd edn (Cambridge University Press, 2012).

Maxwell, P., 'Chile's Recent Copper Driven Prosperity', *Minerals and Energy* 19 (2004) 16.

Maxwell, P., 'Trade in Minerals', in P. Maxwell (ed.), *Australian Mineral Economics: A Survey of Important Issues* (The Australian Institute of Mining and Metallurgy, 2006), pp. 27–34.

Meredith, H., 'Evaluating Movements of Open-Work Glassware', in M. Mundell Mango (ed.), 'Byzantine Trade, 4th–12th Centuries: The Archaeology of Local, Regional and International Exchange', Papers of the Thirty-eighth Spring Symposium of Byzantine Studies (Ashgate, 2004), pp. 191–198.

Michalopoulos, C., 'Trade Policy and Market Access Issues for Developing Countries: Implications for the Millenium Round', The World Bank Policy Research Working Paper No. 2214, 1999, http://elibrary.worldbank.org/content/workingpaper/10.1596/1813-9450-2214 (accessed December 2014).

Mitra, S. and Josling, T., 'Agricultural Export Restrictions: Welfare Implications and Trade Disciplines', IPC Position Paper, Agricultural and Rural Development Policy Series, 2009, www.agritrade.org/documents/ExportRestrictions_final.pdf (accessed May 2014).

MOFCOM, 'China's Rare Earths Reserve Will be Used Up in 20 Years', http://news.xinhuanet.com/mil/2010–10/17/C_12668271 (accessed December 2014).

MOFCOM, 'Comments by the Head of the Chinese Ministry for Commerce, Department of Treaty and Law on US, EU and Japan Requests of Consultations on China', 15 May 2012, http://english.mofcom.gov.cn/aarticle/newsrelease/policyreleasing/201203/20120308016675.html (accessed December 2014).

Morrison, W.M. and Tang, R., 'China's Rare Earths Industry and Export Regime: Economic and Trade Implications for the United States', CRS Report for Congress, 30 April 2012, www.fas.org/sgp/crs/row/R42510.pdf (accessed December 2014).

National Cotton Council, 'Influence of India's Export Restrictions on the World Cotton Market', 2010, www.cotton.org/issues/2010/upload/NCC-Report_India-and-World-Cotton-Market_October-26-2010_FINAL-3.pdf (accessed November 2014).

Nogués, J., 'The Domestic Impact of Export Restrictions: The Case of Argentina', IPC Position Paper, Agricultural and Rural Development Policy Series, 2008, www.agritrade.org/documents/ArgentineExportRestrictions.pdf (accessed December 2014).

OECD, 'Analysis of Non-Tariff Measures: The Case of Export Duties', 31 January 2003, http://search.oecd.org/officialdocuments/displaydocumentpdf/?doc-language=en&cote=td/tc/wp(2002)54/final (accessed December 2014).

OECD, 'Analysis of Non-Tariff Measures: The Case of Export Restrictions', 4 April 2003, http://search.oecd.org/officialdocuments/displaydocumentpdf/?doclanguage=en&cote=TD/TC/WP(2003)7/FINAL (accessed December 2014).

OECD, *The Economic Impact of Export Restrictions on Raw Materials* (OECD Publishing, 2010).

OECD, *Export Restrictions in Raw Materials Trade: Facts, Fallacies and Better Practices* (OECD Publishing, 2014).

OECD, *Glossary of Statistical Terms* (OECD Publishing, 2008).

OECD, 'Minerals and Pro-Poor Growth: The Economics and Politics', DAC Guidelines and Reference Series (OECD Publishing, 2008).

OECD Inventory, 'Restrictions on Exports of Raw Materials', http://qdd.oecd.org/subject.aspx?subject=8F4CFFA0-3A25-43F2-A778-E8FEE81D89E2# (accessed December 2014).

Peeling, G., Stothart, P. and Toms, B., 'Increasing Demand for and Restricted Supply of Raw Materials', in OECD, *The Economic Impact of Export Restrictions on Raw Materials* (OECD Publishing, 2010).

Piermartini, R., 'The Role of Export Taxes in the Field of Primary Commodities', WTO Discussion Paper, Economic Research and Statistics Division, 2004, www.wto.org/english/res_e/booksp_e/discussion_papers4_e.pdf (accessed May 2014).

Piérola, F., *The Challenge of Safeguards in the WTO* (Cambridge University Press, 2014).

Plumer, B., 'Why Oil Prices Keep Falling – and Throwing the World into Turmoil', *Vox*, 23 January 2015, www.vox.com/2014/12/16/7401705/oil-prices-falling (accessed January 2015).

Pogoretskyy, V., 'Energy Dual Pricing in International Trade: Subsidies and Anti-dumping Perspectives', in Y. Selivanova (ed.), *Regulation of Energy in International Trade Law: WTO, NAFTA and the Energy Charter* (Kluwer Law International, 2012), pp. 181–223.

Prestowitz, C.V., 'Export Restraints: The Key to Getting Rich', *Foreign Policy Magazine*, 7 (July 2011).

Price, A. H. and Scott Nance, D., 'Export Barriers and the Steel Industry', in OECD, *The Economic Impact of Export Restrictions on Raw Materials* (OECD Publishing, 2010).

Puissochet, J.P., *L'Etat souverain dans le monde d'ajourd'hui* (Pedone, 2008).

Qin, J.Y., 'The Challenge of Interpreting "WTO-Plus" Provisions', *Journal of World Trade* 44 (2010) 127.

Qin, J.Y., 'Reforming WTO Discipline on Export Duties: Sovereignty over Natural Resources, Economic Development and Environmental Protection', *Journal of World Trade* 46 (2012) 1147.

Qin, J.Y., Speech given at the International Centre for Trade and Sustainable Development (ICTSD), Talking Disputes No. 10: The *China–Rare Earths* Disputes, Geneva, World Trade Organization, 8 October 2014.

Qin, J.Y., '"WTO-Plus" Obligations and their Implications for the World Trade Organization Legal System – An Appraisal of the China Accession Protocol', *Journal of World Trade* 37 (2003) 483.

Radetzki, M., *A Handbook of Primary Commodities in the Global Economy* (Cambridge University Press, 2010).

Rajan, M.S., *Sovereignty over Natural Resources* (Radiant Publishers, 1978).

Ratner, M., Parfomak, P.W., Luther, L., and Fergusson, I.F., 'U.S. Natural Gas Exports: New Opportunities, Uncertain Outcomes', CRS Report for Congress, 8 April 2013, www.fas.org/sgp/crs/misc/R42074.pdf (accessed December 2014).

Raval, A., 'Oil Prices Slide to Five-Year Low', *Financial Times*, 8 December 2014.

Reinert, E.S., *How Rich Countries Got Rich . . . and Why Poor Countries Stay Poor* (Constable & Robinson Ltd, 2008).

Reubens, E., 'Commodity Trade, Export Taxes and Economic Development', *The Academy of Political Science* 71 (1956) 42.

Roessler, F., 'GATT and Access to Supplies', *Journal of World Trade* 9 (1975) 25.

Ruta, M. and Venables, A.J., 'International Trade in Natural Resources: Practice and Policy', *Annual Review of Resource Economics* 4 (2012) 331.

Sacerdoti, G., 'Havana Charter (1948)', in R. Wolfrum (ed.), *Max Planck Encyclopedia of Public International Law* (Oxford University Press, 2013), Vol. IV, pp. 718–725.

Sacerdoti, G., 'La disciplina del commercio internazionale e la protezione dell'ambiente', in P. Fois (ed.), *Il principio dello sviluppo sostenibile nel diritto internazionale ed europeo dell'ambiente* (Editoriale Scientifica, 2007), pp. 63–83.

Sacerdoti, G., 'New International Economic Order (NIEO)', in R. Wolfrum (ed.), *Max Planck Encyclopedia of Public International Law* (Oxford University Press, 2013), Vol. VII, pp. 659–668.

Sachs, J. and Warner, A., 'Natural Resources and Economic Development: The Curse of Natural Resources', *European Economic Review* 45 (2001) 827.

Salzman, J. and Wu, M., 'The Next Generation of Trade and Environment Conflicts: The Rise of Green Industrial Policy', *Northwestern University Law Review*, 108 (2014) 401.

Satake, M., 'Trade Conflict between Japan and the United States over Market Access: The Case of Automobiles and Automotives Parts', Pacific Economic Papers No. 310, APSEG, 2000, https://crawford.anu.edu.au/pdf/pep/pep-310.pdf (accessed November 2014).

Schrijver, N.J., *The Evolution of Sustainable Development in International Law: Inception, Meaning and Status* (Martinus Nijhoff Publishers, 2008).

Schrijver, N.J., *Sovereignty over Natural Resources: Balancing Rights and Duties*, 2nd edn (Cambridge University Press, 2008).

Selivanova, Y., *Energy Dual Pricing in WTO Law: Analysis and Prospects in the Context of Russia's Accession to the World Trade Organization* (London: Cameron May, 2008).

Shan, W., Simons, P. and Singh, D. (eds.), *Redefining Sovereignty in International Economic Law* (Hart Publishing, 2008).

Sharma, R., 'Food Export Restrictions: Review of the 2007–2010 Experience and Considerations for Disciplining Restrictive Measures', FAO Commodity and Trade Policy Research Working Paper No. 32, 2011, www.fao.org/fileadmin/templates/est/PUBLICATIONS/Comm_Working_Papers/EST-WP32.pdf (accessed November 2014).

Sieminski, A., 'Implications of the US Shale Revolution', Presentation given at the Canada Energy Summit, Chicago, 17 October 2014, www.eia.gov/pressroom/presentations/sieminski_10172014.pdf (accessed January 2015).

Sinocast Metals, Mining and Chemicals Beat, 'China Cancels 2015 Tungsten, Molybdenum Export Quotas', *Kitco News* 30 December 2014, www.kitco.com/news/2014-12-30/China-Cancels-2015-Tungsten-Molybdenum-Export-Quotas.html (accessed December 2014).

Solleder, O., 'Trade Effects of Export Taxes', Graduate Institute Working Paper No. 8/2013. http://graduateinstitute.ch/webdav/site/international_economics/shared/international_economics/publications/working%20papers/2013/HEIDWP08-2013.pdf (accessed January 2015).

Staiger, R.W., 'Non-Tariff Measures and the WTO', WTO Economic Research and Statistics Division, Staff Working Paper ERSD-2012-01, www.wto.org/english/res_e/reser_e/ersd201201_e.pdf (accessed December 2014).

Stewart, T.P., Drake, E. J. and Gong, P., 'WTO Rules against China's Export Restraints on Rare Earths: A Step Forward in Leveling the Playing Field', 26 March 2014, Stewart and Stewart, www.stewartlaw.com/Article/ViewArticle/989 (accessed December 2014).

Tanzi, V. (ed.), *Fiscal Policy in Open Developing Economies* (International Monetary Fund, 1992).

Taylor, M., *International Competition Law: A New Dimension for the WTO?* (Cambridge University Press, 2006).

The World Bank, 'Commodity Markets Outlook: Global Economic Prospects', July 2014, www.worldbank.org/content/dam/Worldbank/GEP/GEPcommodities/commodity_markets_outlook_2014_july.pdf (accessed December 2014).

The World Bank, 'Commodity Markets Outlook: Global Economic Prospects', October 2014, www.worldbank.org/content/dam/Worldbank/GEP/GEPcommodities/commodity_markets_outlook_2014_october.pdf (accessed December 2014).

UNCTAD, 'State of Commodity Dependence 2012', www.10.iadb.org/intal/intalcdi/PE/2012/12065.pdf (accessed January 2015).

UNCTAD, 'Trade Agreements, Petroleum and Supply Policies' (United Nations, 2002).

UNEP, 'Recycling Rates of Metals: A Status Report', 2011, www.unep.org/resourcepanel/Portals/24102/PDFs/Metals_Recycling_Rates_110412-1.pdf (accessed December 2014).

USGS, *Mineral Commodity Summaries 2014* (Washington, DC: US Government Printing Office, 2014).

UN General Assembly, 'Declaration on the Right to Development, adopted by the General Assembly by means of Resolution 41/28', 4 December 1986.

UN General Assembly, 'Resolution 626 (VII), Right to Freely Exploit Natural Wealth and Resources', 21 December 1952.

UN General Assembly, 'Resolution 1803 (XVII), Permanent Sovereignty over Natural Resources', 14 December 1962.

UN General Assembly, 'Resolution 2016 (XXVII), Permanent Sovereignty over Natural Resources of Developing Countries', 18 December 1972.

UN General Assembly, 'Resolution 2158 (XXI), Permanent Sovereignty over Natural Resources', 19 December 1968.

UN General Assembly, 'Resolution 2386 (XXIII), Permanent Sovereignty over Natural Resources', 19 December 1968.

UN General Assembly, 'Resolution 2692 (XXV), Permanent Sovereignty over Natural Resources of Developing Countries and Expansion of Domestic Sources of Accumulation for Economic Development', 11 December 1970.

UN General Assembly, 'Resolution 3171 (XXVII), Permanent Sovereignty over Natural Resources', 17 December 1973.

UN General Assembly, 'Resolution 32/176, Multilateral Development Assistance for the Exploration of Natural Resources', 19 December 1977.

UN General Assembly, 'Resolution 3201 (S-VI), Declaration on the Establishment of a New International Economic Order', 1 May 1974.

UN General Assembly, 'Resolution 3202 (S-VI), Programme of Action on the Establishment of a New International Economic Order', 1 May 1974.

UN General Assembly, 'Resolution 3281 (XXIX), Charter of Economic Rights and Duties of States', 12 December 1974.

UN General Assembly, 'Resolution 33/194, Multilateral Development Assistance for the Exploration of Natural Resources', 29 January 1979.

UN HLTF, 'Comprehensive Framework for Action' (United Nations, 2008), www.un.org/en/issues/food/taskforce/Documentation/CFA%20Web.pdf (accessed November 2014).

UN World Commission on Environment and Development, 'Our Common Future' (United Nations, 1987), www.un-documents.net/our-common-future-pdf (accessed December 2014).

US Department of Energy, 'Critical Materials Strategy', 2010, http://energy.gov/sites/prod/files/edg/news/documents/criticalmaterialsstrategy.pdf (accessed December 2014).

US Department of Energy, 'Critical Materials Strategy', 2011, http://energy.gov/sites/prod/files/DOE_CMS2011_FINAL_Full.pdf (accessed December 2014).

US Energy Information Administration (EIA), 'Countries Profiles: China', 2014, www.eia.gov/countries/analysisbriefs/China/china.pdf (accessed December 2014).

US Energy Information Administration (EIA), 'What Drives Crude Oil Prices?' 2014, www.eia.gov/finance/markets/supply-opec.cfm (accessed November 2014).

US National Research Council of the National Academies, *Minerals, Critical Minerals and the US Economy* (The National Academies Press, 2008), www.nma.org/pdf/101606_nrc_study.pdf (accessed December 2014).

Van Damme, I., *Treaty Interpretation by the WTO Appellate Body* (Oxford University Press, 2009).

Van den Bossche, P. and Zdouc, W., *The Law and Policy of the World Trade Organization*, 3rd edn (Cambridge University Press, 2013).

Viner, J., 'National Monopolies of Raw Material', *Council on Foreign Affairs* 4 (1926) 585.

von Braun, J. and Díaz-Bonilla, E., 'Globalization of Agriculture and Food: Causes, Consequences and Policy Implications', in J. von Braun and E. Díaz-Bonilla (eds.), *Globalization of Food and Agriculture and the Poor* (Cambridge University Press, 2006), pp. 1–46.

Waart, P.J.I.M., 'Permanent Sovereignty over Natural Resources as a Cornerstone for International Economic Rights and Duties', *Netherlands International Law Review* 24 (1977) 304.

WCO, 'HS Nomenclature 2012 Edition', www.wcoomd.org/en/topics/nomenclature/instrument-and-tools/hs_nomenclature_2012/hs_nomenclature_table_2012.aspx (accessed December 2014).

Weslowsky, T., 'China Tightens Control on the World's Supply of Rare Earths in 2015', Interview with Jack Lifton (InvestorIntel), 10 January 2015, http://investorintel.com/rare-earth-intel/china-tightens-control-worlds-supply-rare-earths-2015/ (accessed January 2015).

Wilson, J., Otsuki, T. and Sewadeth, M., 'Dirty Exports and Environmental Regulation: Do Standards Matters to Trade?', The World Bank Policy Research Working Paper No. 2806, 2002, http://elibrary.worldbank.org/doi/abs/10.1596/1813-9450-2806 (accessed November 2014).

Wohn, A.J.H., 'Towards GATT Integration: Circumventing Quantitative Restrictions on Textile and Apparel under the Multi-Fibre Agreement', *University of Pennsylvania Journal of International Economic Law* 22 (2001) 375.

Won-Mog, C., *'Like Products' in International Trade Law: Towards a Consistent GATT/WTO Jurisprudence* (Cambridge University Press, 2003).

World Mining Congress, 'World Mining Data Report 2014', www.wmc.org.pl/sites/default/files/WMD2014.pdf (accessed December 2014).

Wright, G. and Czelusta, J., 'The Myth of the Resource Curse', *Challenge* 47 (2004) 6.

WTO, 'Agriculture Negotiations: Phase 1 (2000–2001)', www.wto.org/english/tratop_e/agric_e/negoti_ph1_e.htm (accessed November 2014).

WTO, 'Current Situation of Schedules of WTO Members', www.wto.org/english/tratop_e/schedules_e/goods_schedules_table_e.htm (accessed January 2015).

WTO, 'Goods Schedule: Members Commitments', www.wto.org/english/tratop_e/schedules_e/goods_schedules_e.htm (accessed January 2015).

WTO, 'Interaction between Trade and Competition Policy', www.wto.org/english/tratop_e/comp_e/comp_e.htm (accessed January 2015).

WTO, 'Market Access: Special Agricultural Safeguards (SSGs)', www.wto.org/english/tratop_e/agric_e/negs_bkgrnd11_ssg_e.htm (accessed January 2015).

WTO, 'Ninth Ministerial Conference: Ministerial Declaration and Decisions (Bali, 3–6 December 2013)', http://wto.org/english/thewto_e/minist_e/mc9_e/bali_texts_combined_e.pdf (accessed November 2014).

WTO, 'Non-Agricultural Market Access Negotiations News Archive', www.wto.org/english/news_e/archive_e/nama_arc_e.htm (accessed January 2015).

WTO, 'The Bali Decision on Stockholding for Food Security in Developing Countries', https://www.wto.org/english/tratop_e/agric_e/factsheet_agng_e.htm (accessed November 2014).

WTO, 'Trade Monitoring', www.wto.org/english/news_e/archive_e/trdev_arc_e.htm (accessed December 2014).

WTO, 'Trade Monitoring Database', http://tmdb.wto.org (accessed December 2014).

WTO, 'Trade Policy Review', www.wto.org/english/tratop_e/tpr_e/tpr_e.htm (accessed December 2014).

WTO, 'World Trade Report 2010: Trade in Natural Resources', www.wto.org/english/res_e/booksp_e/anrep_e/world_trade_report10_e.pdf (accessed December 2014).

WTO, 'World Trade Report 2012: Trade and Public Policies: A Closer Look at Non-Tariff Measures in the 21st Century', www.wto.org/english/res_e/booksp_e/anrep_e/world_trade_report12_e.pdf (accessed December 2014).

WTO, 'World Trade Report 2013: Factors Shaping the Future of World Trade', 2010, www.wto.org/english/res_e/booksp_e/world_trade_report13_e.pdf (accessed November 2014).

Wu, X., 'No Longer Outside, Not Yet Equal: Rethinking China's Membership in the World Trade Organization', *Chinese Journal of International Law* 10 (2011) 227.

Yang, Y. and Zhong, C., 'China's Textile and Clothing Exports in a Changing World Economy', *The Developing Economies* 36 (1998) 3.

Ying, L., 'The Applicability of Environmental Protection Exceptions to WTO-Plus Obligations: In View of the *China–Raw Materials* and *China–Rare Earths* Cases', *Leiden Journal of International Law* 27 (2014) 113.

Zakariya, H.S., 'Sovereignty over Natural Resources and the Search for a New International Economic Order', in K. Hossain (ed.), *Legal Aspects of a New International Economic Order* (Bloomsbury, 1980), pp. 208–219.

Zambrano, V., *Il principio di sovranità permanente sulle risorse naturali tra vecchie e nuove violazioni* (Giuffrè, 2009).

INDEX

access to markets
  included in reassessment of GATT provisions, 232
access to resources
  challenge of our era, 290–1
Accession protocols
  China's WTO-plus obligations, 150
  post-1994, 199–202
  post-1994 and WTO-plus obligations, 194
  post-1994 rules on export duties, 202–4
  provisions, and 'objective link' as precondition, 194–202
Action Plan on Food Price Volatility (2011) and G20 Agricultural Ministers, 20
ad valorem export taxes, 74
Ad-hoc Working Group (EU), 53, 55
administered pricing, 96
administrative measures and export restrictions, 68
Afghanistan
  export duties on waste and scrap forms of minerals and metals, 82
  export licensing schemes, 95
African continent, developing nations and commodity dependence, 243
African countries imposing export duties, 75
Agreement on Export Taxes
  new, and a plurilateral agreement, 290
  proposed, beyond implementing GATT disciplines, 289
Agreement on Safeguards (WTO), 139
Agreement on Trade Facilitation, Bali (2013), 165, 190, 269
agricultural sector
  export bans and food crises, 87
  factors affecting, 12
  food security concerns, 11–22
  trade-facilitating measures, 10
Algeria
  export licensing schemes, 95
anti-competitive business practices
  unregulated by WTO, 253
AoA. *See* WTO Agreement on Agriculture
Aquila Food Security Initiative, G8 Summit (2009), 19
Argentina
  emerging economy and export taxes, 114
  energy exporter and export duties, 25
  export bans on copper waste and scrap, 88
  export duties and arguments in favour of, 234
  export duties or taxes, 78, 80
  export duties, proliferation of, 77, 168
  export quotas, 89
  export restrictions on critical minerals and metals, 99
  minimum export prices, 97
  Ministry of the Economy and Infrastructure, 80
  producer of borates and lithium, 59
  resource-rich emerging economy, 15
  revenues from export taxes, 118

Argentina (cont.)
waste and scrap forms of minerals and metals, 82
Argentina–Hides and Leather dispute, 171, 172
Article 2 ASCM, export tax, required to be specific, 251
Article 5 GATT, proposed WTO Agreement on Export Taxes, 262
Article I:1 GATT
most-favoured nation (MFN) clause, 143
Article II GATT, 131, 132
Article II:1(a) GATT, 132, 133, 134, 135, 136, 137, 140, 166
Article II:1(b) GATT, 131, 132, 133, 136, 140
Article II:1(c) GATT, 133
Article II:2(a) GATT, 140
Article II:7 GATT, 131
Article VIII GATT, 143
Article X GATT
export taxes, 163
quantitative restrictions on exports, 188
transparency as main objective, 163
Article XI GATT, 169–75
Article XI:1 and Article VIII GATT, combined reading, 141
Article XI:1 GATT, 141, 142, 166, 169–75, 283
exceptions to, 179–88
Article XI:2 GATT, 180–4
Article XI:2(a), 191
Article XI:2(a) GATT, 222, 242
Article XI:2(b) GATT, 184
Article XII GATT
economic exceptions, 261
other measures to achieve economic gains, 283
safeguarding domestic balance of payments, 283
Article XIII:1 GATT, quotes from, 176
Article XIX GATT, 139, 140, 166
export safeguard mechanism, 284
Article XIX:1(a) GATT, 139
Article XVII:1(b) GATT, 179
Article XVIII GATT, 137, 138, 139, 140
economic exceptions, 261
export taxes and economic development, 283
other measures to achieve economic gains, 283
Article XVIII:1(*bis*) GATT, 132
Article XVIII:2 GATT, 137, 138
Article XVIII:3 GATT, 138
Article XVIII:4(a) GATT, 138
Article XVIII:5 GATT, 139
Article XVIII:7 GATT, 138
Article XVIII:7(a) GATT, 138
Article XVIII:8 GATT, 185
Article XVIII:14 GATT, 187
Article XVIII(*bis*) GATT, 138
Article XVIII(c) GATT, 186
Article XX GATT, 183
chapeau of, 223–5
exceptions to WTO-plus obligations, 194
export duties, 193–208
general exceptions, 289
general exceptions for fourth-tier WTO members, 281
industrial policy exceptions, 221–3
security exception, 261
violations of China's Accession Protocol, 207
WTO case law, recent, 208–25
Article XX(b) GATT
justification for export restrictions, 209
Article XX(g) GATT, 214–21
sovereignty, permanent, over natural resources, 240
sustainable development and export safeguards, 234, 242
Article XX(j) GATT, 222
Article XXVIII GATT, 133, 136, 137, 138, 140, 163, 166, 205
adjustability of export duty concessions, 279
export duty concessions, 162

modification of GATT Schedules, 135
modifying or withdrawing concessions, 289
proposed WTO Agreement on Export Taxes, 260
Article XXVIII:1 GATT
renegotiating tariff bindings, 135
Article XXVIII:2 GATT, 136
Article XXVIII:2(a)(*bis*) GATT, 131
Article XXVIII:3 GATT, 136
Article XXVIII:4 GATT, 136
Article XXVIII:5 GATT, 136
Article XXVIII(*bis*) GATT, 137, 166
ASCM. *See* WTO Agreement on Subsidies and Countervailing Measures
Asia, developing countries and demands for energy and metals, 22
Asia/Pacific region, export duties or taxes and increased use of, 75
Australia
export duty concessions, 133, 278
imbalances in export duty obligations, 204
resources versus amounts of production, 61
Australia Group, 87
automobiles, Japan's export of, 6

balance of payment provisions
Committee on Balance of Payments Restrictions and transparency, 284
balance of payments (BoP), 185–8
exception to GATT rules, 191
exception, to address difficulties, 185
Bali Package on food security, 20
Basel Convention on the Control of Transboundary Movements of Hazardous Wastes and their Disposal, 87
bauxite, 'carrier' of gallium, 61
beggar-thy-neighbour
export restrictions, 231, 234
instruments as export restrictions, 245–54
natural resources policies, 238
policy goals and distortion of trade, 33
Belarus
export licensing of semi-processed products, 95
export quotas, 89
benchmarks for export prices, 250
benchmarks, cross-border, for export prices, 251
BIS. *See* US Bureau of Industry and Security
Bolivia and surtaxes on exports, 85
boom-bust price cycles and primary commodities, 186
Brazil
export licensing of semi-processed products, 94
Brazil–Retreaded Tyres dispute, 211
Article XX GATT and analysis of discrimination, 224
British Empire, 1

Canada
Canada–Herring and Salmon dispute, 170, 184
export quotas on certain wood products, 89
producer of minerals and metals, 59
resources versus amounts of production, 61
Canada–Periodicals dispute, 170
Canada–Wheat dispute, 179
capital costs of mining
uncertainty caused by export restrictions, 108
carbon steel, China's export restrictions and prices, 120
CDDCs. *See* commodity-dependent developing countries
Central Registry of Notifications, 164
Chile and enhanced transparency in export licensing, 266

China, 82, *See also China–Rare Earths* dispute; *China–Rare Materials* dispute
  Accession Protocol, 195–9, 200–1
  coke, export restrictions on, 120
  demand for energy and metals, 22
  dismantling of minimum export prices, 97
  distribution of reserves, 61
  economic slowdown, 23
  emerging economy and export restraints, 114
  energy commodities and reducing import dependence, 26
  energy consumer and importer, 26
  environmental grounds for export laws, 117
  environmental policy, 210
  expanding mine production, 64
  export duties eliminated on some minerals and metals, 83
  export duties or taxes, 78, 79
  export license requirements, strict, 96
  export licensing requirements, 94
  export licensing requirements for ores and concentrates, 94
  export quotas, 89
  export quotas on mineral resources, 191
  export restrictions on minerals and metals, 99
  export taxes on critical minerals and metals, 78
  Foreign Trade Law, 90
  interim export duties, 79
  mine production of critical minerals, 58
  minimum export prices, 97
  mining production, 79
  National Five-Year Plan (2011–2015), 114
  non-automatic export licences, 95
  obligation to eliminate export duties, 205
  reduction of VAT rebates to discourage exports, 85
  Regulation on the Administration of the Import and Export of Goods, 90
  resource-rich nation, 15
  scope and coverage of export duty obligations, 161
  sovereignty over natural resources and export disputes, 239
  surtaxes on exports, 85
  table of mine production, 60
  Tariff Implementation Plan, 79
  WTO-inconsistent export duties, 167
  WTO-plus obligations, 84, 150–2
China's Accession Protocol
  China and export duties, elimination of, 195
  post-1994, 195–9, 200
China–Audiovisuals dispute, 195, 196
China–Rare Earths dispute, 100, 117, 151, 171, 184, 193, 194, 195, 197, 202, 207, 291
  Article XX(g) exceptions to obligations, 214
  environmental exceptions, 209
  export quotas as 'safeguard tools', 218
  future of export restrictions, 273
  health and environmental concerns, 213
  imbalances in WTO obligations, 205
  in imposition of restrictions, 221
  obligations of newly acceded WTO members, 225
  panel's analysis on export restrictions, 223
  protection of domestic industry versus non-discrimination principle, 222
  ruling, 92
  sovereignty over natural resources, 239

China–Rare Earths Report, 83
China–Raw Materials
  Article XX(g) exceptions to obligations, 214
  health and environmental concerns, 212
China–Raw Materials dispute, 92, 99, 117, 151, 171, 173, 180, 184, 193, 194, 195, 198, 199, 202, 204, 291
  accession protocols, 238
  environmental exceptions, 209
  future of export restrictions, 273
  imbalances in WTO obligations, 205
  protection of domestic industry versus non-discrimination principle, 222
  shortages and absence of word 'critical', 222
  sovereignty over natural resources, 239
Chinese Ministry of Commerce (MOFCOM), 26, 90
cobalt and growth in demand for, 63
Colombia
  export licensing of semi-processed products, 95
  taxes on exports, 84
Colombia–Ports of Entry dispute, 175
colour televisions, Japan's export of, 6
Committee on Agriculture and reports on food security and restrictions, 19
Committee on Balance of Payments (BoP) Restrictions
  model for new WTO committee, 277
  transparency, 284
Committee on Market Access and reporting on existing measures, 267
commodities
  price spike of 1972–1975, 4
  versus raw materials, 43
commodities boom, of 2000s, 1
commodities, hard
  2008 price spike, 8
commodity boom of 2003–2008, 7, 8
commodity dependence
  versus resource-dependent economies, 186
  vulnerability to price volatility, 243–4
commodity-dependent developing countries (CDDCs), 11, 235
  export restrictions as development tools, 33
competition
  standard economic theory, 103–5
conservation
  broad definition of, 215
  definition of term in Article XX(g), 239
  goal of, and connection to export restrictions, 219
  limits on 'stretching', 222
  natural resources, exhaustible, and export restrictions, 209
  -related export restrictions, 220
  sustainable economic development, 240
conservation of natural resources
  GATT exceptions to export rules, 182
consumption
  volume restrictions on, 220
Convention on International Trade in Endangered Species of Wild Fauna and Flora, 87
Convention on the Harmonized Commodity Description and Coding System, 48
copper
  'carrier' of tellurium, 61
  export taxes, number of countries using, 78
Costa Rica and enhanced transparency in export licensing, 266
Council for Trade in Goods, 189, 264, 285
  Committee on Export Restrictions, 276
  submission of notifications of export taxes and licensing schemes, 276
countries, developing
  high growth performances, 63

countries, industrialized versus developing, 64
country-specific obligations
  export duties, 145–63
coupled elements, 61
Cournot-Nash oligopoly model, 105, 107
critical, 31–4
critical minerals and metals
  China's WTO-plus obligations, 150
  demand surges and vulnerable countries, 241
  distinctive features, 56–63
  effects of export restrictions, 119–22
  export duties or taxes, 77–84
  list of, 54–6
  recycling of, 63
  substitutability as problem for, 62
  trade patterns, international, 63–5
  unrestricted trade and importers of resources, 65
critical minerals and metals, 73
critical minerals, list of, and Ad-hoc Working Group (EU), 55
critical shortage of essential products
  exceptions to GATT rules, 191
critical shortage of supply clause
  GATT exception to export restrictions, 180–4
critical shortages of foodstuffs or other products
  GATT exceptions to export rules, 180
Croatia and WTO-plus obligations, 149
crude oil and natural gas and US export barriers, 27
Customs Union (Russian Federation, Belarus and Kazakhstan), 81

DAF Brest, 26
Decision on Notification Procedures for Quantitative Restrictions, 189
Decision-Making Procedures, 282
Declaration on Trade Measures Taken for Balance of Payment Purposes (1979), 186
deconsolidation procedure and renegotiating tariff bindings, 135
deforestation
  reduced through export laws, 116
demand and supply of mineral raw materials, 48–51
demand growth for minerals and metals, 63
demand surges
  countries vulnerable to supply disruptions, 241
demand, expansion, and export surges, 241
Democratic Republic of the Congo (DRC)
  export duties, proliferation of, 168
  export restrictions, 81
developed countries
  likely exclusion from export safeguards, 285
developing countries
  commodity dependence, 243–4
  commodity-dependent and restrictions in support of development, 234
  import restrictions, 186
  incentives for reallocation of profits, 110
development
  economic, and export restrictions, 33
development concerns
  resource-endowed countries, 234
developmental tools
  arguments in favour of export restrictions, 234–44
Director General of the WTO, 71
Directorate-General of Trade, European Commission, 10
  Raw Materials Initiative and export restrictions, 71
Doha Agricultural Negotiations, 16
Doha Ministerial Declaration, 257–8

lack of consensus on mandate on export restrictions, 286
Doha Negotiating Committee on Agriculture, 18
Doha negotiations
Agriculture Negotiations of the WTO, 285
analysis of tabled proposals, 269
GATT adjustment procedures applied to imports and exports, 283
Doha negotiations on export restrictions
general stalemate of, 265
Doha proposals on export restrictions, 255, 257
Doha Round
negotiations on export taxes, 277
proposals on enhanced transparency, 275
domestic industries, 248
development goals versus trade liberalization, 246
export restrictions to protect resources, 241
minerals, 107
restrictions on, in symmetry with export tariffs, 214, 219
WTO members and limits on export quantitative restrictions, 187
domestic market obligations, 97
domestic prices of raw materials
reduced through export restrictions, 247
domestic production or consumption
versus export restrictions, Article XX(g) GATT, 214
Dominican Republic
export licensing requirements, 94
taxes on exports, 84
downstream industries
developing nations, 284
domestic supply of raw materials, 241
export duties as instrument to promote, 238
export taxes and benefits for, 250
protected by Article XX GATT, 221
protection of, versus competing with foreign buyers, 284
DRC
export taxes on specific minerals and metals, 84
Dutch disease
dependency on mineral export earnings, 112
versus resource-led development, 33
vulnerability of countries to trade shocks, 186
duties, import and export
asymmetries in commitments, 135–40

EC–Asbestos dispute, 170
economic development goals
low-income, commodity-dependent countries, 192
economic development tools
export restrictions, 101, 111
economic diversification
export duties as instrument to promote, 238
set back by price volatility, 284
economic effects of export restraints, 122–3
economic emergency exception (GATT), 139
economic models, theoretical, and effects of export restrictions, 103–5
economic self-determination and sovereignty over natural resources, 236
economies, emerging, and support of development, 234
economy, resource-based, 186
EEC–Minimum Import Price dispute, 174
elasticity of demand and global welfare, 109
elasticity of supply and demand and global welfare, 108
emerging economies, 113
energy

energy (cont.)
  clean technology and 'critical' materials, 54
energy commodities and export restrictions, 22–8
energy markets, international, 23
energy security and tension with fuel export-dependent countries, 24
environmental concerns
  justifications for export restrictions, 209
environmental considerations
  use of clean technologies, 57
environmental protection, 30
environmental standards
  Article XX(b) GATT, 213
EU. *See also* European Union *entries*
EU and 'criticality' of minerals and metals. *See* European Union
EU economy
  'criticality' of minerals and metals, 52
EU Revised Submission on Export Taxes, 268
European Commission
  Raw Materials Initiative on criticality of resources, 51
  Report of the Directorate-General for Trade, 10
  Report on Potentially Trade-Restrictive Measures, 247
  trade measures and industrial policy, 114
European Union
  Ad-hoc Working Group, 51, 52
  battles over export restrictions, 279
  importer of minerals and metals, 64
  proposal for enhanced transparency, 276
  proposal on new Agreement on Export Taxes, 258–66
  revised submission, proposal on export taxes, 262–6
exchange rates, floating, 185
export barriers
  reserving minerals and metals for a country's own industry, 33
export duties, 165–8
  Article XX GATT, 193–208
  as economic development tools, 239
  ban-equivalent, 141
  critical minerals and metals, 77–84
  GATT provisions, 143–5
  non-discriminatory for WTO members, 143
  post-1994 accession protocols, 202
  proliferation of, 74–7, 165
  proposal of elimination of, for non-agricultural products, 259
  systemic imbalances in WTO obligations, 204–8
  versus import tariffs, 77
  versus prohibitions or restrictions on exports, 169
  weakness of notification requirements under WTO law, 168
  WTO accession regime, 233
export duty commitments
  classified by WTO, 134
  General Agreement on Tariffs and Trade (GATT), 131–5
  non-tariff measures, 130
  scheduled, 133
  WTO members, 132
export duty concessions, 138
  negotiations among WTO Members, 282
  product-by-product basis, 280
  renegotiation procedure for WTO members, 140
export duty rates compared to processing levels, 83
export licensing
  definition, 267
  proposed protocol on enhanced transparency, 266
  requirements, 93–6
  schemes, as frequently reported, 191
export prohibitions or bans, 86–8
export quantitative restrictions
  GATT rules regarding, 169–75
  transparency and notification requirements, 188
export quotas
  distortions of market, compared to export taxes, 105

effect, more negative, than export taxes on global welfare, 108
interchangeable with export taxes, 102, 104
representative period of three years, 177
'signalling' function, 217

export restrictions, 97
access to supplies versus development needs, 231–4
addressed in Doha Ministerial Declaration, 257–8
Article XX GATT, chapeau of, 223–5
as beggar-thy-neighbour instruments, 245
as developmental tools, 234–44
categories of, 68
debate over critical minerals and metals, 31–4
definition of raw materials, 41–4
determination of 'arbitrary or unjustifiable discrimination', 223
disciplined, to avoid beggar-thy-neighbour policies, 254
effects on critical minerals and metals, 119–22
energy commodities, 24
enhanced transparency in reporting, 274–7
environmental concerns, 116–18
GATT avenues for symmetry with import restrictions, 192
GATT provisions, relevant to quantitative restrictions, 175–9
government imposed, at-the-border, 67–9
historical trends, 1–8
increased cost to foreign consumers versus decreased cost for domestic users, 227
increasing in 2000s, 8
increasing numbers of, and responses to, 67–9
industrial policies, revival of, 28–31
Non-discriminatory Administration of Quantitative Restrictions, 176
overview of, 98–100
peak, 2010–2011, 9
present day, 8–11
quantitative, and GATT exceptions to, 179–88
recent trends in, 86–98
safeguard measures to conserve resources, 241–3
state trading enterprises separate from, 179
study of, in this book, 34–7
subsidy-equivalent effects of, 247–52
welfare of countries and redistributional effects, 105–8
WTO case law, recent, 208–25
WTO legal framework, traits of, 254
WTO members and requirements to report on, 178

export restrictions, 'wave of', 1
export royalty, 84
export safeguard mechanism, 284–6
export surges
protection against, in safeguard mechanisms, 284
restrictions to conserve resources, 241

export tax
beggar-thy-neighbour policy, 106

export tax applied by 'large' country
effect, negative, on global welfare, 106

export taxes
distortive effects on global commodity trade, 263
economic effects on large and small nations, 104
energy exporters, 24
new Agreement on Export Taxes, 289
primary cause of tension between countries, 254

export taxes or duties, defined, 72
exporters and access to markets
multilateral trading system, 127–31

extraction and production quotas
compliance with obligations of WTO members, 220

extractive sector
low prices and decline in investment, 22

FAO. *See* Food and Agricultural Organization
financial crisis of, 1930s
  origins of GATT, 128
fiscal tax on exports, 84
Five-year Rare Earths Industry Development Plan (China), 115
Food and Agricultural Organization (FAO)
  survey of countries and commodities, 13
food commodities, 244
food crisis of 2007–2008, 7, 15
food prices
  increase, 2010–2011, 7
  non-commercial humanitarian purposes, 19
  volatility, in 2000s, 13
food security
  agricultural export restrictions, 11
  public stockpiles, 21
foreign and domestic prices
  price gap and unfilled export quotas, 224
Foreign Trade Law China, 90
future of export restrictions and WTO legal system, 273–4

G20 countries
  Agricultural Ministers, 20
  refraining from new export restrictions, 69
  Vice Ministers of Agriculture (May 2012 meeting), 20
G8 Summit (2009), 19
Gabon
  export duties on manganese, 82
  export duties, proliferation of, 168
  export taxes on specific minerals and metals, 84
gallium
  extracted in addition to bauxite, 61
  growth in demand for, 63
  produced in Ukraine, 59
GATS. *See* General Agreement on Trade in Services
GATT. *See also* General Agreement on Tariffs and Trade
  origins of, in 1930s financial crisis, 128
  Schedules of Concessions, 163, 264, 278
  Schedules of Concessions and export duties, 212
GATT Analytical Index, 73
GATT and export restrictions, current era, 231–4
GATT exceptions to restrictions on exports, 179–88
GATT flexibilities, expansion of, 282–4
GATT principle of non-discrimination, 222
GATT provisions for negotiating export duty concessions, 278
General Agreement on Tariffs and Trade (GATT), 5, *See also individual articles in agreement*
  Agreement on Agriculture, 16
General Agreement on Trade in Services (GATS), 162
general elimination obligation, 169
General Elimination of Quantitative Restrictions, 169–75
geoeconomic changes
  trade patterns, international, 63–5
geographic concentrations of minerals and metals, 65
geographic factors and critical minerals, 58
germanium, 61
Germany and economic slowdown, 23
global financial and economic crisis (2008–2009), 8, 10, 29, 30, 78, 99, 122
  export restrictions as reaction to, 69
global value chains (GVCs), 236
  disruptions caused by export restrictions, 245–7
  globalization of production, 246
global welfare
  effects of export quotas, 108

effects of export taxes in producing countries, 107
elasticity of demand and effects on, 109
export tax applied by 'large' country and negative effect on, 106
profit-shifting effect, 107
world prices of minerals and metals, 108
Governmental Assistance to Economic Development, 137
governmental financial contribution requirement for WTO members, 249
governmental stabilization plans, 221
Grenada
export licensing requirements for ores and concentrates, 94
Group of 20. *See* G20
Group of 8. *See* G8
Group on the Sector Approach
Tokyo Round negotiations, 233
Guyana
export licensing schemes, 95
GVCs. *See* global value chains

Harmonized System 6-digit level, 29
Harmonized System Nomenclature (HS), 47
Havana Charter, 128
heavy rare earth elements (HREE), 55
High Level Task Force on the Global Food Security Crisis (HLTF), 15
HREEs and growth in demand for, 63
HS. *See* Harmonized System Nomenclature

import restrictions
imbalances caused by excess imports/exports, 186
targeted by GATT exceptions, 185–8
import-dependent countries
economic effects of export restrictions, 101–3
product-by-product basis for export duty concessions, 280
workable compromise on export taxes, 274
imported products
protection of competitive opportunities, 172
India, 114
agriculture and minimum export prices, 97
captive mining, 98
chromium ore and export tax, 121
congestion surcharge, 85
environmental grounds for export laws, 117
export bans on certain wood products, 87
export duties and arguments in favour of, 234
export duties, proliferation of, 77, 168
export licensing requirements for ores and concentrates, 94
export restrictions, 81
export restrictions on critical minerals and metals, 99
resource-rich emerging economy, 15
India–Autos dispute, 170, 171
India–Quantitative Restrictions dispute, 170
Indicative List of Notifiable Measures
Uruguay Round Ministerial Decision on Notification Procedures (1993), 164, 189
indium, 61
indium, growth in demand for, 63
Indonesia
domestic processing protected by export restrictions, 115
emerging economy and export restraints, 114
environmental grounds for export laws, 117
export bans on certain wood products, 87
export duties, proliferation of, 168
export license requirements, 96
export licensing of semi-processed products, 94, 95
export restrictions, 81
export restrictions and domestic processing, 100

Indonesia (cont.)
export restrictions on critical minerals and metals, 99
export taxes on specific minerals and metals, 84
minimum export prices, 97
mining and domestic processing, 98
Ministry of Energy and Mineral Resources, 88
resource-rich emerging economy, 15
industrial policies
export restrictions, 101
trade measures as part of, 114
industrial policy exceptions
Article XX GATT, 221–3
industrial policy instruments
advantages for domestic industry, 254
industrial policy needs
WTO provisions for protections, 218
industrialization, 280
industrialized countries
concerns about export restrictions, 114
dependence on importing minerals and metals, 31
infant industry protection purposes
export duties and, 283
international law
sovereignty over natural resources, 215
international market prices
undercut by domestic supply and production, 249
International Trade Organization (ITO), 128
iron and steel
manufacturing sectors, recovery of, 63

Jamaica
export licensing schemes, 95
Japan
and United States, trade relations, 6
cotton textiles, and restrictions by US, 4
export licensing of semi-processed products, 95
proposed protocol on enhanced transparency, 266
Japan–Semiconductors dispute, 171, 172, 173, 174, 188

Kazakhstan, 26
energy exporter and export duties, 24
export duties on waste and scrap forms of minerals and metals, 82
resource-rich emerging economy, 15

Lao People's Democratic Republic (Lao PDR)
WTO-plus obligations, 159
Latin America
exports, restrictions on, 2
Latvia
export duties eliminated on certain products, 161
obligation to eliminate export duties, 205
obligation to eliminate export tariffs, 203
WTO-plus obligations, 148
least developed countries (LDCs)
export duties, proliferation of, 75
scheduling of export taxes, 280
least developed countries and GATT schedules, 265
light rare earth elements (LREE), 55
light rare earth elements (LREEs)
growth in demand for, 63

machine tools, Japan's export of, 6
magnesium metal, growth in demand for, 63
Malaysia
energy exporter, 25
environmental grounds for export laws, 117
export duties, proliferation of, 168

export licensing of semi-processed products, 94, 94
export licensing requirements, 94
export licensing requirements for ores and concentrates, 94
export restrictions, 81
export taxes on specific minerals and metals, 84
manufacturing sectors
China, 114
dependence on raw materials, 30
iron and steel and recovery of sectors, 63
mineral raw materials, 46
primary products and commodities, 43
Western economies and dominance of, 122
Market Access for Non-agricultural Products, 257–8
market benchmark
Article 1.1(b) ASCM, 250
Marrakesh Agreement, 195, 199–202, 215, 286
Article XII
1, 146
ratification of new agreement, 289
sustainable development, 239
Marrakesh Protocol, 137
materials, raw, and access to, 7
Mauritius, 95, 96
MEAs. *See* multilateral environmental agreements
MEP. *See* minimum export price
metal extraction, 44
metal prices
between 2005 and 2008, 22
metals
classification of, 46–8
metals for technology
export restrictions, 99
Mexico
energy exporter, 25
producer of silver, fluorspar, and molybdenum, 59
MFN. *See* most favoured nation (MFN) clause
Middle East, and security of supply problems, 23
mineral 'criticality'
defined, 51–4
mineral industry, domestic, 107
mineral raw materials
classification of, 46–8
definition of, 41–4
definition of 'criticality', 51–4
demand and supply determinants, 48–51
mineral resources
adequate domestic supply and exceptions to export rules, 180
conservation of, through export laws, 116
minerals and metals
export restrictions, present day, 9
factors, sector-specific, and export restrictions, 108–10
minerals and metals sector
economic effects of export restrictions, 101
export duties, 127
export licensing requirements, 94
minerals sector
need for enhanced transparency in reporting export taxes, 275
minimum export price (MEP), 96–8
mining
Asian deposits and increasing demand, 63
Australia, and increasing demand, 63
Canada, and increasing demand, 63
captive, 98
China, and critical minerals, 58
defined, 45
deposits for ecological recovery, 213
mineral resources, defined, 89
processing companies and domestic requirements, 98
Rwanda and royalties on exports, 84
uncertainty of global prices caused by restrictions, 108
United States, and increasing demand, 63
volume restrictions and Article XX(b) GATT, 213

Ministry of Energy and Mineral Resources (Indonesia), 88
Missile Technology Control Regime, 87
Modification of GATT Schedules, 137
molybdenum products
  China and export restrictions, 120
Mongolia
  export duties eliminated on certain products, 161
  obligation to eliminate export duties, 203, 205
  WTO-plus obligations, 148
mono-economy
  dominance of mineral exports, 113
Montenegro
  obligation to eliminate export duties, 203, 205
  scope and coverage of export duty obligations, 161
  WTO-plus obligations, 159
Montreal Protocol on Substances That Deplete the Ozone Layer, 87
Morocco and export licensing schemes, 95
most favoured nation (MFN)
  export duties or taxes, 77
most favoured nation (MFN) clause, 143
  and export duties, 143–5
most favoured nation treatment
  customs duties and export fees, 73
multilateral environmental agreements (MEAs), 87
multilateral negotiations
  engagement of WTO members, 291
Multilateral Trade Agreements
  China's Accession Protocol, post-1994, 200
  deal on transparency, 287
  Marrakesh Agreement, 195
  rights and obligations of WTO members, 200
multilateralized export tariff bindings
  benefits to WTO members, 288

National Five-Year Plan (2011–2015) (China), 114
nations, developing
  demands for energy and metals, 22–8
natural disasters
  Article XX(j) GATT, 222
natural gas
  price levels, 24
  US exports, 27
natural resource, exhaustible
  Article XX(g), 214
natural resources
  defined as raw materials, 42
  definition of, 43
  fixed and immobile versus tradable, 43
  four categories, 42
  scarcity, temporary, addressed in Article XX(j) GATT, 222
  sovereignty of nations over, 235–7
  subject to WTO law, once extracted, 219
  UN General Assembly Resolution 626, 236
natural resources, exhaustible
  conservation as justification for export restrictions, 209
Negotiating Group on Market Access, 266
Negotiating Group on Non-agricultural Market Access (NGMA), 257, 258, 259
  limited progress in, 266
  NAMA Negotiating Group, 279
  Textual Report of the NGMA Chairman, 265, 268
negotiation of export duty concessions
  WTO Members' GATT schedules, 277–82
net food-importing developing countries (NFIDCs), 17
new international economic order (NIEO)
  sovereignty over natural resources, 236
NFIDC. *See* net food-importing developing countries

NGMA. *See* Negotiating Group on Non-agricultural Market Access
NGMA and first EU submission on export taxes, 283
NIEO. *See* new international economic order
Nigeria
  energy exporter, 25
  export bans on certain wood products, 87
niobium, growth in demand for, 63
non-agricultural products and WTO Agreement on Export Taxes, 259
non-discrimination
  GATT principle, 222
Non-discriminatory Administration of Quantitative Restrictions, 176
non-tariff barriers (NTBs), 130
non-tariff border restrictions
  prohibited by GATT, 191
Non-Tariff Measure Inventory (1968), 130
notification requirements
  WTO members' state trading, 263
NTBs. *See* non-tariff barriers
Nuclear Suppliers Group, 87

OECD. *See* Organisation for Economic Co-operation and Development
  study on export quotas, 89
  survey of export taxes (2009), 74
OECD Inventory
  export licensing requirements, 94
OECD inventory of industrial raw materials, 72
OECD list of strategic raw materials, 52
OECD Trade Committee
  inventory of export restrictions, 71
oil prices
  between 2005 and 2008, 22
  embargo by OPEC, 4
  Mideast wars and sanctions, 23
  world demand and, 27
oil prices and the shale revolution, 23
oligopolistic (quantity-setting) firms, 105
oligopoly, international
  effects of export taxes, 107
options in reforming WTO disciplines, 286–90
ores
  resource taxes on, 213
Organization of the Petroleum Exporting Countries (OPEC), 4
  production quotas, 27
  reliance of US on Arab oil, 26
  supply management policies, 23
Organisation for Economic Co-operation and Development (OECD), 42
  'criticality' of raw materials, 51
ownership of raw materials, 109

Pakistan
  rice and minimum export prices, 97
panic-buying and export surges, 241
Paraguay
  export licensing of semi-processed products, 95
Peru
  producer of silver, copper, and molybdenum, 59
Philippines
  export bans on PGM ores and concentrates, 88
  export licensing of semi-processed products, 94
  export licensing requirements, 94
  export licensing requirements for ores and concentrates, 94
  producer of nickel, 59
policy objectives
  achieved by export restrictions, 118–19
  export measures and reactions of other producers, 122
pollution
  reduced through export laws, 116
pollution tax
  Article XX(b) GATT, 213

precious metals
  minimum export prices, 97
price elasticity
  minerals and metals sector, 252
price or tax measures on exports, 68, 84–6
price spikes
  critical raw materials, and world demand, 285
  energy sector, in 2000s, 24
  mid-2000s, 22
primary products
  versus raw materials, 43
primary resources
  developing nations and dependence on, 243–4
primary versus secondary materials, 48
private suppliers
  compliance with government WTO-plus obligations, 249
processing of ores and concentrates, 45
production and consumption
  effects on, caused by export restrictions, 105–8
products scheduled and bound in Members' GATT schedules
  export safeguards for agriculture, 285
products, processed
  subsidized, to assist a nation's workforce, 112
profit-shifting effect
  export taxes and global welfare, 107
protection of life and health
  justification for violations of WTO obligations, 210
Protocol on Transparency in Export Licensing to the GATT, 266–9
public health
  protections provided by export laws, 116
public policy goals
  justifications for export restrictions, 208
Publication and Availability of Information
  provision on notification of restrictions by WTO members, 190
qualified exporters lists
  Indonesia, 97
quantitative export restrictions
  GATT exceptions to, 179–88
  limits on, for WTO members, 187
quantity measures
  export restrictions, 68
quotas for export, as restriction measure, 68

Radetzki, 22
rare earth elements
  China's export regime, 121
  clean energy technologies, 54
  used in clean energy technology, 57
  uses for, 57
raw materials
  access to, and beggar-thy-neighbour policies, 245–54
  agricultural versus industrial, 44
  categories of, 41–4
  classification of minerals and metals, 46–8
  domestic prices reduced through export restrictions, 247
  resurgence of export restrictions, 34–7
Raw Materials Initiative, 71
raw materials of mineral origin, 44–6
recycling infrastructures
  Article XX(b) GATT, 213
Regulation on the Administration of the Import and Export of Goods (China), 90
Regulations on Import and Export Tariffs, 79
renegotiation of concessions in GATT schedules, 135
rent-seeking induced by export restrictions, 252–4
Report on Potentially Trade-Restrictive Measures, 247
Republic of Korea
  support for enhanced transparency in export licensing, 266
resource conservation, 30
resource curse theory, 112
resource-based economies, 186

resource-endowed developing countries
justifications for export taxes, 274
resources
extracted versus cultivated, 44
restrictions
defined by WTO case law, 170
restrictions on industrial raw materials, 8
revenues generated by export duties or taxes, 118
Revised Submission on Export Taxes, 257
risk factors
'criticality' of mineral raw materials, 52
supply disruptions, 51
supply risk, defined, 53
Rounds of multilateral negotiations
concessions exchanged by WTO Members, 287
Russia
export duties, proliferation of, 77
Russia's Export Duty Schedule, 157
Russian Federation, 97
emerging economy and export restraints, 114
energy exporter and export duties, 24, 25
export duties of specific products, 167
export duties or taxes, increasing use of, 81
export duty concessions, 278
export licensing requirements, 94
export licensing requirements for ores and concentrates, 94
export quotas on certain wood products, 89
export restrictions on critical minerals and metals, 99
export taxes on critical minerals and metals, 78
incorporation of export duty commitments in GATT Schedule, 278
minimum export prices, 97
non-automatic export licences, 95
relaxation of export restrictions, 100
resource-rich emerging economy, 15
tariff lines in Schedule of Concessions, 161
waste and scrap forms of minerals and metals, 82
WTO-plus obligations, 84, 156–9
Rwanda and mining royalties on exports, 84

Sacerdoti, Giorgio, xiv, xv
safeguard measures taken by WTO Members, 185–8
Saudi Arabia
export duties eliminated on certain products, 161
obligation to eliminate export duties, 203, 205
WTO-plus obligations, 152
scarcity
factor in value of raw materials, 44
feature of natural resources, 43
Schedules of Concessions, 142
export duty concessions, 140
tariff bindings, 131
WTO members, 132, 134
Schedules of Concessions (GATT), 131
scrap metals, 48, 63
Second World War, 1
shortages, severe, caused by, 222
tariffs, import, and financial crisis of 1930s, 128
Secretariats of the WTO, the OECD and UNCTAD
Joint Report on export restrictions, 69
semiconductor producers
export restrictions placed on, 172
semi-conductors, Japan's export of, 6
Separate Customs Territory of Taiwan, Penghu, Kinmen and Matsu
support for enhanced transparency in export licensing, 266
shale revolution
effect on oil prices, 23

shortages, temporary versus permanent
    and GATT exceptions to export rules, 181
single sellers, exporting firms and higher prices of restricted products, 252
SITC. *See* Standard International Trade Classification
SITC categories of minerals and metals, 48
social costs, shifting of, with export taxes, 106
South Africa
    chromite, exports of, 121
    emerging economy and export restraints, 114
    environmental grounds for export laws, 117
    export license requirements, 96
    export licensing of semi-processed products, 94, 94
    export licensing requirements for ores and concentrates, 94
    export restrictions and domestic processing, 100
    non-automatic export licences, 95
    resource-rich emerging economy, 15
sovereignty
    over natural resources, 232
    over natural resources and GATT reassessment, 232
sovereignty over natural resources, xiii, 7, 32, 33, 238, 290
    international law, 215
    permanent, 235
    principle of non-discrimination, 255
    right of a nation to domestic minerals and metals, 33
    sustainable development concerns, 239
    WTO accession regime, 237–9
special agricultural safeguards (SSGs), 285
Sri Lanka and export licensing schemes, 95
SSG. *See* special agricultural safeguards
Standard International Trade Classification (SITC), 47
state trading enterprises (STEs), 98, 175, 178
    export restrictions separate from, 179
    export tax associated with sales or purchases, 249
steel-making and export restrictions, 99
STEs. *See* state trading enterprises
stockpiling strategies as cause of export surges, 241
subsidizing national sectors, 112
subsidy-equivalent effects of export restrictions, 247–52
substitution
    inadequate, due to properties of critical materials, 62
substitution process
    export quotas to replace export duties, 207
supplies, access to, versus developmental needs, 231–4
supply and demand in mineral raw materials, 48–51
supply and use of minerals and metals
    contraction of, to conserve resources, 117
supply distortions, 101
supply risk factor, determined by four factors, 53
supply shortages and export surges, 241

Tajikistan
    elimination of export licensing, 96
    export licensing schemes, 95
    obligation to eliminate export duties, 203, 205
    scope and coverage of export duty obligations, 161
    WTO-plus obligations, 160
tariff binding and bargaining among WTO members, 129
tariff binding commitments
    Schedules of Concessions, 131
tariff bindings
    China, 161
    deconsolidation procedure, 135
    developing countries and LDCs, 261

multilateralized and benefits to WTO Members, 288
Russia's Export Duty Schedule, 157
Tajikistan, 161
tariff concessions
WTO members and flexibility in, 166
tariffs, import versus export (GATT/WTO), 127–31
technologies, environmentally friendly, and Article XX(b) GATT, 213
technology, cutting-edge
effects of export restrictions, 246
tellurium, 61
Tokyo Round
export restrictions on raw materials, 232
negotiations on export duty concessions, 277
TPRB. *See* Trade Policy Review Body
TPRs. *See also* Trade Policy Reviews
data on export duties or taxes, 74
definition of export taxes and duties, 73
trade liberalization
preferences for domestic industrial development, 246
trade measures
part of industrial policies, 114
Trade Policy Review Body (TPRB), 70
Trade Policy Reviews (TPRs), 71
trade shocks
vulnerability of resource-based economies, 186
trade-distortive measures
quantitative restrictions versus tariffs, 207
trading system, multilateral
origins of, 128
transparency
making a deal on, among WTO Members, 286
stricter provisions on, 263
transparency and notification requirements
for WTO members, 188–90
transparency in export licensing
proposed protocol, 266–9
transparency, enhanced
export restrictions, 274–7
Trinidad and Tobago
export licensing schemes, 95
Turkey
export licenses and stringent requirements, 95
Turkey–Textiles dispute, 190
Turkmenistan
export licensing schemes, 95

Ukraine
export duties of specific products, 167
export duties phased down on certain products, 161
export duty rates on non-ferrous scrap, 83
export licensing of semi-processed products, 95
export quotas, 89
export quotas on copper ash and residues, 89
export restrictions on critical minerals and metals, 99
imbalances in WTO obligations, 205
obligation to eliminate export duties, 203
producer of gallium and germanium, 59
relaxation of export restrictions, 100
resource-rich emerging economy, 15
support for enhanced transparency in export licensing, 266
WTO-plus obligations, 84, 155
UN General Assembly Resolution 626
sovereignty over natural resources, 236
Understanding on Balance of Payment Provisions of the GATT 1994, 186
Understanding on the Interpretation of Article XVII GATT, 269
Negotiating Group on Non-agricultural Market Access, 263
proposal by EU for enhanced transparency, 276
United Nations Statistics Division, 47
United States. *See also* US *entries*

United States. (cont.)
  as an energy exporter, 26
  energy commodities and reducing import dependence, 26
  expansion of mining activities, 65
  export bans on certain wood products, 87
  oil and natural gas producer, 23
  producer of minerals and metals, 59
  proposed protocol on enhanced transparency, 266
  resources versus amounts of production, 61
  restrictions on Japan's exports, 4
  trade relations with Japan, 6
Uruguay
  export bans on copper waste and scrap, 88
Uruguay Round, 130
Uruguay Round Ministerial Decision on Notification Procedures (1993), 164, 164, 189
US Bureau of Industry and Security (BIS), 26
US Department of Energy (DoE), 27
  'criticality' of minerals and metals, 54
  critical minerals and uses of, 57
  effects of export restrictions on environmentally-friendly products, 246
US economy
  substitutability rate, defined, 52
US National Research Council, 51, 54
Use of Quantitative Restrictions for Protective and Other Commercial Purposes, The Working Party on (1950), 223
US–Export Restraints dispute, 69, 248
US–Gasoline dispute
  Article XX(g), 214
  chapeau of Article XX GATT, 223
  discriminatory treatment versus methods for avoiding, 224
  disguised restriction, 225
  even-handedness in imposition of restrictions, 220
US–Poultry (China) dispute, 176
US–Shrimp dispute, 170, 208
  clause on 'arbitrary or unjustifialbe discrimination' abuse of GATT XX exceptions, 223
  rational connection test, 216

value added tax (VAT) and denying rebates, to reduce exports, 85
value chains
  effects of export restrictions, 114
value-added mining products, 45
VCLT. *See* Vienna Convention on the Law of Treaties
Venezuela
  export bans on ash and residues, 88
VERs. *See* voluntary export restraints
Vienna Convention on the Law of Treaties (VCLT), 195
Vietnam
  export duty commitments in Accession Protocol, 234
  export duties, 77, 80, 83, 161, 167
  export taxes on critical minerals and metals, 78
  imbalances in WTO obligations, 205
  obligation to eliminate export duties, 203
  producer of titanium, 59
  relaxation of export restrictions, 100
  resource-rich emerging economy, 15
  rice and, 97
  Vietnam's Export Duty Schedule, 154
  waste and scrap forms of minerals and metals, 82
  WTO-plus obligations, 84, 153–5
voluntary export restraints (VERs), 4–8

Wassenaar Arrangement, 87
WCO. *See* World Customs Organization
WFP. *See* World Food Programme
Working Party Reports
  accession protocols and export duties, 233, 236
  section on Export Regulation, 146

Working Party Reports on accession of WTO members and export duties, 145
World Customs Organization (WCO), 48
world demand
  critical materials, 63
  developing Asian nations and China, 63
  expansion, and lack of export safeguards, 187
  imports by China and pressure on world trade, 64
  oil prices, 27
  price spikes of critical raw materials, 285
  rise in world price, 65
  sharp increases, for critical minerals and metals, 63
  surge, and global macroeconomic expansion of mid-2000s, 22
World Food Programme (WFP), 20
world market prices and theoretical economic models, 104
world prices of minerals and metals and global welfare, 108
World Summit on Food Security (2009), 19
world trade
  limiting effect of non-automatic licensing schemes, 191
World Trade Organization
  Agreement on Agriculture (AoA), 285
  competition rules, absence of, 252–4
  food crises of 2007–2008 and 2010–2011, 15
  GATT and expansion of flexibilities in GATT, 282
  imbalances in export duty obligations, 204–8
  minerals and metals sector and restrictions, 30
  reform of export restraints, 31
  tariff concessions, 127–31
  Tokyo Round, 6
  Uruguay Round (1986–1994), 7
World Trade Report on Trade in Natural Resources (2010), 43
WTO. *See also* World Trade Organization
WTO accession regime. *See also* WTO-plus obligations
  by-product of failure of WTO multilateral negotiations, 233
  export duties in violation of WTO commitments, 83
  sovereignty over natural resources, 237
WTO Agreement. *See also* WTO-plus obligations
  accession protocols, 148, 162
  Article 5 on GATT exceptions to export taxes, 260
  Article XII:1
    Marrakesh Agreement, 146
  developed countries versus developing countries and LDCs, 260
  encompassing Marrakesh Agreement and annexed Multilateral Trade Agreements, 200
  inclusion on sovereignty and environmental values, 32
  Multilateral Trade Agreements in Annex 1A, 164
  original WTO Members bound by GATT, 147
  Plurilateral Trade Agreements in Annex 4, 164
  questions regarding Article XX GATT (1994), 197
  versus 'self-contained' agreements in accession regime, 206
WTO Agreement on Agriculture (AoA), 16
WTO Agreement on Export Taxes, 257, 261
  Article 7, proposed, 276
  EU proposals to Doha round, 258–66
  non-agricultural products, 259
  proposed, 268, 270

WTO Agreement on Import Licensing, 267
WTO Agreement on Import Licensing Procedures
  Preamble to, and affirmation of transparency in export licensing, 267
WTO Agreement on Safeguards, 7, 187
WTO Agreement on Subsidies and Countervailing Measures (ASCM), 248
WTO case law
  Article XI:1 GATT on inconsistent 'restrictions', 171
  Article XI GATT, 142
  Article XX GATT and justification of export restrictions, 208–25
  Article XX GATT and post-1994 accession protocols, 193
  Article XX GATT as exception to obligations of WTO members, 194
  assumption of government control of private suppliers, 249
  definition of 'other measures', 171
  definition of 'restrictions', 170
  legal test imposed by chapeau of Article XX GATT, 223
  non-automatic licensing schemes, 191
WTO jurisprudence
  beggar-thy-neighbour policies, 238
  cross-border benchmarks for export prices, 251
  export duties, non-scheduled, 141
  natural resources, once extracted and in commerce, 219
WTO legal treatment
  asymmetries in import and export duties, 140
  asymmetry in import and export duties, 254
  restrictions on imports and exports, 255
  Russia's WTO obligations, 158
  trade measures and asymmetries in import and export duties, 242
WTO Members
  economic effects of export restrictions, 101–3
  export duties or taxes, proliferation of, 74
  export prohibitions or bans, 87
  export quotas, 153
  four tiers of, 277
  GATT schedules and negotiations of concessions, 277
  net exporting, and restrictions for sake of development, 234
  net importing and fair access to supplies, 234
  net importing, and elimination of export taxes, 252
  obligations and exceptions under Article XX(g), 214
  policies versus conservation programmes, 240
  price or tax measures to restrict exports, 84–6
  proposed Agreement on Export Taxes, 289
  rights and obligations in all Multilateral Trade Agreements, 200
  second tier, and negotiated export duty concessions, 280
  third and fourth tiers, and negotiated export duty concessions, 281
WTO Members' GATT schedules
  and export taxes, 264
WTO Multilateral Trade Agreements, 199–202
WTO multilateral trading system
  integrity of, and imbalances in nations' obligations, 207
  negotiations and accession regime, 233
WTO regulation
  export restrictions, 68
WTO-plus obligations
  argument against, in favour of development, 234

export duties, limited flexibility of, 161–3
export duties, post-1994 accession protocols, 194
GATT schedules, 281
incumbent members, 146
newly acceded members, 281
origins of, 127
versus sovereignty over natural resources, 237
WTO-extra obligations, 146
WTO-plus obligations of various nations, 145–63

Yom Kippur War and effects on oil exports, 5

Zambia
export duties on copper, 82
export duties, proliferation of, 168
export licensing schemes, 95
export taxes on specific minerals and metals, 84
semi-processed products, 95
Zimbabwe
export bans on ores and concentrates, 88
export duties on chromium, 82
zinc as 'carrier' of germanium and indium, 61

www.ingramcontent.com/pod-product-compliance
Lightning Source LLC
LaVergne TN
LVHW011530200926
848484LV00001B/214
9781107085961